BIOCHEMISTRY OF FOOD PROTEINS

W0107224

BIOCHEMISTRY OF FOOD PROTEINS

Edited by

B. J. F. Hudson

*Honorary Research Fellow and Consultant,
Department of Food Science and Technology,
University of Reading, UK*

SPRINGER-SCIENCE+BUSINESS MEDIA, B.V.

First published 1992 by Elsevier Applied Science
Reprinted 1996 by Chapman & Hall

ISBN 978-1-4684-9897-4   ISBN 978-1-4684-9895-0 (eBook)
DOI 10.1007/978-1-4684-9895-0

© 1992 Springer Science+Business Media Dordrecht
Originally published by Champman & Hall in 1992
Softcover reprint of the hardcover 1st edition 1992

Typeset in by Enset (Photosetting) Ltd, Bath

A catalogue record for this book is available from the British Library

Library of Congress Catalog Card Number: 92-3541 CIP

PREFACE

Developments in the understanding of food protein structure, behaviour and applications continue apace. Many of these have, in the past decade, been reported and evaluated in the series *'Developments in Food Proteins'*, comprising seven volumes, with a total of 55 chapters.

The time has now come to re-assess many of the topics reviewed in that series and to add certain others. However, instead of assembling, somewhat at random, food protein topics from quite disparate fields in individual volumes, we have decided to bring together homogeneous groups of topics, each representing a specific sector of the subject. Under the general theme of *'Progress in Food Proteins'* the first of these groups covers 'Biochemistry'.

Readers will note that, though six of the topics reviewed in this volume are new, five of them have already featured in *'Developments in Food Proteins'*. These last are in active research fields in which new developments have been of special significance. In this sense, therefore, they are welcome updates.

Proteins, being complex heterogeneous polymers, with a variety of functional groups leading to a wide range of physical and chemical properties, interact readily with several other classes of food components. Chapters 1–6 address this theme, beginning with protein–water and protein–protein associations. Interactions with carbohydrates, both starches and reducing sugars (Maillard Reaction) follow. Attention then turns to metal–protein interaction, followed by a more detailed consideration of the special case of haemoproteins. All these interactions have profound effects on protein behaviour in the context of foods, having a bearing on functionality, acceptability, safety and nutritional status.

The modification of proteins by planned reactions can take several forms, beginning with classical chemical reactions aimed mainly at changing functional properties (Chapter 7) and following with enzymatic methods having a similar objective (Chapter 8). An alternative approach, also having both functional and nutritional implications, is manifest in the Plastein Reaction (Chapter 9).

Enzymes themselves comprise an important segment of the total food protein picture. Their production and applications, from an industrial point of view, are reviewed in Chapter 10. Finally, in Chapter 11, the contribution of proteins to food flavour is discussed.

In this text every effort has been made to stress the most recent developments in food protein chemistry and biochemistry. Readers cannot fail to be impressed with the proliferating methodology, from nuclear magnetic resonance to genetic engineering, now in use by protein investigators. For this reason, as well as for authoritative discussions and comment on the 'state of the art', it can be confidently expected that food protein specialists will find much of value in these eleven chapters.

My sincere thanks are due to the fourteen specialists who have cooperated with me in the preparation of this volume and to the publishers for their encouragement.

B. J. F. Hudson

CONTENTS

LIST OF CONTRIBUTORS

J. M. AMES
Department of Food Science and Technology, University of Reading, P.O. Box 226, Whiteknights, Reading RG6 2AP, UK

S. ARAI
Department of Agricultural Chemistry, University of Tokyo, Bunkyo-Ku, Tokyo, Japan

J. CHRASTIL
US Department of Agriculture Southern Regional Research Center, PO Box 19687, New Orleans, Louisiana 70179, USA

G. M. FROST
21 Spring Walk, Brayton, Selby, North Yorkshire YO8 9DS, UK

R. D. GILLARD
Department of Chemistry and Applied Chemistry, University of Wales, P.O. Box 912, Cardiff CF1 3TB, UK

J. S. HAMADA
U.S. Department of Agriculture, Southern Regional Research Center, PO Box 19687, New Orleans, Louisiana 70179, USA

N. K. HOWELL
School of Biological Sciences, University of Surrey, Guildford GU2
5XH, UK

S. H. LAURIE
School of Chemistry, Leicester Polytechnic, Leicester LE1 9BH, UK

D. A. LEDWARD
Department of Food and Agricultural Chemistry, The Queen's University of Belfast, Newforge Lane, Belfast BT9 5PX, UK

W. E. MARSHALL
US Department of Agriculture, Southern Regional Research Center,
PO Box 19687, New Orleans, Louisiana 70179, USA

M. I. SCHNEPF
Department of Nutritional Science and Hospitality Management, University of Nebraska, 316 Ruth Leverton Hall, Lincoln, Nebraska 68583-0806, USA

F. F. SHIH
US Department of Agriculture, Southern Regional Research Center,
PO Box 19687, New Orleans, Louisiana 70179, USA

M. WATANABE
Food Science Laboratory, Faculty of Education, Tokyo Gakugai University, Koganei-shi, Tokyo 184, Japan

G. S. D. WEIR
Brook Bond Foods Ltd, Development Department, Leon House, High Street, Croydon CR9 1JQ, UK

Chapter 1

PROTEIN-WATER INTERACTIONS

MARILYNN I. SCHNEPF

*Department of Nutritional Science and Hospitality Management,
University of Nebraska, Lincoln, Nebraska, USA*

INTRODUCTION

Water is one of the most important molecules necessary to sustain life. It is the only substance found in all three physical states. The interaction of water with other substances is critical to most chemical and biological reactions. Proteins also play an important role in many life processes. They provide structure to living cells and regulate many cellular functions. The interaction of these two important molecules plays an important role in determining the functional properties of many food ingredients. The focus of this chapter is to discuss the types of bonding which occur between water and protein molecules and to relate the interaction between water and protein molecules to the functional properties of proteins in foods.

THE WATER MOLECULE

A study of the physical constants of water and a comparison of these values with those of molecules of similar molecular weight and atomic structure (CH_4, NH_3, HF, H_2S, H_2Se and H_2Te) show that the water molecule has larger values for heat capacity, melting point, boiling point, surface tension, and heats of fusion, vaporization, and sublimation than would be expected from its size and composition.[1] The unusual properties of water suggest the existence of strong attractive forces within the water molecule itself and with other water molecules. Water is composed of two hydrogen atoms held by covalent bonds to one oxygen atom, with a bond angle of 104·5°. This almost perfect tetrahedral angle causes the water

1

molecule to form a V-like structure which yields an unsymmetrical charge distribution. Each water molecule has two areas of partial positive charge (hydrogen-bond donor sites) and an equal area of partial negative charge (hydrogen-bond acceptor sites). This tetrahedral arrangement allows each water molecule to hydrogen bond with four other water molecules. The extensive three dimensional network of water molecules accounts for the unusual physical properties of water. The hydrogen bonds found in water exist for only a short time compared to those found in ice, which accounts for the fluid nature of water.

The momentary nature of these hydrogen bonds gives rise to different theories on the exact nature of the structure of water. Fennema[2] has divided these theories into two major categories: (1) continuum theories, which are also called uniform or homogeneous theories; and (2) mixture theories. Those who espouse continuum theories propose that all molecules in cold liquid are hydrogen-bonded (four bonds/molecule), but that the bonds differ in angle, length, and energy. Supporters of mixture theories contend that water consists of two or more distinguishable species which exist in dynamic equilibrium and differ in the degree of hydrogen bonding. An example of a mixture theory is the 'interstitial model'. Proponents of this model believe that liquid water exists partly in the form of bulky framework structures that evolve from a 4-coordinated, approximately tetrahedral arrangement of water molecules. Within the cavities formed by this framework, entrapped single water molecules exist.

The structure of water is further complicated by the fact that it is composed of more than simple HOH molecules. It contains more than 33 chemical variants in trace amounts. These include, as well as the common isotopes of ^{16}O and ^{1}H, ^{17}O, ^{18}O, ^{2}H (deuterium) and ^{3}H (tritium). Fortunately, the isotopic variants occur only in small amounts and can usually be ignored.[1]

TYPES OF BOUND WATER

The association of water with macromolecules is difficult to determine, due to a lack of consensus by researchers about the types of bound water. Because water is held in different degrees to macromolecules, few quantitative ways to distinguish among the types of bound water exist. In one sense, all water found in tissue is bound water because of its resistance to flow. Chou & Morr[3] have identified six types of bound water in protein.

The first, structural water, is bound directly to the protein molecule by hydrogen bonding. Structural water stabilizes the native conformation of the protein molecule and is found inside the macromolecule. The water molecules are engaged in two or more hydrogen bonds and are not available for chemical reactions. Only a small portion of the total water associated with the protein molecule is structural, but this water is very important in determining the three-dimensional conformation of the protein.

The second type of bound water, monolayer water, is bound to the surface of the protein by hydrogen bonding or dipole interaction. This type of water makes up about 4–9% of the water associated with the protein. It is not available as a solvent, but may be available for certain reactions.

The third type of bound water is unfreezable water which is the water clustered around each polar group. This water may include both structural and monolayer water and is dependent upon the amino acid composition of the protein and the number of available polar side chains. One gram of protein may have associated with it 0·3–0·5 g of water which is unfreezable. This corresponds to a water activity (a_w) of up to 0·9, so some chemical and biological reactivity is possible.[4]

The remaining three types of bound water, as described by Chou & Morr,[3] are not as well defined as the first three types. The fourth type of water is associated with the protein molecule via hydrophobic hydration. This type of water has been described as clathrate-type or ice-like structured water, but the real nature of this water is not clear. The fifth type, imbibition or capillary water, is held physically or by surface forces. This water is available for chemical reactions and acts as a solvent, but can only be removed by force. Much of the water found in cheese and meat is an example of imbibition or capillary water. The sixth type, hydrodynamic water, is transported along with the protein molecule. This water is independent of a_w and has the normal physical properties of water.

Other researchers have grouped these types into three broad categories—constitutional, interfacial (vicinal and multilayer) and bulk phase water (free and entrapped).[1,2,5] Constitutional water is described as that water which is located in the interstitial sites of proteins. It would correspond to structural water, and in a high moisture food (90% water) would make up less than 0·03% of the total water. It is unfreezable and exhibits essentially zero water activity.

Interfacial water is composed of vicinal water and multilayer water. Vicinal water is the first one or two water molecules adjacent to the protein and multilayer water is the next few layers. Vicinal water strongly interacts with specific hydrophilic sites of proteins by water-ion and water-dipole associations. It also includes water in microcapillaries and corresponds to an a_w of 0·2–0·3. In a high modest food, vicinal water would compose 0·1–0·9% of the total water. Multilayer water is the water that occupies the remaining first-layer sites and forms several additional layers around the hydrophilic groups. This water is held by water–water and water-solute hydrogen bonds and corresponds to an a_w of up to 0·6. Of the total water in a high moisture food, multilayer water represents 1–5% of the water.

Bulk phase water is divided into free and entrapped water. Both of these types compose the water that occupies positions farthest removed from non-aqueous constituents. Water–water hydrogen bonds predominate and the water has properties similar to water in a dilute salt solution. Bulk phase water comprises about 96% of the total water in a high moisture food and corresponds to an a_w of greater than 0·6. The principal difference between free and entrapped water is that the flow of entrapped water is impeded by the matrix of a gel or tissue.

The definitions used to describe the types of bound water may differ, but types of water should be seen as a continuum, with boundaries that may be indistinct. The types of bound water may change somewhat with different products and at different temperatures.

PROTEIN CONFORMATION

Proteins are complex macromolecules whose primary structure is determined by the amino acid sequence, and is unique for each protein. The primary structure gives rise to secondary, tertiary and possibly quaternary structures. The number of different conformations a protein may assume is statistically very large. The final conformation is the result of the interaction of the amino acid side chains with each other and with water molecules.

The net free energy of stabilization of native protein is equivalent to no more than three to five hydrogen bonds per molecule.[6] The low net free energy involved in protein conformation demonstrates that water plays a primary, not a secondary role, in determining the shape of the protein molecule.

Three types of bonds, hydrogen bonding, hydrophobic interaction, and salt linkages, determine the pattern of folding of the protein.[7] Hydrogen bonds occur between the oxygen atoms of carbonyl and hydrogen atoms of amide groups. Hydrophobic interactions are a result of the association of the non-polar amino acid side chains. In the typical protein, valine, leucine, isoleucine and phenylalanine comprise 20–30% of the protein. If proline, alanine and tryptophan are included the percentage rises to 35–45%. Salt linkages, or electrostatic interactions, occur between the charged side chains of the exposed carboxyl (aspartic and glutamic acids) and amino (lysine, arginine, and histidine) groups.

Water molecules can influence each of these types of bonds. The effect of water on hydrogen and hydrophobic bonds is readily apparent. In the presence of separated electrostatic groups, the surrounding water molecules are highly compressed. When these oppositely charged groups come into contact, the solvent molecules are not as strongly oriented and are less compressed. A large increase in entropy and volume results. Both salt linkages and hydrophobic bonds are stabilized by entropy effects rather than by energy effects. The addition of electrolytes will strengthen hydrophobic bonds but will weaken salt linkages, while the addition of non-polar solvents will have the opposite effect. Because of this relationship, solvents such as alcohols and acetone will denature proteins, while salts act as inhibitors of denaturation.

The amounts of energy associated with different types of bonds differ greatly, ranging from 30 to 100 kcal/mol for covalent, to 10–20 for ionic, 2–10 for hydrogen, and 1–3 for hydrophobic bonding.[8] While each of these types of bonds contributes to the native structure of proteins, hydrophobic energy was considered to be essential in the folding of the amino acid chain, with the non-polar groups buried on the inside of the molecule. The loss of entropy on folding was compensated by a gain in 'hydrophobic energy', resulting from a reduction in the number of contacts between the non-polar groups and water. Polar groups were considered to contribute little to the molecule's stability.

With the elucidation of more protein structures, a more complex view of protein conformation is emerging. After a review of the literature on protein structure, Finney[9] made the following generalizations about protein structure and the role of water:

1. Formal charges on protein side groups are almost invariably exposed to the solvent and where they are not exposed, salt bridges are usually formed.
2. Polar groups tend to be more evenly distributed between internal and external positions. Polar groups which are buried are usually involved in hydrogen bonding, with efficiency generally over 80%.

3. As many as 40–50% of the apolar groups may be accessible to the solvent. The simple picture of buried hydrophic side chains may be an oversimplification. The actual protein conformation may be more complex.

With the use of X-ray and neutron scattering techniques it is possible to identify discrete water molecules within crystalline protein structures. Most of the water molecules located by these methods are found linking main chain C=O groups to the NH groups that are too far separated from one another for direct hydrogen bonds. The water molecules may be bonded to three or four different residues. Other water molecules are found linking main chain atoms to side chain hydrogen bond donors and acceptors. Phenylalanine, serine, methionine, and tryptophan would participate in this type of interaction. Water molecules may stabilize reverse beta-turns in regions where few intrapeptide hydrogen bonds occur. Water bridges also may form between amino acid residues of different protein molecules in a crystal or different subunits of the same protein.[6]

Water molecules may play a role in the stabilization of the polypeptide backbone.[10] Peptides form regular hexagonal or honeycomb patterns. The distance between peptide oxygens is about 0·48 nm, which is the same as the 'second-neighbor' oxygen distances in the ice lattice. The ice-like lattice could lie above a polypeptide layer. Exact colinear hydrogen bonds could be formed at each position to satisfy all of the bonding requirements of the peptide and water, thus stabilizing the polypeptide backbone.

Water molecules also form hydrogen-bond bridges and salt bridges between polar groups and ionic groups too far apart to otherwise link. Small groups of water molecules found near internal charged groups may stabilize the protein by spreading the buried charge through a region of higher dielectric constants. Each water molecule found internally will make three or four hydrogen bond contacts.[9]

Polar groups at the surface of proteins are solvated as expected. The water molecules often form short surface bridges between polar groups in the same molecule. 'Clathrate cage' structures around non-polar groups cannot be detected by using X-ray and neutron scattering techniques.[9] Water molecules have been found in the hydrophobic clefts of some proteins such as trypsin and myoglobin. The water molecules may be released as enzymes bind to substrates, thus playing an important role in enzyme activity.[9]

TECHNIQUES USED TO STUDY PROTEIN–WATER INTERACTIONS

Protein–water interactions can be studied by thermodynamic, kinetic, spectroscopic, and diffraction techniques.[3] Each of these techniques provides different information about the nature of this interaction. Other researchers who have described techniques to characterize protein–water interactions include Bull & Breese,[11] Labuza,[12] and Franks.[13]

Thermodynamic techniques measure properties related to enthalpy (H), entropy (S), free energy (G), water activity (a_w), the freezing point, and boiling point. The simplest of the thermodynamic techniques is the determination of water sorption-desorption isotherms. This measures the degree of hydration as a function of a_w. When the protein is exposed to increasing relative humidities or water activities at equilibrium, a sigmoidal sorption isotherm is usually obtained. This reflects the progressive increase in the amount of water associated with the protein.[14] Isotherm equations correlate with the number of water binding sites on the protein molecule and the strength of the binding. The most serious difficulty with adsorption isotherms is that changes in the protein surface are ignored. Protein conformational changes are associated with the degree of hydration of the protein. The same problem exists when measuring enthalpy, entropy, and heat capacity. When dry, the protein may be trapped in an unfavorable energy state and any measurement taken will reflect such conformational changes.[5]

Thermodynamic techniques are used to determine the water holding capacity (WHC) and water binding capacity (WBC) of proteins. While these two terms are often used interchangeably, WHC refers to the amount of water a protein can retain and WBC refers to the amount of water a protein can absorb. The techniques and equipment used to measure WHC and WBC are numerous. Chen et al.[15] divided the methods into four types. The first group applies some kind of external stress such as compression, centrifugation or suction to determine how much water is retained by the protein. The second group of methods determines water absorption by a dry protein and provides sorption isotherms. The third group relates to the swelling of protein as it takes up water by diffusion into the capillary system. The fourth group of methods is based on the determination of a colligative property of water, such as a change in the freezing point. Chen et al.[15] concluded their discussion on water uptake by protein with the caution that results from the different types of tests may yield conflicting data. WBC will strongly depend on the physicochemical properties and composition of the test materials as well as on the experimental conditions.

The second group of techniques, as described by Chou & Morr,[3] measures kinetic properties or the change in the mobility of water as it associates with protein, and changes in the protein molecule as affected by its interaction with water. Specific techniques used to study this interaction include: nuclear magnetic resonance (NMR), dielectric dispersion, laser light scattering, and intrinsic viscosity. Three types of water environments can be identified by kinetic studies. These include 'bulk', 'bound', and 'irrotationally bound', with approximate rotational correlation times in the ps, ns, and μs range, respectively.

The third group of techniques are spectroscopic methods, which measure the nature and strength of hydrogen bonds. These techniques include infrared (IR) and Raman spectroscopy as well as NMR which can determine unfreezable water.

The fourth group of techniques are diffraction methods. These provide information on the average position and orientation of water molecules with respect to each other and to the protein molecule. These methods include light scattering and small angle X-ray scattering, as well as high resolution X-ray and neutron diffraction. These methods are used to locate the highly ordered water molecules located within the protein crystal, and in some cases the water molecules strongly bound at the protein molecular surface.[9]

From the array of methods used to determine the nature of protein–water interactions, conflicting data are bound to result. Chou & Morr[3] conclude their list of techniques by pointing out that since the techniques measure different properties of protein–water interaction, the conflicting data should be viewed as giving supporting information about a complex interaction. No one method can be used to explain fully the different types and degrees of protein–water interactions.

PROTEIN–WATER INTERACTION AND AMINO ACID COMPOSITION

The amount of water bound by a protein is a function of the amino acid composition and conformation of the protein. The number of polar amino acids, cationic, anionic or nonionic, affects the amount of water bound to the protein.[3] Kuntz[5] developed the following formula to predict the amount of water bound to the protein molecule:

$$A = f_e + 0·4f_p + 0·2f_n$$

where A is the amount of bound water in g water/g protein, f_e is the fraction of charged side chains, f_p is the fraction of polar side chains, and f_n is the fraction of non-polar side chains. One problem with the formula was overestimation of the extent of hydration, since all side chains are counted, even those buried in the native protein structure.

Not all sorption sites have equal degrees of hydrophilicity. Carboxyl and amino groups seem to be mainly responsible for the binding of water (Table 1) with other groups showing less of an ability to bind water.[16]

The amount of bound water changes with the increased availability of water. At relative humidities over 65% there is a sharper increase in the actual amount of water bound than would be expected from calculated values.[16] At higher humidities when multilayer formation becomes more evident, the structure of the protein may determine the extent of water uptake. Soluble proteins lack cohesive forces which restrict swelling, while for insoluble proteins, sorption is restricted by swelling constraints.

Anderson & Witter[17] studied the water binding capacity of 22 L-amino acids at an a_w of 0·90, and reported that the amino acids with the highest WBC were γ-aminobutyric acid (1·699 g H_2O/g solid) and proline (1·629 g H_2O/g solid). Glutamic acid and aspartic acid bound 0·004 and 0·001 g H_2O/g solid, respectively. However, WHC increased dramatically when the potassium or sodium salts of these amino acids were tested.

Kuntz[18] determined the amount of water bound to polypeptides of specific amino acids at different pHs and temperatures (Table 2). Aspartic and glutamic acid polypeptides tended to bind the most water. The amount of water bound by the polypeptides was related to pH. Ionized proteins tend to bind more water. When the pH of the protein is at its isoelectric point with no net charge, minimal hydration and swelling can occur. This reduction in water binding at lower pHs is very evident for the polypeptides of aspartic and glutamic acids. Lower temperatures also reduced the amount of water bound.

The effect of denaturation on the hydration of proteins depends to some extent on the degree of denaturation and the type of protein involved. Most researchers report a slight increase in hydration upon denaturation and a slight decrease upon hydrophobic aggregation.[5] Using proton magnetic resonance, Kuntz & Brassfield[19] reported a 10% increase in protein hydration as bovine serum albumin (BSA) was denatured using urea. Water binding was occurring in the regions of the pro-

TABLE 1

WATER ASSOCIATED WITH HYDROPHILIC GROUPS IN PROTEINS

Sorption site	Moles H_2O per mole of sorption site at RH of:						
	5%	10%	20%	35%	50%	65%	80%
Carboxyl (—COOH)	0–7	0–92	1–2	1–63	2–0	2–3	2–5
Amino (—NH₂)	0–6	0–83	1–2	1–63	2–1	2–4	2–7
Guanidino (—NH—C—NH₂) $\overset{\parallel}{NH}$							
Aliphatic hydroxyl (OH)	0–50	0–09	0–17	0–27	0–34	0–46	0–60
Phenolic hydroxyl (—OH)	0–16	0–25	0–5	0–75	1–0	1–3	1–8
Peptide (—CO—NH—)							
Amide (—CO—NH₂)	0–04	0–06	0–11	0–17	0–25	0–36	0–56
Heterocyclic imino (—NH—)							

Reference: Leeder & Watt.[16]

TABLE 2
POLYPEPTIDE HYDRATION

	pH	Moles H_2O per mole of amino acid		
		−25°C	−35°C	−45°C
Polypeptide				
L-Glu	7–12	8·3	7·7	6·3
L-Glu	4–5		1·8	
L-Asp	8–12	8·1	6·0	4·8
L-Asp	4·5	2·1		
L-Tyr	11·5–12		8·5	6·5
L-Tyr	11·3		5·5	5·1
DL- or L-Lys	3–9	5·0	4·3	3·8
DL- or L-Lys	10–12	5·0	4·5	3·7
L-Orn	1·5–9	4·0	3·4	3·5
L-Orn	10–12	4·5	3·7	3·5
L-Arg	3–8	3·1	2·7	
L-Arg	10	3·0		
L-Pro		3·1	2·8	
L-Asn		2·0		
DL-Ala		1·4		
L-Val		0·9		
Gly		0·9		
Polymers				
Lys⁴⁰Glu⁶⁰	2–4	2·5	2·4	
Lys⁴⁰Glu⁶⁰	11–12	7·8	7·5	
Lys⁵⁰Phe⁵⁰	2–9	2·6	3·8	
Lys⁵⁰Phe⁵⁰		1·2		

Reference: Kuntz.[18]

tein that were exposed upon unfolding. Denaturing agents, such as urea or guanidine HCl, may affect protein hydration by direct interaction with the protein or by indirectly affecting the structure of water.[20]

IMPLICATIONS OF PROTEIN–WATER INTERACTIONS FOR FOOD PROTEINS

Protein–water interactions are involved in determining many of the functional properties of proteins. Physico-chemical properties of the protein molecule which influence functional properties include hydrophilic,

TABLE 3

FUNCTIONAL PROPERTIES PERFORMED BY SOY PROTEIN PREPARATIONS IN ACTUAL FOOD SYSTEMS

Functional property	Mode of action	Food system
Solubility	Protein solvation, pH dependent	Beverages
Water absorption and binding	Hydrogen-bonding of HOH, entrapment of HOH, no drip	Meats, sausages, breads, cakes
Viscosity	Thickening, HOH binding	Soups, gravies
Gelation	Protein matrix formation and setting	Meats, curds, cheese
Cohesion-adhesion	Protein acts as adhesive material	Meats, sausages, baked goods, pasta products
Elasticity	Disulfide links in gels deformable	Meats, baked goods
Emulsification	Formation and stabilization of fat emulsions	Sausages, bologna, soup, cakes
Fat absorption	Binding of free fat	Meats, sausages, donuts
Flavor-binding	Adsorption, entrapment, release	Simulated meats, baked goods
Foaming	Forms stable films to entrap gas	Whipped toppings, chiffon desserts, angel cakes
Color control	Bleaching of lipoxygenase	Breads

Reference: Kinsella.[82]

interphasic, intermolecular and organoleptic. Specific properties which are related to each of these characteristics of the protein molecule are listed in Table 3. Many of these properties are related to the interaction of the protein molecule with water.

One of the problems encountered in the study of protein functionality in complex food systems is the many factors that will influence their functional properties. Often food proteins will not behave as predicted by the study of model proteins. Factors that will influence functional properties include: the protein source and variety, extraction conditions, temperature, ionic history, impurities present, and storage conditions.

Solubility

The thermodynamic definition of solubility includes: (1) well defined initial solid and final solution states; and (2) the establishment of equilibrium between these two states. Under these conditions, the solubility of a protein at a given temperature and pressure is the concentration of the sample in solution.[21] The amount of protein in solution is independent of the path used to solubilize the protein.

Because most proteins found in foods are complex heterogeneous mixtures of many proteins, a more practical definition is needed. Operationally, solubility denotes the amount of protein in a sample that goes into solution or into colloidal dispersion under specified conditions and is not sedimented by moderate centrifugal forces.[14] Solubility is markedly affected by environmental conditions, pH, temperature and ions. The pH affects charge and electrostatic balance within and between proteins. Solubility tends to be at a minimum at the isoelectric point of a protein when the net charge is zero. Attractive forces predominate and molecules tend to associate. Away from the isoelectric point, the protein molecules are charged, and dispersibility and solubility are enhanced. In the acidic pH range the proteins have a net positive charge, while in the basic pH range the proteins have a net negative charge. Temperature progressively disorders both the proteins and the solvent. Increasing temperature may destabilize the protein, causing unfolding which may enhance protein–protein interaction and result in aggregation and precipitation. At higher temperatures, hydrophobic interactions may be strengthened as water is forced into closer contact with the hydrophobic core of the protein. Bingham,[22] when studying the solubility of casein proteins, reported that solubility increased as the temperature decreased. The effect of ions on solubility may be due to the ability of the ions to affect both the structure of water and to bind to the protein itself. The effect of ions will be discussed in detail in a later section.

When studying the solubility of soy isolates, Shen[23] found that soy isolates seemed to have no thermodynamically defined solubility concentration. Normally the percentage of protein in solution will remain constant at 100% until the saturation limit is reached, and then the percentage will drop off and approach zero as more and more protein is added. In contrast, the percentage of protein in solution for the soy isolates remained constant as the amount of added protein increased. The isolates behaved as if they constituted an insoluble fraction. This behavior was observed up to the highest concentration of 18%. The distribution of the soluble fraction and the insoluble fraction were not well defined. It depended upon the previous history of the sample, such as the manner of extraction, precipitation, solvent treatment and drying, as well as the method used to dissolve the proteins. Because of this lack of clearly defined solubility states, the comparison of results from one study to another is difficult.

Complex proteins composed of subunits may associate and disassociate as environmental conditions change. This causes changes in solubility. German et al.[24] studied the thermal dissociation of soy protein fractions. They reported that the 11S basic subunits will interact with any 7S proteins present and prevent thermal aggregation by forming a soluble complex.

Bigelow[25] suggested that two structural features of the protein molecule, charge frequency and hydrophobicity, are the major determinants of solubility. The higher the charge frequency and the lower the hydrophobicity, the higher would be the solubility. Any factor which would affect the conformation of the protein would then influence its solubility.

Solubility and Neutral Salts

One of the factors which influences both charge frequency and hydrophobicity is the presence of ions or neutral salts. This three-way interaction of protein, water, and ions has been studied by many researches.[26-28] The major effects of ions have been termed 'salting in' and 'salting out'. 'Salting in' refers to an increase in solubility and usually occurs at lower ion concentrations. 'Salting out' refers to a decrease in solubility, which results in precipitation and usually occurs at ion concentrations of greater than 0.1 M.

Whether the interaction between the protein and the ion will cause an increase or a decrease in solubility will be determined by both the conformation of the protein and the concentration of the ions. Since proteins are colloidal particles, they are surrounded by an electric double layer.[29] Hydrophobic colloids tend to be compact in shape. Neutral electrolytes, at a concentration below 0.1 mol dm^{-3}, increase the thickness of the double layer, which causes an increase in the stability of the protein. Hydrophilic colloids tend to be longer and thread-like in shape and reactions with electrolytes are more complex. Low electrolyte concentrations increase the electric double layer and cause conformational changes in the protein molecule. The conformational changes may be due to a repulsion between adjacent charged groups which may result in the exposure of previously buried hydrophobic groups.[30] At concentrations above 0.1 mol dm^{-3} the electric double layer is suppressed and further changes in conformation occur. At these higher concentrations more solvent–ion interactions occur. Therefore, there may be an exchange between the ions in the solvent and the ions more firmly held by the protein molecule.

Melander & Horvath[31] have proposed a single theory which would account for the effect of electrolytes on charge frequency and hydrophobicity of proteins. Shen[21] simplified their theory by explaining the effect of ions on proteins by the following equation:

$$\ln(W/Wo) = (\Delta F_{electrostatic})/RT + (\Delta F_{cavity})/RT + Const$$

where W is the weight of the protein soluble in 1 liter of salt solution of a given concentration and Wo is the weight of protein soluble in 1 liter of H_2O. The electrostatic term ($\Delta F_{electrostatic}$) is the change in electrostatic free energy when the protein goes from the crystalline state to the solution state. This term is always positive and is responsible for the 'salting in' term. The hydrophobic term (ΔF_{cavity}) is the free energy required to create a cavity in the bulk solvent necessary to house the hydrophobic groups that are exposed when a protein goes from a crystalline state to a solution. This term accounts for the 'salting out' term and takes into account the increase in hydrophobic surface area, the surface tension of the salt solution, and the surface tension of water.

Melander & Horvath's[31] theory, explains the behavior of some proteins such as carboxyhemoglobin and fibrinogen, but more heterogeneous proteins do not behave as expected. Shen[21] found that soy protein does not become more soluble at low salt concentrations when the electrostatic term is more dominant. This 'salting in' should be followed by 'salting out', when the cavity term becomes more dominant. Instead soy protein shows a decrease would result in minimum solubility.

While there is little experimental evidence to support these suppositions, Matsudomi et al.[32] did find that surface hydrophobicity of 11S soy globulin was higher at an ionic strength of 0·01 compared with ionic strengths of 0·1 and 0·5. The maximum value of surface hydrophobicity decreased as the ionic strength of the protein was increased.

The effect of ions on proteins is related to the effect of the specific ion as well as the type of ion. Cations and anions have a different effect on the orientation of water molecules which surround them. In the hydration shell of a cation, the hydrogen atoms of water are directed out, whereas with the anion they are directed in.[28] This different orientation of charges causes anions to decrease the polarity of water to a greater extent than cations. The degree to which the water structure is affected depends on the size and charge density of the ions. Both cations and anions tend to arrange themselves in a series known as the Hofmeister, or chaotropic or lyotropic series. In general for anions, the extent of structure breaking effect on water progressively follows the order of $F^- < CH_3COO^- < SO_4^= < Cl^- < Br^- < I^- < NO_3^- < ClO_4^- < SCN^- < Cl_3CCOO^-$. With cations the series is as follows $Li^+ < Na^+ < K^+ < Rb^+ < Cs^+$.

Ions with the largest ionic radius and lowest charge density have the greatest ability to break the hydrogen bonded structure of water.

Since hydrophobic interactions have an important role to play in the conformation of many proteins, the destabilizing effect of ions on water structure would have an effect on protein conformation. Damodaran & Kinsella[28] studied the changes in the ability of bovine serum albumin to bind 2-nonanone in the presence of various anions. The anions studied followed the classical Hofmeister series. These researchers concluded that the effect observed on the 2-nonanone binding to bovine serum albumin at 0·15 M concentration of various anions may be attributed to changes in the hydrophobic regions in the protein, mediated via changes in water structure.

Ions may also affect the protein by preferentially binding to the protein molecule itself. Bull & Breese[27] studied the binding of ions to egg albumin. They reported that the size of the cation had little effect on the binding of the ion to the protein. About 5·7 mol of LiCl, RbCl, or CsCl were bound per mole of egg albumin. Very little NaCl or KCl were bound. In contrast to the cations, the anions bound to the protein according to the size of the ionic radii. The larger monovalent anion the greater the tendency to bind to the protein and the greater was its dehydrating effect on the protein.

Schnepf & Satterlee[33] reported that iron ($FeSO_4$) was bound to soy protein isolate in different degrees. Some of the surface bound iron was easily removed by increasing the NaCl concentration or by using chelating agents such as ethylenediamine tetraacetic acid (EDTA) or 1,10-phenanthroline. Some of the iron was more difficult to remove and may have been located on the interior of the molecule and thus partially responsible for protein aggregation.

Because most food systems contain protein, water, and solutes the interactions of these components have many practical implications. Hardy & Steinberg[34] studied the interaction of sodium chloride and p-casein as a function of water sorption. These researchers reported that as the salt concentration increased, the amount of interacted salt also increased. Salt tended to partition itself between the protein and water. At a lower water content, the salt tended to interact with the protein, reaching a constant level at 0·7 g salt/g p-casein. At a higher a_w, the salt tended to bind water and interact less with the protein. These interactions are important in cheese production, since the binding of salt will influence the growth of microorganisms which, in turn, will influence the hydrolysis of protein during ripening. This hydrolysis will determine many of the rheological and textural properties of the final product.

Konstance & Strange[35] studied the effect of temperature, pH, salt type and salt concentration on the solubility and viscosity of various commercial caseins and caseinates. Sodium phosphate had the greatest impact on solubility and viscosity for most of the caseinates, with the exception of the normally insoluble rennet casein. Calcium chloride was able to solubilize the caseinates at their isoelectric points, at the appropriate concentration and temperature. These researchers demonstrated the complexity of the interaction of many factors on solubility and viscosity of proteins.

In meat products, salts are used for their functional properties to obtain acceptable products. Salts influence the solubility of meat proteins and play an important role in water holding capacity. The effect of sodium chloride on water holding capacity of meat is dependent on pH.[36] At a pH below the isoelectric point, sodium chloride (2%) decreases water holding capacity, while above the isoelectric point, water holding capacity is greatly enhanced. These results can be explained by an understanding of the effect of anions and cations on water and protein. Anions have a greater effect on the structure of water and have a greater ability to bind to the protein molecule itself. Hamm[36] reported that the isoelectric point of the protein was lowered in the presence of salt. This effect would be due to the greater ability of the chloride ion to bind to the protein. As the net positive charge was neutralized by the chloride ion, the protein matrix would shrink and water holding capacity would be reduced. Above the isoelectric point the sodium ions were absorbed poorly, some chloride ions neutralized the positive charges that still existed and the net negative charge increased. As this happened the swelling of the matrix occurred and water holding capacity increased.

The hydration of milk proteins also demonstrates the importance of ions. Berlin[37] concluded that the ions in milk played a greater role in controlling water binding than the status of either the casein or whey proteins when milk was denatured. The loss of protein solubility through thermal denaturing apparently had little effect on the capacity of the protein to bind water.

German et al.,[24] when studying the thermal dissociation of soy protein fractions, reported a significant negative interaction between pH and ionic strength. Protein solubility of the basic subunits of the 11S protein was reduced by the neutralization of charged groups upon the addition of salt.

Viscosity

Viscosity is that property of a fluid which gives rise to forces that resist the relative movement of adjacent layers in the fluid.[38] Knowledge of the viscosity of fluids is important in the processing of many foods and will affect the desired texture and appearance of products. Protein and water interactions obviously have a great effect on viscosity. The problem with much of the information on viscosity as a thermodynamic principle is the inability to apply this information to actual food systems. The viscosity of solutions of food proteins is usually too high to exhibit equilibrium behavior and may be affected by shear rate, as well as time. The term 'apparent viscosity' is usually applied to the measurement of viscosity in most food systems. Kinsella[14] cautions that this term is often misused. This term applies to a value obtained when a non-Newtonian fluid is subjected to a constant rate of shear as with a Brookfield viscometer. The Brookfield equipped with a T-bar or disc-type fixture is useful for the measurement of Newtonian fluids in which viscosity is independent of shear rate, but cannot provide meaningful data in terms of flow behavior of more complex rheological systems. Data obtained in this manner should be referred to as Brookfield consistency.

A study by Richardson et al.[39] illustrated the problem of studying viscosity of actual foods. These researchers determined the viscosity of wheat flour suspensions at various concentrations. At low concentrations, between 0 and 30%, wheat flour exhibited Newtonian flow characteristics, between 30 and 85%, pseudoplastic, and above 85%, viscoelastic properties as gluten develops. NMR response was directly related to apparent viscosity only within each region, not for all concentrations studied.

In a review of research on soy protein and viscosity, Shen[21] reported that the viscosity of soy isolates could be decreased drastically as total shear was increased. Time was also a factor in measuring viscosities. If the concentrated slurries (12%) were allowed to stand, the viscosities increased slowly and reverted to a much higher value. The slurries were not at equilibrium. Temperature also has a dramatic effect on viscosity. As soy proteins were heated, there was a large decrease in viscosity followed by a large increase as the slurry was cooled. These changes in viscosity can be related to the ability of the soy proteins to form first a progel and then a gel. Gel formation will be discussed in a later section. Protein concentration was shown to cause an exponential increase in viscosity, but different soy isolates exhibited different behavior. For the soy isolates studied, viscosity was not necessarily correlated with solubility.

In a model system, viscosity should be inversely related to solubility. Because many additional factors such as conformation, hydration, exposure of hydrophobic groups, and charge distribution affect viscosity, a direct correlation with solubility may be difficult to obtain. However, within a series of similarly processed isolates, a more direct correlation between solubility and viscosity may be found.[21]

Gelation

The ability of proteins to form gels is important in the structure and texture of many foods. The gelling ability of proteins found in meat, milk, fish, eggs, and soy is necessary for the structure of products such as yogurt,[40] processed meats,[41] omelets[42] and surimi.[43]

Protein gels are defined as the three-dimensional network in which polymer–polymer and polymer–solvent interactions occur in an ordered manner, resulting in the immobilization of large amounts of water by a small proportion of protein.[44] Gelation is related to but differs from denaturation, aggregation and coagulation. Denaturation refers to any process which causes a change in the three-dimensional structure of the native protein without involving the rupture of peptide bonds. Protein-solvent interactions may be involved as well as changes in the physical properties of the protein. Denaturation is often the first step in the gelation process. Aggregation refers to protein–protein interactions which result in the formation of complexes of higher molecular weight. Coagulation is the more random aggregation of already denatured protein molecules in which polymer–polymer interactions are formed in preference to polymer–solvent interactions. Gelation differs from coagulation and aggregation in that gelation involves a well ordered three-dimensional matrix with a balance between repulsive and attractive forces. Aggregation and coagulation are more random complexes.[45]

In general, gelation is thought to proceed by a two-step mechanism. The first step involves the denaturation of the protein. The protein begins to change its native conformation and unfold. The second step, which often proceeds more slowly, involves the reorientation of the denatured proteins into a three-dimensional network.[44,45] The second step is the critical step involved in gel formation and will determine whether a gel, a coagulum, or an aggregation is formed. The rate of this step is also important. If the second step is slow, the protein polymer may form a fine network and the gel may be less opaque and more elastic, and may exhibit less syneresis. If step two is fast, a coarser network will be set up and the gel will be opaque, with more solvent expressed.[45]

Hamm[46] divided gels into two classes, thermo-reversible secondary valence gels and principal valence gels. In thermo-reversible gels the macromolecules are cross-linked by hydrogen bonds. In principal valence gels the macromolecules are bound together by multivalent cations or salt bridges. Syneresis occurs if macromolecules are too close and colloidal solutions occur if intermolecular cohesion becomes too weak.

Beveridge et al.[47] found that egg white and whey protein concentrate gels consisted of a network of spherical particles apparently adhering together. They divided the gelation process of these proteins into three steps following the initial denaturation and unfolding of the native protein. The first step is turbidity development, which occurs during the first 3–10 min of heating. This results in the formation of spherical aggregates and is probably directed by hydrophobic interactions. The second step is sulfhydryl–disulfide interchange and sulfhydryl oxidation. This step results in the stiffening of the preformed aggregates and perhaps enhances interaggregate adherence. The third step is the sudden, large increase in elasticity that occurs upon cooling. This increase in elasticity could be due to the rapid formation of multiple hydrogen bonds causing a marked increase in the rigidity of the aggregates.

The types of forces that hold a gel together will depend on the characteristics of the native protein. These include hydrophobic interaction, hydrogen bonding, electrostatic interaction, and disulfide cross-links or thio-disulfide interchange.[44]

In gels formed from gelatin, Labuza[12] reported that the principal bonding of the protein was due to hydrogen bonds between the $C{=}O$ and NH groups of the peptide linkages, not electrostatic bonds. Chemical modification of polar groups did not greatly influence the mechanical properties of the gels.

Electrostatic bonds may have a greater effect in gel formation as the pH of the protein moves away from the isoelectric point (pI). At a pH lower than the pI, there may be too many positively charged groups and above the pI too many negatively charged groups for a gel to form. An increase in salt content may be needed to form a gel under these conditions. Hegg[48] found that as the pH of conalbumin was raised and lowered, aggregation occurred only when the salt content was increased.

For soybean protein gels, non-covalent crosslinks seem to play an essential role. Babamjimopoulos et al.[49] studied gel forming capabilities of soy isolates, 7S, and 11S components and concluded that the major forces involved were hydrogen bonds and Van der Wals attractions. Hydrophobic and electrostatic interactions were negligible in gel formation. Soy gels dissolved in an 8 mol/l urea solution, indicating that non-covalent interactions were essential in gel formation. Urea solutions destabilize hydrophobic interactions but also affect hydrogen bonding. Utsumi & Kinsella[50] reported that 11S gels were formed by electrostatic and disulfide bonds, 7S gels were formed by hydrogen bonds, and soy isolate gels were formed by hydrogen and hydrophobic bonds. Considering the number of sulfhydryl groups that are present in soy protein it may seem surprising that disulfide bonds do not play a major role in gel formation. However, the sulfhydryl groups are not distributed in an even pattern to allow for a continuous covalent network. Disulfide crosslinks may be formed, leading to small aggregates, and could act as multi-functional crosslinks for other molecules or aggregates.[51] O'Riordan et al.[52] concluded that blood plasma protein gels were formed mainly by hydrophobic interactions. The gel network was controlled by a balance of attractive forces between hydrophobic segments of protein molecules, generated by thermal unfolding, and repulsive forces generated by electrostatic charges.

Many factors influence the type of gel that each protein will form. These factors include concentration of the protein, other protein and non-protein components, pH, ionic environment, reducing agents, and heat treatment conditions.[53] In the gelation of food proteins these factors will interact with each other and affect the characteristics of the gel formed. Most proteins can form either a coagulum or a gel depending upon the conditions present. However, proteins tend to prefer one of the two states due to their inherent characteristics. Shimada & Matsushita[54] determined that for some proteins whether a coagulum or a gel is formed depended upon concentration, while for other proteins gel formation was concentration independent. With large proteins (> 60 Kda) such as hemoglobin, egg albumin and catalase, that have a high molar percentage of hydrophobic amino acids ($> 31 \cdot 5\%$), the pH range for gelation is dependent upon concentration. The pH range for gelation is concentration independent for large protein (gelatin, soy conalbumin and prothrombin) with a low molar percentage of hydrophobic amino acids (22–$31 \cdot 5\%$) and for smaller proteins such as β-lactoglobulin. Hydrophobic and disulfide bonds formed at high-protein concentrations can compensate for the repulsive electrostatic forces associated with pH values well removed from the isoelectric point. Proteins showing concentration-dependent coagulation have a larger number of hydrophobic groups than those without concentration dependence.

Van Kleef[51] studied soybean protein and ovalbumin gels at 10–35%

concentrations and found the characteristics of the gels to be highly dependent on pH. These concentrations were higher than the 8% concentration which is needed for gelation of soy proteins. The gels formed at a higher pH exhibited more protein–water interactions and showed less syneresis. At a higher pH, less intramolecular interactions occurred, due to electrostatic repulsion. This may lead to more flexible protein chains that could extend further. Because the protein molecules interacted less, the overall distribution of the protein chains in the gel was more uniform. At a lower pH, more protein–protein aggregation was possible, which resulted in regions of the gel of higher and lower protein concentration. The uneven distribution of the protein in the gel was manifested by weak points in the gels formed at lower pHs.

Much of the above discussion has focused on the structure of gels as determined by the interaction of the protein chains with each other. The interaction of water with the protein chains is important in determining the type of gel that is formed. Substantial protein–water interaction in the system at the time of heating results in a highly hydrated viscoelastic gel. A low degree of protein–water interaction in the system results in aggregation or precipitation due to exclusion of water from the network.

Water may be held in a gel by hydrogen bonding to hydrophilic groups, dipolar interactions with ionic groups, or structured around hydrophobic groups. Protein–water interactions tend to reduce protein–protein interactions.[44]

Since gels can be formed with a concentration as low as 1%, the distance between macromolecules is very large. A tremendous amount of free space is available for water. The properties of water in gels are almost the same as that of pure bulk water. Water does not leak out of the gels, due to either some unusually weak, long range forces, or due to capillary suction in the pores formed between macromolecules.[12]

Goldsmith & Toledo[55] used pulsed NMR spectrometry as a nondestructive measurement of protein–water interaction in egg albumin gels. Spin-lattice relaxation times for proteins (T_1) should be shorter, the greater the degree of water binding present. As the heating temperature of egg albumin increased from 60 to 90°C, T_1 values of the resulting gels became shorter. The shorter T_1 values would indicate an increased degree of water structuring. The T_1 values were highly correlated with the physical strength of the gels. A higher temperature was needed to form a stronger gel because of the dependence of gel structure on hydrophobic interactions.[54]

Labuza & Busk[56] observed the same decrease in T_1 when studying gels composed of gelatin. At higher concentrations, especially above 30%, the T_1 decreased. Once the major network has formed, increased additions of gelatin only enhance the development of the helix content, without affecting the overall network spaces. Since gelatin is a highly hydrophobic molecule the water contained within the pore spaces would become more highly structured through hydrogen bonding induced by the hydrophobic shell.

The relationship between water binding and texture may not always be correlated. In studies using blood plasma proteins, Hermannsson & Lucisano[57] reported that water binding and texture are not always correlated and should be treated separately. Changes in gel structure may affect texture and water binding quite differently. As gels are heated above the gelation temperature, protein–protein interactions increase. Shrinkage will occur if protein–protein interactions are uniform throughout the gel. If the protein has not been distributed evenly in the gel, the stronger protein–protein interactions will cause a partial disruption of the gel network. The gel will become more aggregated and more water will be lost, since it is easily pressed out through larger capillary structures. The increase in these denser regions will contribute to an increase in force during the compression and penetration of these gels.[58] Likewise, the changes in blood plasma gels as the pH was increased was not reflected in water binding properties. The finer and more continuous the structure, the less the tendency toward phase separation and moisture loss, regardless of the mobility of the polymer chains.

On a practical level, the problem of protein–protein and protein–water interactions in gels can be seen in the study of gels in cooked and frozen egg mixtures. The loss of water from these gels is a major problem in the production of high quality products. Albumin has been identified as the component mainly responsible for the percent expressible moisture (%EM) in precooked, frozen and thawed whole egg mixtures.[59] One way to alleviate this problem is to increase the net negative charge in the protein, thereby decreasing protein–protein interactions and increasing protein–water interactions. Gossett & Baker[59] used two methods to increase the charge on the protein, increasing the pH and succinylation. When the pH of the mixture was raised to 9·5 or greater, the %EM decreased. Succinate anions, by binding with two carboxyl groups caused a net gain in negative charge and decreased %EM.

Busk[60] in a review of polymer–water interactions in gelation, concluded that there was little or no correlation between gel macrostructures and the microstructures formed by polymer–water interaction. Macrostructures were assigned to the physical chemistry of the polymer itself.

Foam Formation

Foams are defined as dispersions of gas bubbles in a continuous liquid or semisolid phase that contains a soluble surfactant.[61] The gas dispersed is usually air or carbon dioxide and the continuous phase is an aqueous solution or suspension containing proteins. In many ways foams or interfacial protein films can be considered as thin gel layers.[62] Additional components which must be considered in foam formation are surface pressure (π) and surface concentration (Γ). Film formation involves two steps. First, the protein molecules slowly penetrate into the surface of the interface and then slowly rearrange at the surface. The first step is diffusion controlled with both π and Γ changing. Two different modes of action are possible in the second step. If both π and Γ are changing, a positive energy barrier to both penetration and molecular rearrangement at the surface exists. If π is changing and Γ remains constant, conformational changes are taking place in the adsorbed layer. The different modes of action depend upon the type of protein involved, with β-casein representing the first type and lysozyme the second type.[63]

Molecular characteristics needed for foaming include solubility, the ability of the protein to unfold at the interface, and the possession of substantial surface hydrophobicity. Denaturation improves foaming by enhancing macromolecular flexibility and surface hydrophobicity.[64] The degree to which the protein is denatured at the interface depends on both the concentration of the protein and the characteristics of the native protein. In dilute films, protein molecules are completely unfolded and spread out at the surface so no protein tertiary or secondary structure remains. In concentrated films, native and unfolded molecules coexist at the surface. The type of film formed also depends upon the conformation of the original protein. β-Casein, which is a flexible disordered molecule, forms either trains of amino acid residues at the interface, or layers and tails of residues protruding into the bulk phase. In contrast, globular proteins such as lysozyme and BSA retain elements of their native structure when adsorbed at the interface.[65]

The two main attributes used to characterize foaming proteins include foam power or capacity and foam stability. For a protein to be considered a good foaming protein it must have a high foam volume and exhibit stability. Foam power or capacity is a measure of the increase in foam volume upon the introduction of a gas into the protein solution. Foam stability is the measure of the rate of liquid leakage from the foam, or the rate of a decrease in foam volume with time. Foam power is partially dependent upon the method used to introduce the gas into the protein.

Foam stability depends on the thickness and strength of the adsorbed film at the air–water interface.

Attempts to relate foam characteristics to molecular properties of proteins have not always been successful. Good foaming proteins usually have a mixture of hydrophilic and hydrophobic amino acids.[66] The measurement of hydrophobicity of proteins is difficult, since many of the hydrophobic amino acids are buried on the inside of the protein and not exposed at the interface until denatured. Two measurements of hydrophobicity include average (or total) and surface (or effective) hydrophobicity. Townsend & Nakai[67] were not able to correlate protein surface hydrophobicity with foaming capacity, but were able to correlate total hydrophobicity with foaming capacity. Surface hydrophobicity does increase upon heat denaturation as Kato et al.[68] observed when heating ovalbumin and lysozyme under conditions that did not coagulate the proteins. Total hydrophobicity may relate to foaming capacity better than surface hydrophobicity because of the extensive uncoiling of the protein molecules at the air–water interface. Townsend & Nakai,[67] using normalized regression coefficients, indicated the degree of contribution to foaming capacity to be viscosity, surface hydrophobicity, and solubility in descending order. They also reported a significant negative relationship between foam stability and net charge density. As the net electrical charge on a protein molecule increased the electrical potential barrier to surface adsorption was intensified.

Water Binding and Meat Proteins

Concern about the water binding ability of meat proteins is evident in all stages of meat processing, from slaughtering to final cooking procedures. Quality of meat is often judged by the amount of water associated with the muscle protein. Cross-striated muscle contains about 75% water, most of which is associated with the myofibrillar proteins. Hamm[69] described three types of water associated with muscle tissue, constitutional, interfacial, and bulk or entrapped water. Constitutional water represents about 0·1% of the water in muscle tissue, interfacial 5–15%, and the remaining water bulk or entrapped. The bulk or entrapped water is of importance in meat processing since the other types of water are tightly bound to the protein and would not be affected by most processing treatments. Whether this water behaves as if it were free water or bulk water in dilute salt solution is a matter of debate.[70] The bulk or entrapped water should not be viewed as homogeneous, but rather as a continuum with some of the water strongly immobilized within the tissue and therefore difficult to remove. The water at the other end of the scale is easily squeezed out by very low pressure. Because this bulk water, or as Hamm[69] prefers to call it immobilized water, is not homogeneous, several definitions are needed to describe it. Hamm[69] has defined the following terms:

(1) *Drip loss:* the exudate from meat or meat systems without the application of external force.
(2) *Thawing loss:* the formation of exudate from meat after freezing and thawing without the application of external force.
(3) *Cooking loss:* the release of fluid after heating of meat either with or without the application of external forces such as centrifugation or pressing.
(4) *Expressible juice:* the release of juice from unheated meat or meat systems during application of external forces such as pressing, centrifugation, or suction.

The immobilized water in meat is found in the spatial arrangement formed by the myofibrillar proteins, with myosin being mainly responsible for the water binding. Any factors which affect the attractions between adjacent filaments will influence the amount of water held by the muscle tissue. Swelling of muscle tissue could occur with a decrease in attractive forces holding the filaments in close contact. As the swelling occurs, an increase in the water holding capacity of meat will result. One way to cause a decrease in the attractive forces between adjacent filaments would be increasing the pH above 5, which is the isoelectric point of myosin. This will result in an increase in the net negative charge of the molecule. Another way to decrease the attractive forces is the addition of NaCl at a pH below 5 which will cause the screening of positive protein charges by the preferential binding of Cl^- ions. Pyrophosphate of ATP will cause the dissociation of linkages between myosin heads and thin filaments. The interaction between hydrophobic protein groups and the hydrophobic part of smaller molecules such as lecithin, will cause a weakening of interaction between hydrophobic groups. Cleavages of linkages between z-lines, which occur during aging also increase water binding.[69]

Conversely, any factors which will increase the attractive forces between adjacent molecules will decrease the space available for water and will result in a decrease in water holding capacity. The tightening of the protein network will result from lowering the pH of the protein to pH 5, the isoelectric point of myosin. The association of actin and myosin during *rigor mortis* results in a tighter protein complex. The heat coagulation of protein increases the interaction between hydrophobic groups.[69] Meat processors use many of these techniques to improve the quality of their products.

Salts and phosphates are commonly used by meat processors to improve the water holding capacity of meats and prevent drip loss.[71] Offer & Trinick[72] found that the myofibrils swelled to more than twice their original volume in 0·8 M NaCl solutions. Pyrophosphate had a synergistic effect on water holding when used with NaCl. With 0·3% pyrophosphate, maximum swelling occurred with a NaCl concentration of 0·4 M. Shults et al.[73] concluded that the elevation of pH by phosphates may be one of the most important ways by which phosphates improve water holding capacity.

Trout & Schmidt[74] studied the effect of ionic strength, pH and pyrophosphate on the cooking yield and tensile strength of beef homogenates. They reported that increasing the ionic strength, pH, and pyrophosphate content increased the temperature at which syneresis occurred. The role of salts in preventing water loss from meat gels may be due to their ability to increase the temperature at which proteins aggregate. Protein aggregation, which results in syneresis, usually occurs at approximately 55°C. In the presence of salts, the temperature at which aggregation occurred increased.

Nutritional Implications

Protein–water interactions have important nutritional implications. Detrimental chemical reactions may take place as the amount of water in protein increases. One of the most important of these reactions is non-enzymatic browning. Non-enzymatic browning, which is a function of the amount of free water available, may result in the loss of available lysine. Labuza & Saltmarch[75] reported that a 0·1 water activity (a_w) increase for food doubles the reaction rate constant for most reactions related to non-enzymatic browning. Maximum browning occurs at water activities of 0·3–0·7.[76] At lower a_w, water is tightly bound to surface polar sites and is generally unavailable for reaction. An a_w of 0·2–0·3, which corresponds to the upper limit of the BET monolayer, is the most stable moisture content for dehydrated foods.[77] The decrease in browning at higher a_w is attributed to the dilution of reacting substances and to the law of mass action.

Non-enzymatic browning or Maillard browning can become a major problem during the drying of foods.[78] If reducing sugars are present and react with amino groups of proteins, especially the free amino group of lysine, visible browning may result. When drying foods, it is important that the food does not remain at intermediate a_w levels any longer than necessary. Similar problems can result if humectants that are also reducing compounds are used to control a_w.[77]

Wang & Damodaran[79] reported the loss of cysteine and cystine residues in soy protein gels. At a concentration of 8% soy protein isolate, and at pH 8 and 100°C, these researchers reported an 18% loss of half-cystine. The rate of the loss was decreased by an increase in viscosity of the protein, a decrease in pH, a decrease in temperature, and a higher salt concentration (0·1 M). All of these factors may help stabilize the conformation of the protein. The control of these processing conditions may be used to help preserve the nutritional quality of soy protein. These researchers did not find that lysinoalanine was formed under conditions necessary for gelation. A destruction of 52% of half-cystine was reported by Friedman et al.,[80] when 1% of soy protein solutions were heated at 75°C for 3 h. Lysinoalanine also was formed in these dilute protein solutions. As the moisture content is reduced, the distance between protein chains is diminished, allowing for cross-linkages between adjacent chains to form. These cross-linkages may lead to a tighter network of protein and the water holding capacity and solubility of the protein will be reduced. Thermal stability may be increased at low moisture contents since the bonds formed are tighter and less free to rotate with the addition of energy.[81]

CONCLUSIONS

Protein–water interactions are important for understanding both the structure of the protein molecule and the functional properties of the protein molecule in food systems. Water molecules play a central role in determining the conformation of the protein which in turn determines many of the physical properties of the protein. A number of problems are encountered in the study of these interactions. One is application of findings using model proteins to proteins found in foods. Proteins and water often interact differently when studied in dilute model systems as opposed to more concentrated and complex food systems. Another problem encountered is the correlation of the structure of proteins to functional properties. If the structure of proteins could be correlated with specific functional properties, it would be possible to manipulate protein structure to obtain desirable properties. Because of the many environmental conditions affecting functional properties, more research is needed to correlate protein structure with functional properties.

REFERENCES

1. Fennema, O., Water and Ice. In *Food Chemistry*, ed. O. Fennema. Marcel Dekker, Inc. NY, 1985, pp. 23–67.
2. Fennema, O., Water and protein hydration. In *Food Proteins*, ed. J. Whitaker & S. Tannenbaum. AVI, Westport CT, 1977, pp. 50–90.
3. Chou, D. & Morr, C., Protein-water interactions and functional properties. *J. Am. Oil Chem. Soc.*, **56** (1979) 53A–62A.
4. Bull, H. & Breese K., Protein hydration. II. Specific heat of egg albumin. *Arch. Biochem. Biophys.*, **128** (1968) 497–502.
5. Kuntz, I., The physical properties of water associated with biomolecules. In *Water Relations of Foods*, ed. R Duckworth, Academic Press, NY, (1975), pp. 93–109.
6. Franks, F., Water in aqueous solutions: recent advances. In *Properties of Water in Foods*, ed. D. Simatos & J. Multon. Martinus Nijhoff Publishers, Dordrecht, The Netherlands, 1985, pp. 1–23.
7. Kauzmann, W., Some factors in the interpretation of protein denaturation. *Adv. Protein Chem.*, **14** (1959) 1–63.
8. Pomeranz, Y., *Functional Properties of Food Components*. Academic Press, Orlando, FL, 1985, p. 153.
9. Finney, J., Organization and function of water in protein crystals. In *Water, A Comprehensive Treatise*, ed. F. Franks. Plenum Press, NY, 1979, pp. 47–122.
10. Warner, D., Theoretical studies of water in carbohydrates and proteins. In *Water Activity: Influences on Food Quality*, ed. L. Rockland & G. Steward. Academic Press, NY, 1981, pp. 435–465.
11. Bull, H. & Breese, K., Protein hydration. I. Bind sites. *Arch. Biochem. Biophys.*, **128** (1968) 488–496.
12. Labuza, T., The properties of water in relationship to water binding in foods, a review. *J. Food Proc. Preser.*, **1**(2) (1977) 167–190.
13. Franks, F., Water, ice and solutions of simple molecules in water relations of foods. In *Water Relations in Food*, ed. R. Duckworth. Academic Press, NY, 1975, pp. 3–22.
14. Kinsella, J., Milk proteins: physicochemical and functional properties. *CRC Crit. Rev. Food Sci. Nutr.*, **21**(3) (1984) 197, 262.
15. Chen, J., Piva, M. & Labuza, T., Evaluation of water binding capacity (WBC) of food fiber sources. *J. Food Sci.*, **49** (1984) 59–63.
16. Leeder, J. & Watt, I., The stoichiometry of water sorption by proteins. *J. Colloid Interface Sci.*, **48** (1974) 339–344.
17. Anderson, C. & Witter, L., Water binding capacity of 22 L-amino acids from water activity of 0.33 to 0.95. *J. Food Sci.*, **47** (1982) 1952–1954.
18. Kuntz, I., Hydration of macromolecules. III. Hydration of polypeptides. *J. Am. Chem. Soc.*, **93** (1971) 514–516.

19. Kuntz, I. & Brassfield, T., Hydration of macromolecules. II. Effects of urea on protein hydration. *Arch. Biochem. Biophys.*, **142** (1971) 660–664.
20. Tombs, M., Phase separation in protein-water systems and the formation of structure. In *Properties of Water in Foods*, ed. D. Simatos & J. Multon. Martinus Nijhoff Publishers, Dordrecht, The Netherlands, 1985, pp. 25–36.
21. Shen, J., Solubility and Viscosity. In *Protein functionality in foods*, ed. J. Cherry. Am. Chem. Soc. Washington, D.C., 1981, pp. 89–109.
22. Bingham, E., Influence of temperature and pH on the solubility of αs_1,-β and κ-Casein. *J. Dairy Sci.*, **54** (1971) 1077–1080.
23. Shen, J., Soy protein solubility: The effect of experimental conditions on the solubility of soy protein isolates. *Cereal Chem.* **53**(6) (1976) 902–909.
24. German, B., Damodaran, S. & Kinsella, J., Thermal dissociation and association behaviour of soy proteins. *J. Agric. Food Chem.*, **32**, 807–811.
25. Bigelow, C., On the average hydrophobicity of proteins and the relation between it and protein structure. *J. Theoret. Bio.*, **16** (1967) 187–211.
26. Von Hippel, P. & Wong, K., The effect of ions on the kinetics of formation and stability of the collagen fold, **1** (1962) 664–674.
27. Bull, H. & Breese, K., Water and solute binding by proteins. II. Denaturants. *Arch. Biochem. Biophys.*, **139** (1970) 93–96.
28. Damodaran, S. & Kinsella, J., Effects of ions on protein conformation and functionality. In *Protein Structure Deterioration*, ed. J. Cherry. ACS Publication, Am. Chem. Soc., Washington, DC, 1982, pp. 327–356.
29. Eagland, D., Protein hydration – its role in stabilizing the helix conformation of protein. In *Water Relations of Foods*, ed. R. Duckworth. Academic Press, NY, 1975, pp. 73–92.
30. Eagland, D., Nucleic acids, peptides and proteins. In *Water, A comprehensive Treatise*, ed. F. Franks. Plenum Press, NY, 1975, pp. 305–518.
31. Melander, W. & Horvath, C., Effect of neutral salts on the formation and dissociation of protein aggregates. *J. Solid-Phase Biochem.*, **2**(2) (1977) 141–161.
32. Matsudomi, N., Mori, H., Kato, A. & Kobayashi, K., Emulsifying and foaming properties of heat-denatured soybean 115 globulins in relation to their surface hydrophobicity. *Agric. Biol. Chem.*, **49**(4) (1985) 915–919.
33. Schnepf, M. & Satterlee, L., Partial characterization of an iron soy protein complex. *Nutr. Repts. Int.*, **31** (1985) 371–380.
34. Hardy, J. & Steinberg, M., Interaction between sodium chloride and paracasein as determined by water sorption. *J. Food Sci.*, **49** (1984) 127–131.
35. Konstance, R. & Strange, E., Solubility and viscous properties of casein and caseinates. *J. Food Sci.*, **56** (1991) 556–559.
36. Hamm, R., Water-holding capacity of meat. In *Meat*, ed. D. Cole & R. Lawrie. AVI Publishing Co., Inc. Westport, CT, 1975.
37. Berlin, E., Hydration of milk proteins. In *Water Activity: Influences on Food Quality*, ed. L. Rockland & G. Steward. Academic Press, NY, 1981, pp. 467–488.
38. Geankoplis, C., *Transport Processes and Unit Operations*. Allyn and Bacon, Inc., Boston, 1978, p. 46.
39. Richardson, S., Baiann, I. & Steinberg, M., Relation between oxygen-17 NMR and rheological characteristics of wheat flour suspension. *J. Food Sci.*, **59** (1985) 1148–1151.

40. Schmidt, R. & Morris, H., Gelation properties of milk proteins, soy proteins, and blended protein systems. *Food Technol.*, **38**(5) (1984) 85–96.
41. Ziegler, G. & Acton, J., Mechanisms of gel formation by proteins of muscle tissue. *Food Technol.*, **38**(5) (1984) 77–82.
42. O'Brien, S., Baker, R., Hood, L. & Liboff, M., Water-holding capacity and textural acceptability of precooked frozen white egg omelets. *J. Food Sci.*, **47** (1982) 412–417.
43. Burgarella, J., Lanier, T., Hamann, D. & Wu, M., Gel strength development during heating of surimi in combination with egg white or whey protein concentrate. *J. Food Sci.*, **50** (1985) 1595–1597.
44. Mulvihill, D. & Kinsella, J., Gelation characteristics of whey proteins and β-lactoglobulin. *Food Technol.*, **41**(9) (1987) 102–111.
45. Gossett, P., Rizvi, S. & Baker, R., Quantitative analysis of gelation in egg protein systems. *Food Technol.*, **38**(5) (1984) 67–74, 96.
46. Hamm, R., The water imbibing power of foods. *Rec. Adv. Food Sci.*, **31** (1963) 218.
47. Beveridge, T., Jones, L. & Tung, M., Progel and gel formation and reversibility of gelation of whey, soybean and albumen protein gels. *J. Agric. Food Chem.*, **32** (1984) 307–313.
48. Hegg, P., Conditions for the formation of heat-induced gels of some globular food proteins. *J. Food Sci.*, **42** (1982) 1241–1244.
49. Babajimopoulos, M., Damodaran, S., Rizvi, S. & Kinsella, J., Effects of various anions on the rheological and gelling behaviour of soy proteins: thermodynamic observations. *J. Agric. Food Chem.*, **31** (1983) 1270.
50. Utsumi, S. & Kinsella, J., Forces involved in soy protein gelation: Effects of various reagents on the fermenting hardness and solubility of heat-induced gels made from 7S, 11S, and soy isolate. *J. Food Sci.*, **50** (1985) 1278–1282.
51. Van Kleef, F., Thermally induced protein gelation: gelation and rheological characterization of highly concentrated ovalbumin and soybean protein gel. *Biopolymers*, **25** (1986) 31–59.
52. O'Riordan, D., Morrissey, P., Kinsella, J. & Mulvihill, D., The effects of salts on the rheological properties of plasma protein gels. *Food Chem.*, **34** (1989) 249–259.
53. Schmidt, R., Gelation and coagulation. In *Protein Functionality in Foods*, ed. J. Cherry. Am. Chem. Soc. Washington, D.C., 1981, pp. 132–147.
54. Shimada, K. & Matsushita, S., Relationship between thermocoagulation of proteins and amino acid compositions. *J. Agri. Food Chem.*, **28** (1980) 413–417.
55. Goldsmith, S. & Toledo, R., Studies on egg albumin gelation using nuclear magnetic resonance. *J. Food Sci.*, **50** (1985) 59–62.
56. Labuza, T. & Busk, G., An analysis of the water binding in gels. *J. Food Sci.*, **44** (1979) 1379–1394.
57. Hermannsson, A. & Lucisano, M., Gel characteristics—waterbinding properties of blood plasma gels and methodological aspects on the water-binding of gel systems. *J. Food Sci.*, **47** (1982) 1955–1959.
58. Hermannsson, A. M., Gel characteristics—structure as related to texture and waterbinding of blood plasma gels. *J. Food Sci.*, **47** (1982) 1965–1972.

59. Gossett, P. & Baker, R., Effect of pH and of succinylation on the water retention properties of coagulated frozen and thawed egg albumin. *J. Food Sci.*, **48** (1983) 1391–1394.
60. Busk, G., Polymer-water interactions in gelation. *Food Technol.*, **38**(5) (1984) 59–64.
61. Cheftel, J., Cug, J. & Lorient, D., Amino acids, peptides and proteins. In *Food Chemistry*, ed. O. Fennema. Marcel Dekker, Inc., NY, 1985, pp. 245–369, 303.
62. Graham, D. & Phillips, M., Proteins at liquid interfaces. V. Shear properties. *J. Colloid Interface Sci.*, **76** (1980) 240–250.
63. Graham, D. & Phillips, M., Proteins at liquid interfaces. I. Kinetics of adsorption and surface denaturation. *J. Colloid Interface Sci.*, **70** (1979) 403–414.
64. Halling, P., Protein stabilized foams and emulsions. *CRC Crit. Rev. Food Sci. Nutr.*, **15** (1981) 155–203.
65. Graham, D. & Phillips, M., Proteins at liquid interfaces. III. Molecular structure of adsorbed films. *J. Colloid Interface Sci.*, **70** (1979) 427–439.
66. Nakai, S., Structure-function relationships of food proteins with an emphasis on the importance of protein hydrophobicity. *J. Agri. Food Chem.*, **31** (1983) 676–683.
67. Townsend, A. & Nakai, S., Relationship between hydrophobicity and foaming characteristics of food proteins. *J. Food Sci.*, **48** (1983) 588–594.
68. Kato, A., Takahashi, A., Matsudomi, N. & Kobayashi, K., Determination of foaming properties of proteins by conductivity measurements. *J. Food Sci.*, **48** (1983) 62–65.
69. Hamm, R., The effect of the quality of meat and meat products: problems and research needs. In *Properties of Water in Foods*, ed. D. Simatos & J. Multon. Martinus Nijhoff Publishers, Dordrecht, The Netherlands, 1985, pp. 591–602.
70. Blanchard, J. & Derbyshire, W., Physico-chemical studies of water in meat. In *Water Relations of Foods*, ed. R. Duckworth. Academic Press, NY, 1975, pp. 559–571.
71. Shults, G. & Wierbicki, E., Effects of sodiuim chloride and condensed phosphates on the water-holding capacity, pH and swelling of chicken muscle. *J. Food Sci.*, **38** (1973) 991–994.
72. Offer, G. & Trinick, J., On the mechanism of water holding in meat: The swelling and shrinking of myofibrils. *Meat Sci.*, **8** (1983) 245–281.
73. Shults, G., Russell, D. & Wierbicki, E., Effect of condensed phosphates on pH, swelling and water-holding capacity of beef. *J. Food Sci.*, **32** (1972) 860–864.
74. Trout, G. & Schmidt, G. The effect on cooking temperature on the functional properties of beef proteins: the role of ionic strength, pH, and pyrophosphate. *Meat Sci.*, **20** (1987) 129–147.
75. Labuza, T. & Saltmarch, M., The nonenzymatic browning reactions as affected by water in foods. In *Water Activity: Influences on Food Quality*, ed. L. Rockland & G. Steward. Academic Press, NY, 1981, pp. 605–650.
76. Eichner, K. & Ciner-Doruk, M., Formation and decomposition of browning intermediates and visible sugar-amine browning reactions. In *Water Activity:*

Influences of Food, ed. L. Rockland & G. Steward. Academic Press, NY, 1975, 567–603.

77. Labuza, T. Water binding of humectants. In *Properties of Water in Foods,* ed. D. Simatos & J. Multon. Martinus Nijhoff Publishers, Dordrecht, The Netherlands, 1985, pp. 421–445.

78. Eichner, K., Laible, R. & Wolf, W., The influence of water content and temperature on the formation of Maillard reaction intermediates during drying of plant products. In *Properties of Water in Foods,* ed. D. Simatos & J. Multon. Martinus Nijhoff Publishers, Dordrecht, The Netherlands, 1985, pp. 191–210.

79. Wang, C. & Damodaran, S., Thermal destruction of cysteine and cystine residues of soy protein under conditions of gelation. *J. Food Sci.,* **55** (1990) 1077–1080.

80. Friedman, M., Levin, C. & Noma, A., Factors governing lysinoalanine formation in soy proteins. *J. Food Sci.,* **49** (1984) 1282–1288.

81. Rustad, T. & Nesse, N., Heat treatment and drying of capelin mince. Effect of water binding and soluble protein. *J. Food Sci.,* **48** (1983) 1320–1322, 1347.

82. Kinsella, J., Functional properties of soy proteins. *J. Am. Oil Chem. Soc.,* **56** (1979) 242–258.

Chapter 2

PROTEIN–PROTEIN INTERACTIONS

NAZLIN K. HOWELL

School of Biological Sciences, University of Surrey, Guildford GU2 5XH, UK

INTRODUCTION

Protein–protein interactions occur in many chemical and physical processes which are involved in the complex molecular organisation of plant and animal cells. Most of these interactions affect the processing, storage and behaviour of proteins in food. There are various reasons for studying protein interactions, including acquiring knowledge of structure–function relationships of proteins; optimising the use of product constituents; improving quality, cost reduction and new protein applications.

Studies of protein–protein interactions comprise investigations which have stemmed from observations made during processing; a classic example being the interaction between β-lactoglobulin and κ-casein in milk. Alternatively, protein–protein interactions are studied by deliberately mixing proteins in order to create new products. It was apparent at the outset that studies reported vary considerably in the types of proteins used, and in experimental methods and conditions employed so that these studies are not usually directly comparable. However, most of these investigations fall into two main categories related to functional properties. One is the effect of heating, incorporating viscosity and gelation studies, on protein–protein interactions. The second area is foaming. Within these sections, the main studies reported for selected proteins have been discussed.

The selected proteins are globular proteins and include milk whey, egg albumen, blood plasma and soya globulins. Interactions of these proteins either with each other or with non-globular proteins such as casein or myosin are discussed. However, complexes of proteins with enzymes and antigen–antibody complexes are highly specific interactions and cannot be covered within the limitations of this discussion.

General aspects of intermolecular forces and conformation governing protein interactions are briefly discussed in this chapter. The significance of intermolecular interactions in foods is illustrated by relating the structure of the proteins to their functional behaviour both in isolation and in combination with other proteins. Functional properties affecting the behaviour of proteins, and therefore the texture, include gelation, foaming, emulsification, viscosity and water-binding.[1,2] Where possible, elucidation of the interaction mechanism is highlighted. However, only a modest amount of information exists on the mechanisms of interaction phenomena which are observed in the functional testing of proteins. One of the main difficulties experienced in the study of interactions is the lack of suitable techniques which can provide unequivocal evidence of interactions. The methods used by various workers are therefore critically discussed.

MOLECULAR FORCES GOVERNING PROTEIN INTERACTIONS

It is appropriate to begin with the molecular forces governing protein interactions as a knowledge of these forces enables an understanding of the relationship of the structure of individual proteins to their functional properties, as well as the association of a protein with other protein structures.

The linear association of amino acids which forms the primary structure, subsequent folding and bonding into α-helix and β-sheet secondary structures, and further organisation into tertiary and quaternary structures have been extensively reviewed.[3,4] These conformations are influenced by intramolecular forces, particularly the forces which determine the strength and length of the chemical bond between two atoms, e.g. C—C, the forces which determine angles between adjacent chemical bonds and the torsional forces which control the rotations around chemical bonds. The geometric structure of molecules can be obtained by X-ray diffraction, electron diffraction and microwave spectroscopy.

Molecular forces involved within a protein, and interactions with other proteins, include covalent disulphide linkages and the non-covalent electrostatic van der Waals, hydrogen and hydrophobic forces (Fig. 1).[3,5] Steric and hydration repulsive forces are also reported.[6]

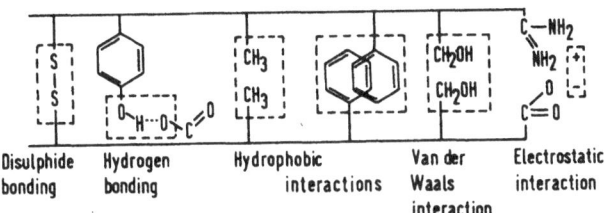

Disulphide   Hydrogen   Hydrophobic   Van der   Electrostatic
bonding   bonding   interactions   Waals   interaction
                                           interaction

FIG. 1. Molecular forces involved in protein interactions.

Covalent Bonds

Covalent bonds include the peptide bond of the primary structure which has a heat of formation value of about 100 kcal/mol.[4] The peptide bond is not broken during protein denaturation, which relates to changes in the secondary and tertiary structures only. However, hydrolysis with strong acid or alkali, or proteolytic enzymes, may be used to break peptide bonds.

More importantly, covalent disulphide bonds are involved in the tertiary structure and are therefore affected by protein denaturation during heat gelation or foaming. Disulphide bonds are formed between two Cys residues; the sites depending on the conformation of the polypeptide chain:[5]

$$
\begin{array}{lcl}
\mid & & \mid \\
NH & & C{=}O \\
\mid & & \mid \\
CH{-}CH_2{-}S{-}S{-}CH_2{-}CH \\
\mid & & \mid \\
C{=}O & & NH \\
\mid & & \mid
\end{array}
$$

The bonds are strong, with a heat of formation of about 50 kcal/mol.[4] They are stable at most pH values but may undergo disulphide–sulphydryl interchange in the presence of oxidising and reducing agents.

Other types of covalent cross-links not commonly found include those of the connective tissue proteins which are formed between lysine residues after oxidative deamination of an amino group, ester linkages and amide linkages in fibrin during blood clotting.[5]

Electrostatic Interactions

Electrostatic interactions occur between two charged particles.[4,5] Protein molecules are zwitterions which carry a net positive or a net negative

charge depending on their pI and the pH of the environment. Electrostatic
interactions are repulsive for like charges and attractive for opposite signs
of the charges. They occur over large distances and vary inversely with
distance.

The change in energy of two point charges, as they approach as a function of their distance of separation, is given by Coulomb's Law modified
to take into account solvent molecules.[5]

$$\Delta E = \frac{Z_A Z_B \varepsilon^2}{D r_{AB}}$$

where r_{AB} is the distance between the two charges A and B,
Z_A and Z_B are their respective unit charges,
ε is one unit of electronic charge, and
D is the dielectric constant of the solvent.

The equation is valid when the average number of solvent molecules
between two interacting charges is large enough so that the properties of
these solvent molecules are similar to those of the bulk solvent. Non-
polar liquids have low dielectric constants (2) whereas polar liquids, e.g.
water have higher values (80). Thus electrostatic interactions are reduced
in polar liquids. In food systems, the charged surfaces of particles are surrounded by diffuse layers of ions to form the electrical double layer. Since
these are of similar charge, repulsion of particles results as described by
the DVLO (Derjaguin, Landau, Verwey & Overbeek) theory.[7]

In addition, electrostatic forces also occur with uncharged but polar
molecules. In these molecules electrons are distributed in such a way that
there is an excess of negative charge in one area and an excess of positive
charge in another.[5] This is known as the dipole moment (μ) which may be
defined as:

$$\mu = zd$$

where z is the amount of excess charge separated in the molecule and d is
the distance between the centres of the excess charge.

Compared with two simple ions the energy of interactions involving
dipoles is complex, as the dipolar nature is influenced by the molecules
with which the protein interacts. Electrostatic bonds are quite strong,
with a heat formation of 10–20 kcal/mol.[4] The surface charge of the protein can be obtained by measuring the zeta-potential, which is related to
the rate at which the charged particles move in an electric field.[6]

Van der Waals or London Dispersion Forces

Van der Waals forces are weak attractive interactive forces between polar or non-polar molecules. They occur over short distances when the positively-charged nuclei and negatively-charged electrons of two approaching molecules affect the electron cloud distribution of each molecule. Dipoles are induced and reversed and the resulting oscillating dipoles reduce the total energy of the interacting molecules, thereby creating an attractive force.[5,8] However, if the molecules come too close, so that their electron clouds .overlap, then a strong repulsive force between the molecules overrides the attractive force, ensuring that the two different molecules cannot occupy the same space.[5]

Hydrogen Bonds

A particular type of electrostatic interaction force between polar molecules is known as the hydrogen bond. The H atom is shared between an acid (proton donor) group, e.g. —OH or —NH, and a base (proton

$$\overset{\displaystyle O}{\underset{\displaystyle \|}{}}$$

acceptor), e.g. —C—. Although is it not clear exactly how the bond is formed it may be considered as an intermediate stage in the transfer of a proton from an acid AH to a base B.[5,8,9]

$$A\text{—}H + B \leftrightarrow {}^{\delta^-}A \cdots {}^{\delta^+}H \text{ - - - - } B \leftrightarrow A^- \cdots H\text{—}B^+$$

A partial bond is formed between the proton and the base. The extent of proton transfer depends on the strengths of the acid and base. Hydrogen bonds with linear geometry are stronger than bent hydrogen bonds. To form H bonds between proteins, hydrogen bonds between protein and water are substituted for bonds between protein atoms. The energy of hydrogen bonds is reported to be 1–6 kcal/mol.[4,5]

Hydrophobic Interactions

In a polar aqueous environment the absence of hydrogen bonding between water and a non-polar solute is referred to as the hydrophobic interaction.[5] Electrostatic, hydrogen and van der Waals forces between protein molecules are not favourable energetically in aqueous solution because the tendency is for the polypeptide to interact with the surrounding water. With a non-polar molecule, only weak van der Waals interactions with water exist. In order to compensate, water molecules interact strongly with each other. This in turn results in strong attractive forces between the non-polar groups causing them to aggregate in water.

In this reaction the Gibbs free energy, ΔG, is negative whereas the entropy, ΔS, and enthalpy, ΔH, are positive, resulting in greater interactions with increasing temperature, except at very high temperatures when ΔS and ΔH become negative and hydrophobic interactions become weaker.[5]

Despite a poor understanding of hydrophobic interactions, hydrophobicity can be measured in a number of ways. Average hydrophobicity can be obtained theoretically by adding the hydrophobicity for individual non-polar amino acids and dividing by the total number of amino acid residues in the proteins.[10] Tanford[11] measured hydrophobicity by calculating the free energy change when 1 mol of amino acid was transferred from an aqueous solution to an alcoholic solution. Side chain hydrophobicity was then determined by subtracting the transfer free energy attributed to glycine.

More recently Nakai and co-workers have correlated surface hydrophobicity, due to the non-polar amino acids on the protein surface to emulsification.[12,13] In addition, exposed hydrophobicity of proteins denatured prior to measurement is reported to correlate well with foaming[14] and gelation.[15] In these studies hydrophobicity was measured using a fluorescent probe, cis-parinaric acid (CPA), according to the method of Sklar et al.[16] cis-Parinaric acid incorporates a linear hydrocarbon chain and is thought to bind aliphatic groups. An alternative fluorescent probe, 1-Anilinonaphthalene-8-sulphonate (ANS), has an aromatic ring structure and is thought to bind to aromatic groups. Kato & Nakai[17] have presented views to suggest that CPA is preferable to ANS. The fluorescent probe method is rapid and simple to perform, offering a distinct advantage over hydrophobic chromatographic methods which have also been used to measure surface hydrophobicity.[18]

Hydration Repulsion Forces

Hydration repulsion forces have been proposed by Israelachvili & Pashley[19] to be of equal importance to electrostatic and hydrophobic interactions. Association of water molecules with protein polar groups causes repulsion between the protein groups when they are separated by a few water molecules. These repulsive forces are due to work required to remove water surrounding the protein polar groups as they approach each other. The repulsion is overcome if the charges on the interacting molecules have an energetically better fit with each other than with the surrounding water molecules.[20]

Steric Repulsion Forces

In addition to the electrostatic and hydration repulsion forces, there is a third type of repulsion force, namely steric repulsion. Steric repulsions occur when proteins are within atomic distances of each other.[21,22] Two factors contribute to steric repulsion. They are an osmotic effect, produced by the high concentration of protein where chains overlap, and a volume restriction effect, due to the loss of possible conformation in the restricted space between two surfaces.[23]

PROTEIN–PROTEIN INTERACTIONS IN HEATED DISPERSIONS AND GELS

Viscous Dispersions

Proteins are complex food ingredients by virtue of both their structure and their presence in large concentrations together with other food ingredients. The behaviour of proteins during heating is therefore governed by aggregation and intermolecular interaction.

Rha & Pradipasena[24] have proposed that since molecular properties have little influence under these conditions, rheology can offer measurements in fundamental units to describe protein behaviour.

Viscosity is affected by the geometry, i.e. size and shape of proteins, which varies with conformational change.[25] In addition, a measure of the viscosity or resistance to flow can be used to monitor protein–protein interactions during heating before a gel is formed. This method has been used by the author[26] to examine the interaction of blood plasma and egg albumen proteins.

Blood Plasma–Egg Albumen Protein Interactions

Plasma proteins. Plasma contains over 100 proteins;[27,28] fibrinogen comprising about 5% of the plasma proteins. The removal of fibrinogen from plasma following blood coagulation results in serum. Serum consists of the main protein albumin as well as α-, β- and γ-globulins, classified according to their electrophoretic mobility. The α- and β-globulins are a complex mixture comprising glyco- and lipo-proteins (Table 1).

Egg albumen proteins. Egg albumen contains some 40 different proteins.[29] The main proteins and their properties are listed in Table 2.

TABLE 1

PHYSICO-CHEMICAL PROPERTIES OF SERUM PROTEINS

Protein	% Serum proteins	pI	Molecular weight
Albumin	59	4·9	69 000
α_1-globulins	5	2·7–4·4	44 000–435 000
α_2-globulins	8	3·6–5·6	41 000–20×10⁶
β-globulins	12	3·0–5·9	80 000–3·2×10⁶
γ-globulins	16	5·8–7·3	100 000–160 000

TABLE 2

PHYSICO-CHEMICAL PROPERTIES OF EGG ALBUMEN PROTEINS

Protein	Amount in egg white (%)	pI	Molecular weight
Ovalbumin	54	4·5	45 000
Ovotransferrin (conalbumin)	12	6·05	76 600
Ovomucoid	11	4·1	28 000
Ovoinhibitor	1·5	5·1	49 000
Ovomucin	3·5	4·5	110 000
Lysozyme	3·4	10·7	14 300
Ovoglycoprotein	1·0	3·9	24 400

Adapted from Osuga & Feeney.[30]

Interactions. The potential for using blood plasma proteins in meat and bakery products is high, for both economic and functional reasons.[31] Howell & Lawrie[26] determined the viscosity of egg albumen and blood plasma proteins using a high ratio cake-type model system (6% (w/w) protein and 45% (w/w) sucrose in distilled water at pH 8·0). The interaction of egg albumen with bovine and porcine blood plasma proteins was examined by heating the solutions from 20 to 79°C for 30 min, cooling to 20°C and measuring the viscosity on a Ferranti-Shirley Cone and Plate Viscometer. Viscosity values for mixtures of porcine and bovine plasma and egg albumen proteins were lower than the additive values calculated for each component in the mixture, indicating negative interaction (Table 3). However positive synergistic interaction occurred at high temperatures of heating (73–79°C) for porcine plasma Fraction I and Fraction III

TABLE 3

VISCOSITY OF PLASMA PROTEINS: EFFECT OF TEMPERATURE ON THE APPARENT VISCOSITY (CENTIPOISES (cps±SD) AT 20ºC, SHEAR RATE 625/s, SHEARING TIME 5 min) OF PORCINE PLASMA AND EGG ALBUMEN PROTEINS IN 45% (w/w) SUCROSE SOLUTION[26]

Protein	Temperature						
	20ºC	40ºC	60ºC	70ºC	73ºC	76ºC	79ºC
Protein (w/w)							
6% plasma cps	70·9±4·5[a]	65·8±5·8	70·9±1·4	81·2±1·4	97·7±1·4	159·4±13·0	443·9±17·1
4% plasma+2%							
egg albumen cps	49·4±2·9	51·4±1·0	55·5±2·9	67·9±5·8	73·0±1·4	121·3±2·9	318·8±14·5
Additive value[b]	58·6	44·6	58·3	80·8	112·0	165·3	364·3
3% plasma+3%							
egg albumen cps	44·2±1·4	45·2±2·9	53·5±2·9	56·6±1·4	67·9±2·9	113·1±2·9	273·5±2·9
Additive value	53·5	49·2	53·0	80·8	120·8	171·0	330
2% plasma+4%							
egg albumen cps	40·1±1·45	42·1±1·45	42·1±1·45	51·4±2·92	65·8±1·0	102·8±1·0	218·0±11·6
Additive value	47·1	43·1	46·5	81·1	127·2	173·3	288·4
6% egg albumen cps	36·0±1·45	32·5±0·57	35·0±2·91	82·2±1·51	144·0±2·0	183·0±20·36	216·0±2·92

[a]Mean values based on three determinations.
[b]Additive value—the value derived from summing the contributions of the component proteins, as measured in isolation, in proportion to the concentration of each in the mixture.

(albumin) and porcine serum with egg albumen proteins. It is quite likely that the interaction at high temperatures is due to forces similar to those experienced in gelation. However as the gel-like dispersions were reversible at temperatures under 73°C, covalent bonds such as disulphide bonds are precluded.

Egg Albumen–Whey Protein Interactions

According to Sato *et al.*[32] heating of the main egg albumin protein, ovalbumin, with κ-casein at 70–80°C for 5 min followed by cooling to 28°C resulted in an increase in viscosity accompanied by turbidity, in contrast to the individual proteins. The viscosity also increased when ovalbumin was heated prior to mixing with κ-casein and holding at 28°C. On the other hand, prior heating of κ-casein before mixing with unheated ovalbumin at 28°C did not result in increased viscosity. Conformational changes in ovalbumin therefore seem to be necessary for interaction. Electrostatic interactions are proposed by Sato *et al.*[32] to contribute to the ovalbumin–κ-casein interaction, as acetylation of the proteins influenced the viscosity. Since blocking agents *N*-ethylmaleimide and *p*-chloromercuribenzoic acid did not affect the viscosity of the mixture disulphide-sulphydryl groups were considered not to be involved in viscous interactions.

It is interesting to note that although interaction, as judged by viscosity measurements, may not be detected on mixing proteins at lower temperatures of, e.g. 25°C, changes do appear to occur when sound velocity measurements are used. Sound velocity is affected by changes in water structure around the protein in solution. The sound velocity was reduced for a mixture of ovalbumin and β-casein compared with values calculated from those of individual proteins. This may be due to conformational changes imparted by β-casein to ovalbumin by reducing the number of SH groups.[32]

Heat Set Gels

For most globular proteins high temperatures of heating lead initially to conformational changes indicated by changes in viscosity, hydrophobicity, circular dichroism and differential scanning calorimetry measurements, and eventually to the formation of gels. Protein gels are composed of three-dimensional networks in which water is either tightly bound to the polar groups of the polypeptide chains or more loosely associated within the network interstitial spaces.

The commonly held view of protein gelation is that described by Ferry.[33] He proposed the two-stage gelation process which consists of an initial unfolding of native protein, thereby exposing reactive groups followed by an association of the long unfolded polypeptide chains by covalent and non-covalent forces. For stable gel formation a correct balance of attractive and repulsive forces is essential.

Later Barbu & Joly[34] and Kratchovil[35,36] proposed that linear aggregation of partially-unfolded globular proteins took place when repulsion was large, whereas random aggregates or clumping occurred when repulsion was small, e.g. at the isoelectric point. Tombs[37] also distinguished between ordered and random aggregation and suggested that protein gelation could be considered in terms of statistical theories developed by Flory[38] for synthetic polymers. However, Bezrukov[39] reported that statistical aggregation is unlikely to occur because protein globules have a surface mosaic structure with sites differing in charge, density and sign, degree of hydrophobicity and the presence of disulphide groups. Howell & Lawrie[40] pointed out that the reactions in protein gelation, those of plasma proteins and egg albumen proteins at least, occur in an orderly sequence resulting in reproducible gels provided that the protein concentration and time and temperature of heating are carefully controlled. The present view[39–41] supports the theory that partially-unfolded globular proteins associate linearly at low temperatures and randomly to form

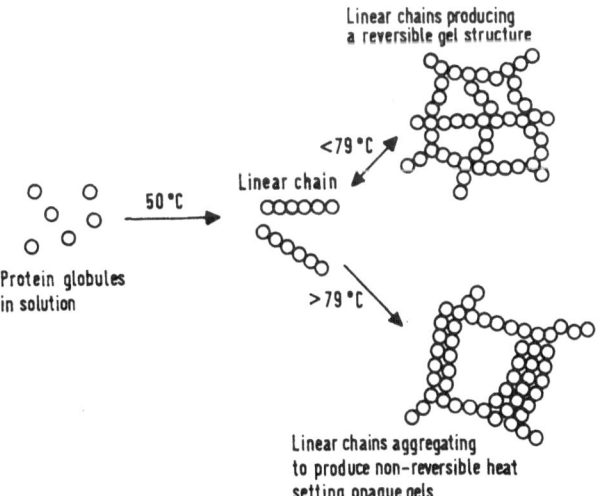

FIG. 2. Proposed mechanism of the gelation of blood plasma proteins.

aggregates at high temperatures or at the pI. A proposed mechanism of the gelation of plasma proteins is suggested in Fig. 2. The presence of non-covalent electrostatic, hydrophobic and hydrogen bonds as well as covalent disulphide bonds in globular protein gels is widely reported.[37,40,42]

Plasma–Egg Albumen Protein Interaction
Major studies on the interaction of plasma proteins with egg albumen proteins include those by Howell & Lawrie[31,40] and Clark & Lee-Tuffnell.[41]

TABLE 4

EFFECT OF HEATING TIME (min) AND OF PROTEIN CONCENTRATION ON THE GEL STRENGTH (g) OF EITHER COMMERCIAL PORCINE PLASMA (P) OR EGG ALBUMEN (E) OR A MIXTURE OF BOTH (P+E), ALL IN 45% SUCROSE SOLUTION AT 85°C[43a]

Time of heating	% Protein (w/w)				
	P 6	P+E 4+2	P+E 3+3	P+E 2+4	E 6
15 min (n)	8	8	8	6	8
Mean gel strength	428	437	411	318	101
SD±	18·0	8·8	21·3	31·6	24·8
CV %	4·2	2·0	5·1	9·9	24·4
Additive value	428	319	264	210	101
Interaction Index		+37	+55	+51	
30 min (n)	8	9	9	8	8
Mean gel strength	582	510	433	338	137
SD±	21·2	23·9	18·0	20·3	11·4
CV %	3·6	4·7	4·1	5·9	8·2
Additive value	582	434	360	285	137
Interaction Index		+17	+20	+18	
60 min (n)	8	8	7	7	8
Mean gel strength	633	534	467	367	172
SD±	34·6	52·8	46·4	17·9	29·8
CV %	5·4	9·8	9·9	4·9	17·3
Additive value	633	422	403	326	172
Interaction Index		+26	+15	+12	

Note: n, number of replicates; SD, standard deviation; CV, coefficient of variation; additive value, the value derived from summing the contributions of the component proteins, as measured in isolation, in proportion to the concentration of each in the mixture;

$$\text{Interaction Index} = \frac{\text{actual value—additive value}}{\text{additive value}} \times 100.$$

Gelation of the proteins of blood plasma, egg albumen and mixtures of the two were examined using the cake-type model system described previously for the measurement of viscosity.[31,40] Small quantities of solutions (7 ml) of the proteins were heated at 80–95°C for 15–60 min in stainless steel tubes (50×16 mm) sealed with rubber bungs at each end.[40,43a] Synergistic interaction was observed at the outset, which led the authors to devise an interaction index, using gel strength or breaking strength values obtained by compression testing on the Instron Universal Testing Machine.

The Interaction Index is based on

$$\frac{\text{actual value} - \text{additive value}}{\text{additive value}} \times 100$$

TABLE 5

EFFECT OF HEATING TIME (min) AND OF PROTEIN CONCENTRATION ON THE BREAKING STRENGTH (g) OF EITHER COMMERCIAL BOVINE PLASMA (P) OR EGG ALBUMEN (E) OR A MIXTURE OF BOTH (P+E), ALL IN 45% SUCROSE SOLUTION AT 95°C[43a]

Time of heating	% Protein (w/w)				
	P 6	$P+E$ 4+2	$P+E$ 3+3	$P+E$ 2+4	E 6
15 min (n)	12	14	14	16	8
Mean breaking strength (g)	1760	1825	1509	928	241
SD±	157·9	242·2	132·5	117·8	35·7
CV %	8·97	13·2	8·7	12·7	14·8
Additive value	1760	1253	1000	747	241
Interaction Index		+45	+50	+24	
30 min (n)	10	15	14	14	13
Mean breaking strength (g)	1931	1790	1657	983	239
SD±	176·9	128·2	112·6	87·0	21·0
CV %	9·1	7·1	6·7	8·8	8·7
Additive value	1931	1366	1085	802	239
Interaction Index		+30	+52	+22	
60 min (n)	11	13	12	13	14
Mean breaking strength (g)	2065	1883	1589	1026	250
SD±	258·1	99·7	101·4	74·2	32·4
CV %	12·4	5·3	6·3	9·2	12·9
Additive value	2065	1459	1157	855	250
Interaction Index		+29	+37	+20	

and provides a quantitative reproducible measurement of interaction. Interaction Index values demonstrate that the gel strength and breaking strength values for mixtures of plasma and egg albumen proteins are higher than the calculated values (additive values) obtained by adding the contributions of the individual plasma and egg albumen protein components as measured in isolation in proportion to the concentration of each in the mixture. Tables 4 and 5 show the gel strength and breaking strength values and interaction indices of porcine and bovine plasma proteins heated with egg albumen proteins. Similarly, the cohesiveness value of the gel, which is represented by the distance the gel can be pressed before it ruptures, is greater for the mixture of porcine plasma and egg albumen proteins than that expected by linear addition. The effectiveness of the use of the Interaction Index is clearly illustrated by the interaction of porcine albumin (Fraction III)[43a] with egg albumen proteins as shown in Fig. 3.

The interaction pattern is similar to that of whole porcine plasma, i.e. greatest synergistic interaction when heated at low temperatures of 80 and 85°C for a short time of 15 min. This is not surprising as albumin constitutes about 50% of the blood plasma proteins and is mainly responsible for the gelation and interaction of whole plasma with egg albumen proteins. As with whole porcine plasma, interaction increased with increased concentration of albumin in the mixture. Other plasma Fractions I and II incorporating fibrinogen, β-globulins and immunoglobulins and α-globulins[43b] also interact with egg albumen, but to a lesser extent.

Interactions also depend on the species and type of protein and vary with time and temperature of heating and protein concentration in the mixture as illustrated for porcine and bovine plasma with egg albumen proteins. For porcine plasma, interaction with egg albumen was high at a short heating time (15 min), low temperature (85°C) and low plasma protein concentration (2%). In contrast, interaction between bovine blood plasma and egg albumen was high at higher temperatures of 90–95°C but also with low plasma protein concentration (2%). The difference in the behaviour of porcine and bovine plasma proteins may be due, for example, to the difference in the N-terminal and C-terminal amino acids of bovine and porcine albumin.[150,151] At present it is not clear how the conformational changes resulting from differences in the primary structure affect gelation. Further work in this area is necessary.

The effect of other proteins on the gelation of plasma proteins highlights the importance of the compatibility of different ingredients in a product. In addition to interaction with egg albumen proteins, a degree of heating of the plasma proteins encourages interaction with other globular proteins such as whey. On the other hand, mixtures of porcine plasma and sodium caseinate, as well as soya isolate proteins, inhibit gelation or lead to phase separation, probably due to steric hindrance by the larger molecular weight and subunit structure of caseinate and soya isolates.

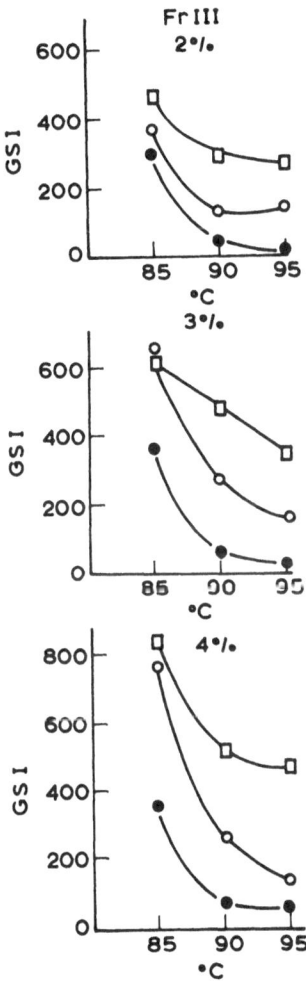

FIG. 3. Variation of interaction index I in terms of gel strength (GS) with temperature (°C) for 2, 3 and 4% (w/w) porcine plasma Fraction III (FR III) protein levels. In all cases the plasma fractions were made up to 6% (w/w) total protein in 45% (w/w) sucrose solution with egg albumen proteins.[43a]

Mechanism of Interaction

If the general theory of protein denaturation and conformational changes of molecules to expose reactive groups is applied then it appears that, at short times and low temperatures of heating, the egg albumen proteins undergo substantial conformational changes as indicated by viscosity increases (Table 3). In this state they appear to interact with the partially unfolded plasma proteins. At higher temperatures, longer times of heating and higher protein concentration, plasma–plasma protein interaction becomes dominant and the interaction value declines.

Elucidation of the actual mechanism of interaction in the gelation process is difficult to achieve. Chromatographic characterisation of protein mixtures,[44,152] differential scanning calorimetry[45] and conformational changes by circular dichroism[46] have been used to some extent to monitor interactions. However, these methods employ dilute protein solutions which may represent foaming or emulsification properties, but not gelation involving large concentrations of proteins. Chemical modification of proteins to block certain reactive groups would provide the means of determining which bonds are involved, e.g. cysteine hydrochloride reduces disulphide bonds. However, the obvious problem here is that it is not possible to ascertain whether for example disulphide groups within the individual proteins are being reduced, thereby affecting gelation, or whether the linkages between the two types of protein are disrupted. In some studies reagents are used to block certain groups in individual proteins prior to mixing, whereas in other studies reagents are added after gelation. It may be possible to use chemical modification in this way providing adequate controls are applied and side reactions minimised.

One way of examining mixed protein gels is by electron microscopy. For blood plasma and egg albumen proteins, and mixtures of the two proteins heated at 85 or 95°C for 15 min, transmission electron microscopy (TEM) indicates a similar type of gel network, showing the compatibility of these proteins in gel formation. The plasma protein gel has a 'close' dense network. In contrast, the egg albumen gel has an 'open' structure and the mixed gel a structure in density between the two (Fig. 4).[40]

Similarly Clark & Lee-Tuffnell[41] have reported compatibility as judged by TEM between bovine serum albumin (BSA) and ovalbumin in gels (10% protein, pH 7·0) containing varying amounts of BSA and ovalbumin. Electron microscopy studies to date do not indicate the individual protein domains in the mixture. Therefore, in an attempt to elucidate the composition of the network and the role of the individual BSA and ovalbumin proteins in the gel, TEM using an immuno-cytochemical technique is currently being examined by the author.

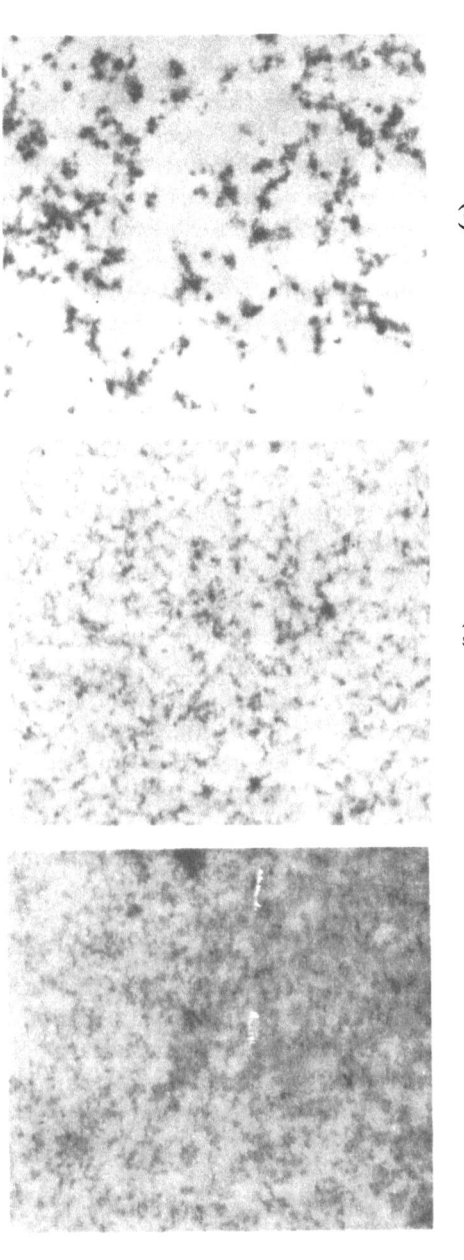

(a) (b) (c)

FIG. 4. Electron micrographs of gels prepared by heating the proteins of (a) 6% (w/w) plasma; (b) a mixture of 3% plasma and 3% (w/w) egg albumen; or (c) 6% (w/w) egg albumen all in 45% (w/w) sucrose solution for 15 min at 85°C.[40]

Difficulties in differentiating the contribution of individual protein in mixtures have also been pointed out by Clark & Lee-Tuffnell.[41] In particular, infra-red spectroscopic data for gel networks made from mixtures of BSA and ovalbumin shows that the β-sheet content varies in an approximately linear way with the protein composition. Similarly, small angle X-ray scattering results show no discontinuities of scattering behaviour with mixtures ranging from all-BSA to all-ovalbumin; the trend is towards less scattering as gels increase in their content of the lower molecular weight component ovalbumin.[41] Thus, little information on the nature of interaction between ovalbumin and BSA has been provided by these two methods.

Interaction of Bovine Serum Albumin with Basic Proteins
The importance of electrostatic interactions and molecular size are apparent in the interaction of bovine serum albumin with basic proteins including clupeine (pI 12), hydrolysed clupeine, arginine and modified β-lactoglobulin (pI 9·5).[47] These basic proteins are considered to interact electrostatically with the negatively-charged BSA at pH 7. A lower concentration of BSA is required for the gelation of mixed BSA-clupeine gels which resemble BSA gels, made at pH 5, when examined by electron microscopy. The interaction involves perturbation of the BSA tertiary structure by clupeine, which reduces the temperature and enthalpy of denaturation of BSA as measured by differential scanning calorimetry.

Molecular size also appears to be important. Hydrolysed clupeine or basic amino acids of arginine added to BSA result in weak gels.[47] Small molecules, therefore, do not appear to take part in cross-linking. On the other hand, very large molecules of soya isolates and sodium caseinate also inhibit gelation by steric hindrance.[40] A compatible molecular size is therefore required for gelation.

Johnson & Zabik[48] have also emphasised the role of electrostatic interactions in egg albumen and have taken the view that the effect of protein–protein interactions should be evaluated in relation to their performance in a food system. They used a custard model system containing 1·27% egg albumen proteins in salt solution, ionic strength 0·275 and pH 8. The temperature of coagulation during heating was measured, and gel strength was determined by penetration using the Instron Universal Testing Instrument.

Their studies indicate that binary mixtures of the egg albumen proteins, lysozyme, globulins, ovomucoid, conalbumin and ovalbumin have significant effects on the coagulation, gel strength and drainage. In particular, ovomucoid increases the coagulation temperature range of the globulins, conalbumin and ovalbumin, and prevents the coagulation of lysozyme. On the other hand, conalbumin and ovalbumin reduce the coagulation temperature and gel strength and increase drainage loss. On the whole, aggregation occurs near the denaturation transition temperature of the least stable protein.

Whey Protein Interactions

Whey proteins comprise 15–25% of skimmed milk proteins and consist of β-lactoglobulin (7–12%), α-lactalbumin (2–5%), immunoglobulins (1·5–2·5%), bovine serum albumin (0·7–1·3%) and protease peptones (2–4%). β-Lactoglobulin is a globular protein with a MW of 36 000 and exists as a dimer of the 18 000 MW monomer unit cross-linked by two disulphide bonds.[49,50] The sulphydryl groups of β-lactoglobulin (1 SH/mol) contribute 90% of the SH groups in milk. It is believed that the proximity of disulphide and sulphydryl groups make β-lactoglobulin susceptible to heat denaturation.[51]

During heating two reactions occur in both β-lactoglobulin variants, A and B.[52,53] Sawyer[50] has reviewed the thermo-denaturation of β-lactoglobulin which may be summarised as follows. In the primary reaction disulphide-bonded aggregates form which migrate slowly during zone electrophoresis at pH 7·5 and have a higher sedimentation coefficient (3·7S) for a sample heated at 97°C for 50 min than the native protein (2·6S). The secondary reaction leads to the formation of a heavy component (29S) which is not stabilized by disulphide bonds. However, the occurrence of the secondary reaction depends on the primary reaction.

It is interesting to note that the kinetics of the reaction between the SH group of β-lactoglobulin and 5,5′-dithiobis-(2-nitrobenzoic acid) (DTNB) have been used to detect the SH/SS interchange reactions occurring during the heating of whey protein isolate dispersions. The method is based on the assumption that the reactivity of the SH group adjacent to the SS bond is low in the presence of SDS. This method indicated that heating a 9% protein dispersion caused the formation of a highly elastic gel at pH 7·5 with a predominance of SS bonds through SH/SS interchange processes and the formation of a non-elastic gel at pH 2·5.[54]

The effect of pH (range 2–9·5) on the texture and viscoelastic properties of β-lactoglobulin gels in distilled water has also been reported.[55] In addition the presence of salts such as NaCl and $CaCl_2$ can markedly affect the rheological properties.[56]

Aggregation due to disulphide bonds also occurs during alkali denaturation[49] and urea denaturation[57] of β-lactoglobulin. The β-lactoglobulin
variants A, B and C are susceptible to denaturation to different extents,
possibly due to the varied reactivities of the thiol group.[50]
α-Lactalbumin is essentially a monomer and has a MW of 14 000 daltons. It contains four disulphide bonds, no sulphydryl groups and has a
pI of 5·1. Hillier et al.[42] and Morr[57] have discussed the importance of
α-lactalbumin denaturation on the functional properties of whey. The
degree of thermodenaturation of α-lactalbumin is greater when heated in
the presence of β-lactoglobulin than when heated by itself, particularly
when pH and temperature are increased. Elfagm & Wheelock[153] concluded that α-lactalbumin interacts with an aggregated form of β-lactoglobulin. The interaction between α-lactoglobulin and β-lactoglobulin is
reported to be due to disulphide links.[154]

β-Lactoglobulin–κ-Casein Interaction

Milk is heat-treated to promote the controlled interaction of denatured
β-lactoglobulin and κ-casein. This heat-induced complex affects the heat
stability of milk and milk products.[58,59] Further, heat treatment of milk
delays the rate of formation of a gel by rennet (chymosin).[60] Rennin hydrolyses κ-casein by cleaving the Phe_{105}–Met_{106} bond to release the
hydrophobic para-κ-casein and the hydrophilic macropeptide which ini-
tiates coagulation.[6] One of the reasons for the delay in gel formation has
been attributed to the interaction of β-lactoglobulin with κ-casein. The
interaction interferes with the primary phase in the action of rennet,
probably by shielding the chymosin-susceptible bond, and also through
the secondary non-enzymic action of rennet, which usually results in random aggregation as described by the von Smoluchowski theory.[61–63]
In order to understand how stability is conferred on milk by the β-
lactoglobulin–κ-casein complex it is necessary to note the structure of the
caseins. The $α_{s1}$, $α_{s2}$, β- and κ-caseins exist as micelles in milk, which are
composed of submicelles associated hydrophobically and by calcium
phosphate linkages. Although a number of models for the casein micelles
have been proposed, the Slattery model[64] is the one most favoured.[6]
According to Walstra & Jenness,[65] κ-casein protrudes as chains on the
surface of the submicelles and prevents association at these sites.
Casein molecules are amphiphilic; that is, they contain a high level of
proline and hydrophobic amino acids, which exist as discrete regions as
well as charged, especially negatively-charged, carboxylic acids, situated
also in clusters. The amphiphilic nature of κ-casein stabilises the micelle

structure so that the hydrophobic residues occupy the interior whereas the hydrophilic molecules are orientated at the surface.[64,66] About 70% of the κ-casein molecules are thought to be at the surface whereas 30% occupy the interior.

Caseins exist in solution as random coils with little ordered structure that can be disrupted by heat,[50,67] especially mild heat treatment during forewarming. However, heating κ-casein at 90°C has been reported to reduce the rennin clotting time by 60%, although its ability to stabilise the casein micelle against precipitation by calcium still exists.[68] In addition, heating increases the turbidity[69] and sedimentation[68] values for κ-casein.

It has been proposed by Dziuba[70] that the hydrophobic interactions stabilising the casein micelle structure are reduced when κ-casein interacts with β-lactoglobulin. The casein proteins therefore interact only weakly with each other, producing a fine network of micelles and consequently firm, stable gels.

The complex formation between β-lactoglobulin and casein was first detected by Tobias, Whitney & Tracy,[71] and Slatter & van Winkle[72] by moving boundary electrophoresis. As was pointed out earlier, it is difficult to demonstrate interactions unequivocally by certain methods. In the above case for example, the interaction was assumed to occur due to the lowering of the β-lactoglobulin peak and an increase in the α-casein peak. However, a β-lactoglobulin solution heated on its own also results in a peak close to that of the α-casein peak.[71–74] Nevertheless, unequivocal evidence of complex formation between β-lactoglobulin and κ-casein has been obtained by Zittle et al.,[68] using electrophoresis with an acid buffer at pH 2·1. In this system a distinct peak of intermediate mobility between heated β-lactoglobulin and κ-casein proteins was identified. Further evidence[75,76] of complex formation using electrophoresis and ultra-centrifuge techniques showed that the complex had a sedimentation velocity value of 44–48S compared with 3–5S for heated β-lactoglobulin and 15–18S for heated κ-casein.

Various groups are implicated in the complex formation. Thiol groups are probably involved, as the complex is not formed in the presence of N-ethylmaleimide or β-mercaptoethanol.[59,75] Purkayastha et al.[77] believe that thiol–disulphide interchange is involved in the formation of the β-lactoglobulin–κ-casein complex based on the following findings: no complex was formed when alkylated β-lactoglobulin and unheated κ-casein were heated together; a limited complex was formed when heated and recooled β-lactoglobulin solution was mixed with κ-casein solution and only slight complex formation resulted when heated β-lactoglobulin was alkylated. The involvement of disulphide–sulphydryl groups is also postulated by Doi et al.[78] based on the findings that the 288-nm peak in the difference spectrum is decreased when the β-lactoglobulin–κ-casein complex is reduced by β-mercaptoethanol. Moreover β-lactoglobulin and S-carboxymethylated κ-casein do not exhibit complex formation when examined by gel filtration.

As pointed out by Sawyer[50,79,80] the above findings do not provide unequivocal evidence that disulphide bonds are involved in complex formation. The addition of blocking agents may simply prevent the primary aggregation of β-lactoglobulin which could subsequently prevent complex formation with κ-casein.

In addition to the presence of disulphide bonds, hydrophobic interactions have also been suggested. Doi et al.[78] examined changes in the ultraviolet spectrum when κ-casein, β-lactoglobulin and the mixtures were heated. They interpreted changes occurring during complex formation to be due to a decrease in the polarity surrounding the tyrosine residue. Since all the tyrosine residues of κ-casein are situated in the para-κ-casein and the tyrosine residues of β-lactoglobulin are located in the hydrophobic regions, they conclude that hydrophobic interactions are involved. On further examination of this hypothesis using the ANS (1 anilino-8-naphthalene sulphonate) fluorescent probe, major increases in the fluorescence intensity and a shift in the emission maximum to a shorter wavelength were detected in the heated mixture as compared with the heated and unheated individual proteins (Fig. 5). Additionally, based

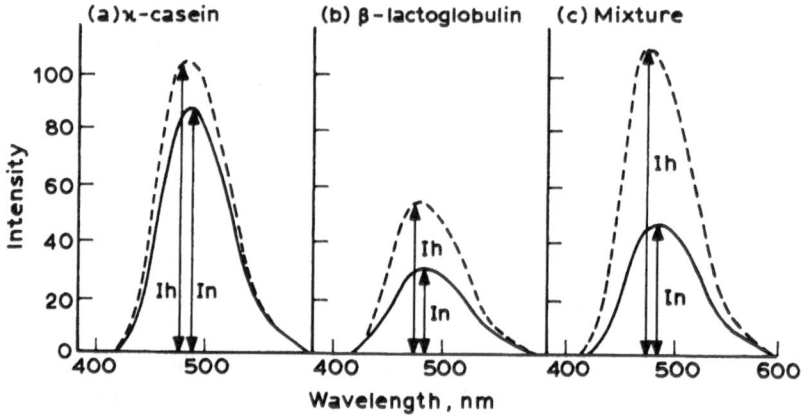

FIG. 5. The fluorescence emission spectra of ANS bound to (– – –) heated and (——) unheated proteins.[78]

on adsorption shifts in the ultraviolet difference spectra, it appears that the presence of carbohydrate in κ-casein is favourable for hydrophobic bonding with β-lactoglobulin.[81,82]

Further investigations indicate the predominant role of hydrophobic interactions in the initial stages prior to covalent bond formation when mixtures of β-lactoglobulin and κ-casein are heated at 70°C at pH 6.8. The apparent hydrophobicity decreased with a concomitant rise in acidic groups.

Various factors affect β-lactoglobulin–κ-casein complex formation, including the concentration of proteins, time and temperature of heating and the presence of salts. For instance, heat treatment of 1 : 1 mixture of β-lactoglobulin and κ-casein results in 3·4% β-lactoglobulin interacting at 69°C, 82·7% at 85°C and 76·7% at 99°C.[76]

Mixing ratios of the two proteins also influence complex formation. An increase in the proportion of β-lactoglobulin results in a higher percentage of the protein interacting at 85°C. However, the mixing ratio has little effect on the molecular size of the complex. Thus at mixing ratios 1 : 1 and 3 : 1 the sedimentation coefficients are 44S and 48S respectively.[76] The size of the complex can be affected by heating the individual proteins prior to complexing.[32]

The presence of inorganic salts affects the heat stability of milk through their action on the β-lactoglobulin–κ-casein complex.[83] The amount of complex during heating increases with added calcium and decreases with disodium phosphate.[84,85] El-Negoumy et al.[85] heated mixtures of 2% each of β-lactoglobulin and κ-casein at 110°C for 30 min in salt solutions containing de-ionised water, 0·02 M calcium, 5% lactose, 0·035% phosphorous, 0·2% citrate, 0·14% chlorine, synthetic ultra-filtrate and milk dialysate. Interaction products assessed by electrophoresis indicated that only 13·8% β-lactoglobulin and 17·1% κ-casein in de-ionised water, but 76·6% β-lactoglobulin with 83·3% κ-casein in the presence of salts, interacted in a milk dialysate. This increased interaction in the presence of salts is attributed to the reduction of electrostatic repulsion between charged κ-casein molecules. Variation in the degree of attraction and repulsion depends on the types of salt present.

In addition to electrophoresis and centrifugation techniques, the β-lactoglobulin–κ-casein complex has been characterised by light scattering and fluorescence depolarisation. These techniques appear to be advantageous as they provide thermodynamic and kinetic information at high temperatures. However the information gained may be anomalous and difficult to interpret, as shown below.

The turbidity of mixtures heated at 90°C shows that the weight average molecular weight of the complex (6.54×10^6) is less than the value of κ-casein alone at 90°C (1.77×10^7) but double the value for κ-casein alone at 30°C.[86] This is in contrast to the sedimentation analysis data which suggest a complex with molecular weight greater than those of aggregates which form when two proteins are heated separately. These discrepancies are believed to be due to the fact that the properties of the complex at 90°C may differ from those after cooling to room temperature. Secondly, light scattering techniques which are sensitive to the presence of insoluble material formed on heating, may lead to anomalous values for the weight average molecular weight.[50]

Fluorescence depolarisation techniques have been used for studying the effect of mixtures containing different concentrations of proteins on the formation of the complex at 25 and 65°C.[87] Difficulty in the interpretation of results reported as relaxation time and average apparent molecular volume was encountered.[50]

Compared with the information that exists on the formation of the β-lactoglobulin–κ-casein complex, little has been reported on its rheological properties. The apparent viscosity of the heated mixture is reported to increase to a greater extent than expected compared with that of a heated κ-casein solution.[78]

α-Lactalbumin–κ-Casein Interaction

In contrast to the numerous studies on the β-lactoglobulin–κ-casein interaction, only a few studies have been reported on the α-lactalbumin–κ-casein complex. Elfagm & Wheelock[88] concluded that α-lactalbumin interacts with the κ-casein–β-lactoglobulin complex. The α-lactalbumin–κ-casein complex formation is sensitive to temperature and the presence of salts. Complex formation was detected after heating at 90°C for 15 min by gel filtration. However, Sedmerova *et al.*[89] observed the formation of a κ-casein–α-lactalbumin complex in 0·1 M NaCl, pH 7, at room temperature, which was destroyed by subsequent heating at 80°C for 10 min. In contrast, heat treatment at 74·5 or 85°C in a milk–salt solution did not show complex formation by electrophoresis.[90] Doi *et al.*[83] revealed that complex formation was observed in certain buffers of high pH. In particular, gels may be formed when α-lactalbumin is heated with κ-casein (5% w/w total protein) at temperatures greater than 75°C in phosphate buffer (pH 7·6) containing 0·4 M NaCl.[91] Neither α-lactalbumin nor κ-casein on their own form gels under the same conditions.[91] The breaking strength of the κ-casein–α-lactalbumin gel is less than that of the κ-casein–β-lactoglobulin gel. Lowering the pH of the mixture from 7·6 to 5·8 prior to heating allows a gel to form at a lower protein concentration of 3% (w/w), probably due to greater insolubility of the protein near the isoelectric point.

Optical studies using fluoresence emission spectra of ANS bound to heated and unheated proteins indicate changes in α-lactalbumin (but not for κ-casein) which are thought to be necessary for complex formation.[83] Hydrophobic interactions have not been detected between the two proteins. However, sulphydryl–disulphide interchange is proposed, based on evidence that carboxymethylation of cysteine residues inhibits complex formation and that β-mercaptoethanol dissociates the complex during SDS-electrophoreses. Furthermore, since the sulphydryl content of the heated mixture is less than the sum of those in heated individual proteins Doi et al.[83] conclude that sulphydryl–disulphide interchange is involved.

Apart from the studies of interactions of α-lactalbumin and β-lactoglobulin with κ-casein, gelation of whey proteins by a dynamic rheological technique which allows the non-destructive testing of gels is reported by Paulsson et al.[92] α-Lactalbumin does not gel on its own, even at 20% (w/w) concentration at pH 6·6, probably due to the lack of sulphydryl groups. In addition, α-lactalbumin delays the onset of the gelation of 2% BSA solution to a higher temperature. β-Lactoglobulin also increases the gelation temperature of the BSA solution and the viscous component of the gel. Interactions of whey proteins with rapeseed,[93] pea flour[94] and peanut proteins[95] have been reported. These observations are of practical significance. However, the mechanism of interaction has not been studied.

Interactions of Soya Proteins

Soya proteins are widely used, mainly for their gelation and water and fat binding properties. Compared with whey, egg albumen and blood plasma proteins, soya proteins consist of heterogeneous complex quaternary structures of high molecular weight.

The major globulins of soya proteins are conglycinin (7S) and glycinin (11S) comprising 70% of the proteins. The remainder consist of 2S and 15S globulins. Conglycinin is a trimeric glycoprotein with molecular weight 140 000–190 000. It is composed of six fractions, each comprising three subunits associated by hydrophobic interactions.[96] Glycinin (11S) has a molecular weight ranging between 300 000 and 400 000 and consists of twelve units packed in two hexagonal rings.[97] Each ring contains three pairs of disulphide-linked acid (MW 35000) and basic (MW 20000) subunits.[98]

Heat gelation of soya proteins involves the dissociation and association of subunits.[99-104] Gelation studies[105-109] suggest that 11S globulins undergo only small conformational changes on heating.

Utsumi & Kinsella[110] examined the effect of various reagents namely NaCl, NaSCN, β-mercaptoethanol, dithiothreitol, propylene glycol, N-ethylmaleimide and urea, on the gelation of 7S and 11S globulins and soya isolate proteins. The reagents were added both prior to and subsequent to gelation. Their results indicate mainly hydrogen bonding in 7S globulin gels, electrostatic interactions and disulphide bonds in 11S globulin gels and both hydrogen bonding and hydrophobic interactions in soya isolate gels.

The presence of non-covalent bonds in 7S globulin gels and disulphide bonds in 11S globulin gels has also been proposed by Kamata et al.[111] based on studies on the effect of heating rate and high temperature holding on the viscosity and thixotropic behaviour of soya globulin gels. Their studies indicated that the 7S globulin gels were able to reform, after shearing, on standing whereas the 11S gels were unable to reform after shearing. Damodaran[112] suggested that 11S globulins at 8% concentration undergo partial refolding to such a considerable extent during the cooling regime of the thermal gelation process that they fail to form a gel. The extent of refolding can be altered by the addition of sodium chloride. Based on these findings it is suggested that by controlling the extent of refolding of proteins during cooling, it is possible to improve the gelation of globular proteins event at suboptimum protein concentration. The rheological properties of 7S and 11S can also be modified by limited proteolysis using trypsin.[113]

Interactions Between 11S and 7S Globulins

It is believed that the basic subunits of the 11S globulins interact with the β-75 globulins in soya isolate gels[99,100,110,155] via electrostatic interactions. The association of basic subunits is by disulphide bonds. In addition, the acidic subunit (AS III) is an essential component in the formation of 11S globulin and soya isolate gels and contributes to the hardness. However acidic subunit IV is considered to be the component which interacts with 7S globulins. With the exception of ASIV, the acidic subunits are associated in the gel by disulphide bonds.[110] In the 7S globulin gel the three subunits contribute uniformly to the gel and are associated by similar types of bonds including disulphide linkages.[110]

Further elucidation of the mechanism of interaction between 7S and 11S globulins during gelation indicates the importance of protein concentration and presence of relative proportions of globulins present.[114] The lowest protein concentration for the formation of self-supporting gels for the individual 7S and 11S globulins were 7·5 and 2·5%, respectively. For the mixture 7·5% protein concentration, but containing 3·75% 11S globulin, is required. Thus the gelation of 11S globulins is suppressed by the 7S globulins.

Gel hardness of the mixed system increases sigmoidally with increased total protein concentration compared with the exponential curves exhibited by the individual 7S and 11S globulin gels. Depending on the total protein concentration, the proteins interact either positively or negatively to give a high gel strength (e.g. at 12·5% protein concentration), or low gel strength (e.g. at 10 and 20% protein concentration) respectively. The presence of 7S globulins also reduces the turbidity of the mixed gels,[114] probably by suppressing the dissociation of the 11S globulins into acidic and basic subunits and the subsequent aggregation of the basic subunits (Fig. 6).

In the mixed gel the gelling time for the formation of a self-supporting gel decreases with increased concentrations (5–10%) of 11S globulins. 7S Globulins on their own do not gel at these concentrations. However, in the mixed system 7S globulins increase the gelling time of 11S globulins at low concentrations and lower the gelation time at high concentrations (Fig. 7) This is due to the effect of the 7S globulins on the soluble aggregates reported to form in the early stages of gelation.[105–108] Soluble aggregates produced by heating mixtures of 11S globulins and 7S globulins may be identified by sucrose density gradient centrifugation.[114] Heating produces an increase in the higher MW faster sedimenting components, especially in the mixture and in the presence of a higher concentration of 7S globulin.[108] Electrophoretic analysis indicates that the soluble aggregates consist of both 7S and 11S globulins. The interactions of 7S globulin with itself, and those between 7S and 11S globulins, are non-covalent. On the other hand, both non-covalent (mainly hydrophobic since electrostatic forces are suppressed by a high ionic strength of 0·5) and covalent disulphide bonding are involved in the interaction of 11S globulins with itself.[114]

Protein–protein interactions by ^{1}H-NMR using ^{17}O-NMR relaxation measurements at 54·2 MHz and 21°C in connection with hydration studies have facilitated quantitation of protein–protein interaction by fitting the ^{1}H-NMR data by virial expansion.[115] Although this method indicates the

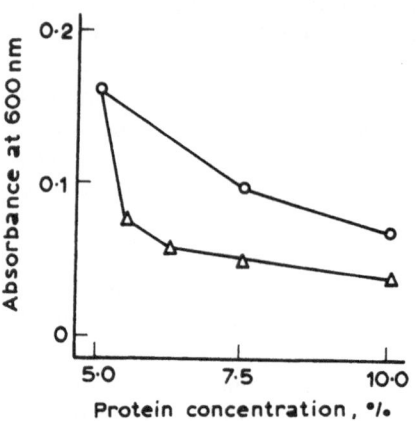

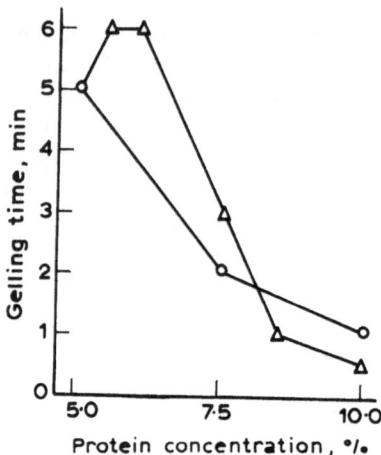

FIG. 6. Effects of protein concentration on gel turbidity.[114] O, 11S
Globulin; △, mixture of 7S and 11S
globulins (5% 11S plus 7S with various
concentrations).

FIG. 7. Effects of protein concentration on gelling time.[114] O, 11S
Globulin; △, mixture of 7S and 11S
globulins (5% 11S plus 7S with various
concentrations).

magnitude of interaction it does not provide precise information on
chemical groups interacting, which can be obtained by high resolution
NMR.

Interaction of Soya Proteins with Muscle Proteins

The interaction of soya proteins with muscle proteins has also been
studied, due to the extensive use of soya proteins in meat products.
Studies are based on simplified meat-type model systems as meat products are very complex.

The structure and properties of muscle proteins have been reviewed
recently.[116] In particular myosin, which constitutes 50–55% of the myofibrillar proteins, is a high molecular weight protein containing two identical polypeptide chains with a high degree of α-helical structure. The two
chains are coiled around each other and terminate in two globular heads.
Myosin can be cleaved by proteolytic enzymes into heavy and light
meromyosin fragments.[1] Myosin is characterised by a high proportion of
basic (17%) and acidic amino acids (18%) and about 42 thiol residues.
Actin, which represents 15% of the meat proteins has been reported to increase the rigidity of myosin gels by the formation of F-actomyosin which
then interacts with free myosin.[116] The gelation of myosin has been

thought to involve the oxidation of SH groups leading to aggregation of the globular head regions and non-covalent interactions between the myosin tail regions.[116]

Soya Protein–Myosin Interaction

Complex formation between 7S globulins and rabbit myosin occurs at temperatures of 75–100°C, and is characterised by an increase in the specific viscosity of the mixture, a decrease in the sedimentation coefficient and a change in the composition of the soya protein component of the aggregate.[117] As with most methods of assessing interaction, it is difficult to ascertain whether the changes are due to changes in individual components or to the formation of a genuine complex. The latter is probably the case at high temperatures as only a single band of the same sedimentation coefficient, containing both myosin and 7S globulins, was obtained using a sucrose density gradient.[117]

In contrast to the above findings, soya-β-conglycinin (7S globulins) interacted with chicken myosin at temperatures between 60 and 100°C, as judged by changes in turbidity and solubility.[118] In addition, SDS-polyacrylamide gel electrophoreses indicated that the presence of β-conglycinin results in diminished aggregation of myosin heavy chains between 50 and 100°C. Differences in the temperatures at which interaction occurs may be attributed to the use of different myosin species (rabbit and chicken) and differences in the pH and method of preparation of β-conglycinin.

Interaction between soya 11S proteins (cold insoluble fraction, CIF) and myosin occurs at 85–100°C[119–120] as shown by gel filtration, solubility, electrophoresis and turbidity measurements. On its own the myosin solution increases in turbidity from 40°C reaching a maximum at 50°C and decreases at higher temperatures. In contrast, soya 11S proteins increase in turbidity at around 70°C producing a precipitate at 80–100°C. The mixture of 11S proteins and myosin exhibit turbidity values similar to myosin from 40 to 80°C. However, at 80–90°C values lie between those of soya 11S proteins and myosin (Fig. 8) indicating that myosin interferes with the aggregation of 11S globulins.

Solubility, electrophoresis and titration methods confirm that the 11S globulins dissociate into acidic and basic subunits and the basic subunits subsequently interact with the myosin heavy chains.[120] A limited amount of denaturation of the basic subunits by preheating from 0 to 4 min leads to an increase in sulphydryl groups. Thus, interaction via disulphide bonding produces firmer gels at 70°C rather than at 100°C.[121]

In contrast, Haga & Ohashi[122] claim that the interaction between 11S globulins and myosin via disulphide bonds occur prior to heating. Their evidence is based on gel filtration of the proteins and mixtures treated with β-mercaptoethanol and guanidine hydrochloride. Scanning electron microscopy also indicates that the mixed protein gel network is different from that produced by the individual proteins prior to heating.

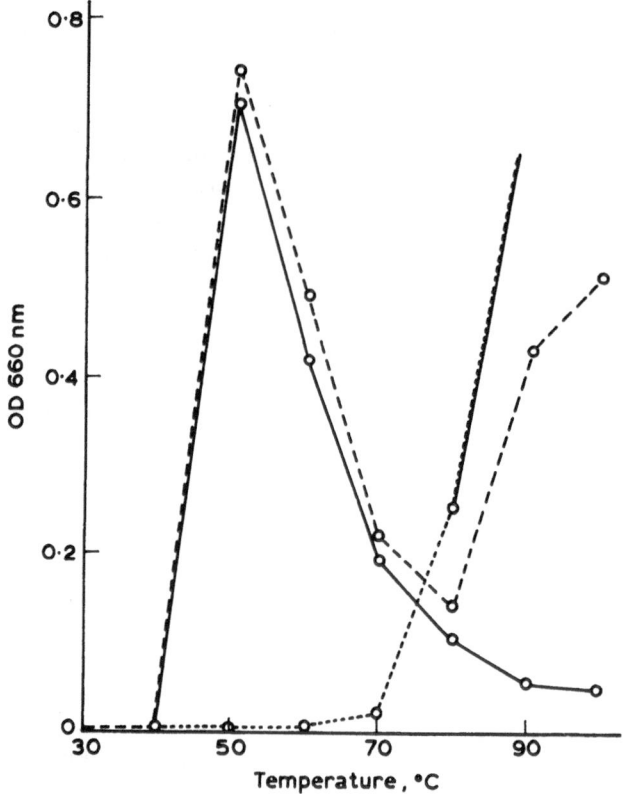

FIG. 8. The turbidity curves of myosin, 11S protein and the mixture of these two proteins as a function of temperatures. ———, Myosin; · · · · ·, 11S protein; – – –, mixture. Heating conditions: myosin (2 mg/ml), 11S protein (2 mg/ml) and the mixture (2 mg/ml each) were suspended in buffer and heated at various temperatures for 30 min. The solid line portion of the 11S protein curve is arbitrary indicating the appearance of aggregates.[120]

Soya Protein—Actin and Alpha-actinin Interaction

Interaction of soya proteins and alpha-actinin with F-actin filaments in sedimentation analysis were assessed by SDS-PAGE.[123] Soya proteins were observed to bind to F-actin filaments at pH 5·5 or pH 7 at 25°C whereas no interaction was observed between alpha-actinin and soya proteins. However alpha-actinin can bind to actin filaments. This binding is not affected by the presence of soya proteins at pH 7 but is reduced at pH 5·5. These findings were confirmed by viscosity studies.

PROTEIN–PROTEIN INTERACTIONS IN FOAMS

Foaming properties of proteins have been extensively reviewed by Halling,[124] Stainsby[125] and Mitchell.[126] Briefly, the formation of foams requires the lowering of surface tension to produce a larger surface area for the creation of new small air bubbles. This is achieved by the adsorption of proteins with an appropriate balance of hydrophilic and hydrophobic groups at the surface. Once formed, the air bubble stability needs to be maintained to prevent coalescence and eventual breakdown of the foam. The stability depends on the ability of the protein to form a viscoelastic film at the bubble surface. These properties are achieved by the partial denaturation and unfolding of the protein followed by electrostatic and hydrophobic interactions. Rheological properties such as viscosity and network formation imparted by the protein can also enhance foam stability.

Interaction of Egg Albumen Proteins

Interaction of ovomucin with other egg albumen proteins is widely reported. In particular the ovomucin–lysozyme complex is related to the thinning of egg white on storage.[127–130] Cotterill & Winter[128] and Robinson & Monsey[131] proposed that the ovomucin–lysozyme complex is a result of the electrostatic interaction between the negative charges of ovomucin molecules with the positively charged lysozyme. Other egg albumen proteins which are thought to exhibit electrostatic interactions with lysozyme are ovalbumin[132] and ovotransferrin (conalbumin).[133]

The electrostatic interaction in the ovomucin–lysozyme complex occurs between the carboxyl group of the terminal sialic acid in ovomucin and the positive charges of the lysyl ε-amino group in lysozyme.[134,135] This has been demonstrated by Kato et al.[134,135] who separated ovomucin into a sialic acid rich, fast-moving component (F-ovomucin) and a sialic acid poor, slow-moving ovomucin (S-ovomucin) by preparative electrophoresis. Lysozyme interacts mainly with the F-ovomucin as shown by turbidity measurements. Furthermore, the ovomucin–lysozyme interaction decreases correspondingly at a rate depending on the enzymatic removal of sialic acid by neuramidase action from ovomucin. The carboxyl group is essential for the interaction whereas the polyhydroxyl group is not essential and may hinder the interaction by partially blocking the carboxyl group.[135] In addition, ovomucin–lysozyme interaction decreases when acetylation, succinylation or carbamylation of the amino groups of lysozyme reduce the positive charges of the lysyl ε-amino groups of lysozyme. Disulphide bonds between lysozyme and ovomucin do not occur, as the complex does not form after the alkylation and reduction of lysozyme and ovomucin.[136]

The ovomucin–lysozyme complex observed in egg white is also important in the foam capacity and foam stability of cakes. Johnson & Zabik[137-139] studied the interactions of egg albumen proteins using a complex angel cake model system. They used response surface methodology which involves experimental formulations derived from factorial arrangements. The extreme vertices design was used to determine the levels of six different proteins incorporated[140] and the data obtained from the cakes (for 57 combinations) were treated by multiple regression analysis.

Foaming was enhanced by ovomucin and its interaction with globulins, probably due to the concentration at the film surface due to increased solution viscosity.[137,139,141] In the presence of lysozyme, which has a limited ability to whip, however, the foaming capacity was depressed but the foam stability and cake volume increased. The lysozyme–ovomucin complex appears to exist as an insoluble complex at the bubble surface which can heat denature and set during baking. In the absence of lysozyme, ovomucin cannot maintain or produce a cohesive layer, probably due to its inability to gel on heating.[139]

Transmission electron micrographs of whipped egg protein solutions show that the film at the bubble surface in globulin, conalbumin and ovalbumin foams appears as a flexible smooth sheet. In contrast, the film in a lysozyme foam is rigid and rough textured. Lysozyme strengthens the film in foams in the presence of ovomucin.[139] Optimum foaming and cake volume may be obtained with mixtures of ovomucin, lysozyme and globulins ranging from 0·2 to 1·0%, 0·0 to 1·8% and 12·2 to 14·8%, respectively.

Interaction of Egg Albumen with Whey, Soya and BSA

In addition to the electrostatic interaction of lysozyme (pI 10·7) with ovomucin, lysozyme exhibits electrostatic interaction with soya 7S and 11S globulins[156] and enhances the foaming properties of other acidic globular proteins such as BSA (pI 4·7) and whey proteins (pI 5·1–5·3) even in the presence of corn oil.[142,143] Various milk proteins including β-lactoglobulin, α-lactalbumin, casein and proteose peptones increase the overrun of egg white foams.[144] However, a lipoprotein associated with casein and whey reduced overrun. In addition, the presence of milk fat globule membrane proteins reduced foam expansion and stability of egg or whey protein concentrate foams.[145] In order to test the importance of electrostatic linkages in foaming, the carboxyl groups of β-lactoglobulin were modified to yield a basic protein pI 9·5. These basic proteins, preferably pI 9·9, also enhanced the foaming properties of acidic proteins in the presence or absence of lipids. However, the conformation of the basic protein appears to be important for providing charged sites to allow interactions to occur.

Bovine serum albumin (BSA) amidated to yield a basic protein enhances the foaming properties of ovalbumin.[142,146]

Interactions of Native and Modified BSA
Taylor[147] and Howell & Taylor[148] have recently found that whilst extensively amidated BSA (pI 9) on its own has poor foaming properties, its combination with native BSA and blood plasma can produce enhanced foam expansion and foam stability particularly when mixed at the 1:1 ratio (Figs 9 and 10).

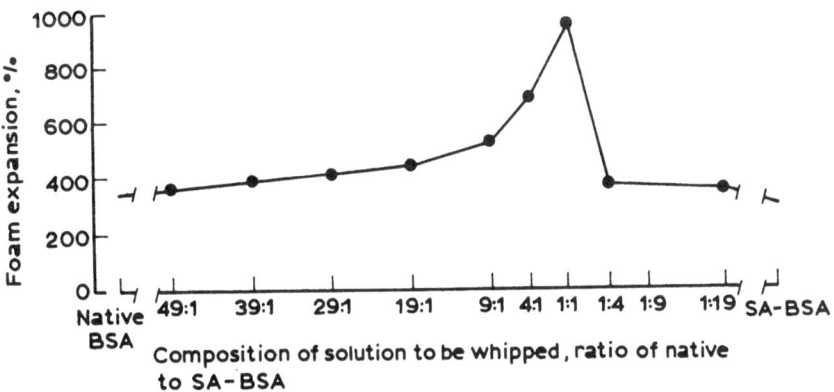

FIG. 9. Interaction of substantially amidated BSA (SA-BSA) with native BSA at pH 7: foam expansion.[146]

A novel method of enhancing foam expansion and foam stability without the addition of other proteins is to partially modify a protein to obtain a range of pI comprising negative and positive components within the same protein. Partial amidation of BSA, using a carbodiimide mediated reaction and ammonium sulphate as the nucleophile, resulted in a product with a pI range 5·3–7·5 and excellent foaming properties at pH 7·0.[146] The interaction is probably due to electrostatic interaction as well as other forces, as judged by the increase in exposed hydrophobicity[146] which reflects conformational changes,[46] and also correlates well with foaming properties.[14]

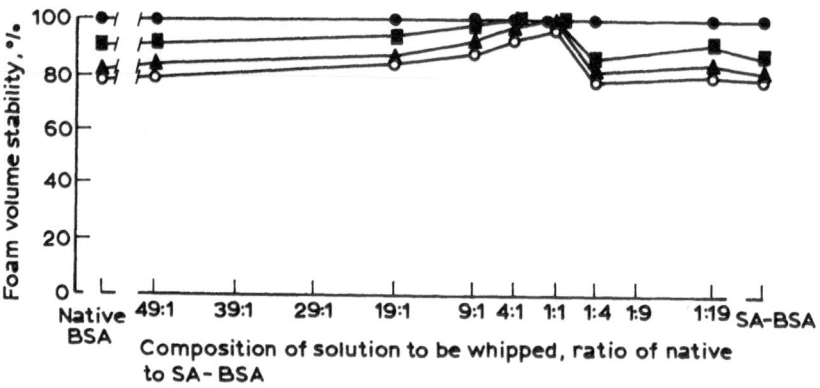

FIG. 10. Interaction of substantially amidated BSA (SA-BSA) with native BSA at pH 7: foam volume stability. Minutes post whipping: 0(●), 5(■), 15(▲), 30(○).[135]

The interactions between differently chemically modified proteins and their gelling and foaming properties in 1:1 mixtures by weight were examined by Murphy & Howell.[149] Two types of negatively charged modified proteins, namely succinylated and thiolated BSA were mixed with either native BSA or a positively charged lysyl-BSA. Increased whipping properties compared with native BSA were shown only by the mixture of native BSA and lysyl-BSA. In general, mixtures which included succinylated and thiolated proteins performed poorly, with the exception of a mixture of thiolated and lysyl-BSA which gelled. Thus the availability of amino groups makes a significant contribution to the mechanism of gelling.

CONCLUSIONS

It is apparent from the above discussion that protein–protein interaction studies range from fundamental physico-chemical studies in simple systems to more complex food-type model systems. Although the latter may not provide a detailed knowledge of the mechanism of interactions they are of immense practical value in food processing.

Unfortunately many of the reported studied cannot be compared directly with each other. It is obvious that the type of protein, species and experimental conditions used affect protein–protein interactions. In this context a parallel can be drawn to the study of functional properties, where a number of tests are employed by different research establishments to study a particular functional property. Consensus relating to methods is often difficult to achieve since, in many cases, in the final analysis the testing of functional properties and of protein–protein interactions has to relate to the food product and appropriate model systems may have to be used. Nevertheless, a concerted effort in establishing the use of suitable methods for studying protein interactions may assist in the elucidation of the mechanisms of protein interactions.

Finding suitable methods to provide unequivocal evidence for interactions is not simple. A number of methods have been cited in this chapter including turbidimetry, viscosity, sedimentation analysis, electrophoresis, electron microscopy, gel permeation chromatography, DSC as well as chemical modification. With many of these methods, particularly chemical modification, it is difficult to ascertain whether the changes denoting interaction are due to modification of the individual proteins or whether they are the result of genuine complex formation. To overcome this problem, the author is currently investigating the use of non-empirical techniques such as high resolution NMR, immunocytochemistry with gold-labelled antibodies, chromatography with radiolabelled protein groups and molecular modelling to elucidate protein–protein interactions. A novel genetic method to detect protein–protein interactions may also be worthy of investigation.[158]

Interactions can be studied effectively, by using several diverse appropriate methods together with stringent controls, to provide unambiguous evidence and precise descriptions of the interactions.

REFERENCES

1. Howell, N., *Int. Flavours and Food Additives*, May/June, 1978, 119.

2. Kinsella, J. E., *CRC Crit. Rev. Food Sci. Nutr.*, **7** (1976) 219.
3. Anglemier, A. F. & Montgomery, M. W., In *Principles of Food Science Part I. Food Chemistry*, ed. O. R. Fennema. Marcel Dekker, New York, 1976.
4. Whitaker, J. R., In *Food Proteins*, ed. J. R. Whitaker & S. R. Tannenbaum. AVI Publishing Company, Westport, Connecticut, 1977, p. 14.
5. Creighton, T. E., *Proteins, Structures and Molecular Properties*. W. H. Freeman and Company, New York, 1983, p. 86.
6. Bringe, N. A. & Kinsella, E., In *Developments in Food Proteins—5*, ed. B. J. F. Hudson. Elsevier Applied Science Publishers, London, 1987, p. 159.
7. Friberg, S., In *Food Emulsions*, ed. S. Friberg. Marcel Dekker, New York, 1976, p. 1.
8. Lifson, S., In *Protein–Protein Interactions*, ed. R. Jaenicke & E. Helmreich. Springer-Verlag, Berlin, 1972, pp. 3–16.
9. Schultz, G. E. & Schirmer, R. H., *Principles of Protein Structure*. Springer-Verlag, NY, 1979.
10. Bigelow, C. C., *J. Theor. Biol.*, **16** (1967) 187.
11. Tanford, C., *J. Am. Chem. Soc.*, **84** (1962) 4240.
12. Nakai, S., Ho, L., Tung, M. A. & Quinn, J. F., *Can. Inst. Food Sci. Technol. J.*, **13** (1980) 14.
13. Nakai, S., *J. Agricultural and Food Chemistry*, **31** (1983) 676.
14. Townsend, A. & Nakai, S., *J. Food Sci.*, **48** (1983) 588.
15. Voutsinas, L. P., Cheung, E. & Nakai, S. J., *J. Food Sci.*, **48** (1983) 26.
16. Sklar, L. A., Hudson, B. S. & Simoni, R. D., *J. Supramol. Struct.*, **4** (1976) 449.
17. Kato, A. & Nakai, S., *Biochim. Biophys. Acta*, **624** (1980) 13.
18. Kesharvarz, E. & Nakai, S., *Biochim. Biophys. Acta*, **576** (1979) 269.
19. Israelachvili, J. N. & Pashley, R. M., In *Biophysics of Water*, eds F. Franks & S. F. Mathias. John Wiley, New York, 1982, p. 183.
20. Parsegian, V. A., *Adv. Colloidal Interface Sci.*, **16** (1982) 49.
21. Barclay, L. & Ottewill, R. H., *Special Discuss. Faraday Soc.*, **1** (1970) 169.
22. Payens, T. A. J., *J. Dairy Res.*, **46** (1979) 291.
23. Overbeek, J. Th. G., In *Colloid Dispersions*, ed. J. W. Goodwin. Royal Society of Chemistry, London, 1982, p. 1.
24. Rha, C. & Pradipasena, P., In *Functional Properties of Food Macromolecules*, ed. J. R. Mitchell & D. A. Ledward. Elsevier Applied Science, London, 1986, p. 79.
25. Rha, C., *Food Technol.*, **32** (1978) 77.
26. Howell, N. & Lawrie, R., *Int. J. Food Technol.*, **22** (1987) 145.
27. Peters T., In *The Plasma Proteins*, Vol. 1, 2nd edn, ed. F. W. Putnam. Academic Press, New York, 1975, p. 133.
28. Putnam, F. W., In *The Plasma Proteins*, Vol. 1, 2nd edn, ed. F. W. Putnam. Academic Press, New York, 1975, p. 57.
29. Vadehra, D. V. & Nath, K. R., *CRC Crit. Rev. Food Tech.*, **4** (1973) 193.
30. Osuga, D. T. & Feeney, R. E., In *Food Proteins*, ed. J. R. Whitaker & S. R. Tannenbaum. AVI Publishing Company, Westport, Connecticut, 1977, p. 220.
31. Howell, N. & Lawrie, R., *J. Food Technol.*, **19** (1984) 289.
32. Sato, Y., Iwatsuki, K. & Hayakawa, M., *Agric. Biol. Chem.*, **41** (1977) 1331.

33. Ferry, J. D., *Adv. Protein Chem.*, **4** (1948) 1.
34. Barbu, E. & Joly, M., *Faraday Discuss. Chem. Soc.*, **13** (1953) 77.
35. Kratchovil, P., Munk, P. & Sedlacek, B., *Coll. Czech. Chem. Commun.*, **27** (1962) 115.
36. Kratchovil, P., Munk, P. & Sedlacek, B., *Coll. Czech. Chem. Commun.*, **27** (1962) 788.
37. Tombs, M. P., In *Proteins as Human Food*, ed. R. A. Lawrie. Butterworths, London, 1970, p. 126.
38. Flory, P. J., *J. Phys. Chem.*, **46** (1942) 132.
39. Bezrukov, M. G., *Angew Chem.* (Engl. Ed.), **18** (1979) 599.
40. Howell, N. K. & Lawrie, R. A., *J. Food Technol.*, **20** (1985) 489.
41. Clark, A. H. & Lee-Tuffnell, C. D., In *Functional Properties of Food Macromolecules*, ed. J. R. Mitchell & D. A. Ledward. Elsevier Applied Science, London, 1986, p. 203.
42. Hillier, R. M., Lyster, R. L. & Cheeseman, G. C., *J. Sci. Food Agric.*, **31** (1980) 1152.
43*a*. Howell, N. K. & Lawrie, R. A., *J. Food Technol.*, **19** (1984) 297.
43*b*. Howell, N. K. & Lawrie, R. A., *J. Food Technol.*, **18** (1983) 747.
44. Moore, W. E. & Carter, J. L., *J. Texture Studies*, **5** (1974) 77.
45. Ledward, D. A., *Meat Sci.*, **2** (1978) 241.
46. Kato, A., Fujimoto, K., Matsudomi, N. & Kobayashi, K., *Agric. Biol. Chem.*, **50** (1986) 417.
47. Poole, S., West, S. I. & Fry, G. C., *Food Hydrocolloids*, **1** (1987) 301.
48. Johnson, T. M. & Zabik, M. E., *Poultry Sci.*, **60** (1981) 2071.
49. McKenzie, H. A. & Sawyer, W. H., *Nature*, **212** (1966) 161.
50. Sawyer, W. H., *J. Dairy Sci.*, **52** (1969) 1347.
51. Schmidt, R. H. & Morris, H. A., *Food Technol.*, **38** (1984) 85.
52. Briggs, D. R. & Hull, R., *J. Am. Chem. Soc.*, **67** (1945) 2007.
53. Gough, P. & Jenness, R., *J. Dairy Sci.*, **45** (1962) 1033.
54. Stading, M. & Hermansson, A.-M., *Food Hydrocolloids*, **4**(2) (1990) 121.
55. Mulvihill, D. M. & Kinsella, J. E., *J. Food Sci.*, **25**(1) (1988) 231.
56. Haque, Z. & Kinsella, J. E., *J. Dairy Res.*, **55**(1) (1988) 67.
57. Morr, C. V., *J. Dairy Sci.*, **50** (1967) 1752.
58. Rose, D., *Dairy Sci. Abstr.*, **25** (1963) 45.
59. Trautman, J. C. & Swanson, A. M., *J. Dairy Sci.*, **41** (1958) 715.
60. Wilson, G. A. & Wheelock, J. V., *J. Dairy Res.*, **39** (1972) 413.
61. Dalgleish, D. G., *J. Dairy Res.*, **46** (1979) 653.
62. Green, M. L. & Morant, S. V., *J. Dairy Res.*, **48** (1981) 57.
63. Fox, P. F., In *Developments in Food Proteins—3*, ed. B. J. F. Hudson. Elsevier Applied Sciences, London, 1984, p. 69.
64. Slattery, G. W., *J. Dairy Res.*, **46** (1979) 253.
65. Walstra, P. & Jenness, R., (ed.), *Dairy Chemistry and Physics*. John Wiley, New York, 1984.
66. Schmidt, D. G., *Neth. Milk Dairy J.*, **34** (1980) 42.
67. Herskovits, T. T., *Biochemistry*, **5** (1966) 1018.
68. Zittlè, C. A., Thompson, M. P., Custer, J. H. & Cerbulis, J., *J. Dairy Sci.*, **45** (1962) 807.

72 NAZLIN K. HOWELL

69. Kresheck, G. C., Van Winkle, Q. & Gould, I. A., *J. Dairy Sci.*, **47** (1964) 117.
70. Dziuba, J., *Acta Aliment. Pol.*, **5** (1979) 97.
71. Tobias, J., Whitney, R. M. & Tracy, P. H., *J. Dairy Sci.*, **35** (1952) 1036.
72. Slatter, W. L. & van Winkle, Q., *J. Dairy Sci.*, **35** (1952) 1083.
73. McGugan, W. A., Zehren, V. F., Zehren, V. L. & Swanson, A. M., *Science*, **120** (1954) 435.
74. Della, M. E. S., Custor, J. H. & Zittle, C. A., *J. Dairy Sci.*, **41** (1985) 465.
75. Sawyer, W. H., Coulter, S. T. & Jenness, R., *J. Dairy Sci.*, **46** (1963) 564.
76. Long, J. E., Van Winkle, Q. & Gould, I. A., *J. Dairy Sci.*, **46** (1963) 1329.
77. Purkayastha, R., Tessier, H. & Rose, D., *J. Dairy Sci.*, **50** (1967) 764.
78. Doi, H., Ideno, S., Ibuki, F. & Kanamori, M., *Agric. Biol. Chem.*, **47** (1983) 407.
79. Harper, W. J., *J. Dairy Sci.*, **64** (1981) 1028.
80. Brunner, J. R., *J. Dairy Sci.*, **64** (1981) 1038.
81. Doi, H., Ibuki, F. & Kanamori, M., *J. Dairy Sci.*, **62** (1979) 195.
82. Doi, H., Ideno, S., Ibuki, F. & Kamori, M., *Agric. Biol. Chem.*, **45** (1981) 2351.
83. Doi, H., Tokuyama, T., Kuo, F.-H., Ibuki, F. & Kanamori, M., *Agric. Biol. Chem.*, **47** (1985) 2817.
84. Rao, P. S., *Abstrs.*, **21** (1961) 3251.
85. El-Negoumy, A. M., *J. Dairy Sci.*, **57** (1974) 1302.
86. Kresheck, G. C., van Winkle, Q. & Gould, I. A., *J. Dairy Sci.*, **47** (1964) 126.
87. Morr, C. V., van Winkle, Q. & Gould, I. A., *J. Dairy Sci.*, **45** (1962) 817.
88. Elfagm, A. A. & Wheelock, J. V., *J. Dairy Res.*, **44** (1977) 367.
89. Sedmerova, V., Helesicova, H. & Sicho, V., *Milchwissenschaft*, **27** (1972) 481.
90. Hartman, G. H. & Swanson, A. M., *J. Dairy Sci.*, **48** (1965) 1161.
91. Doi, H., Hiramatsu, M., Ibuki, F. & Kanamori, M., *J. Nutr. Sci., Vitaminol.*, **31** (1985) 77.
92. Paulsson, M., Hegg, P.-O. & Castberg, H. B., *J. Food Sci.*, **51** (1986) 87.
93. Thompson, L. U., *Can. Inst. Food Sci. Technol. J.*, **10** (1977) 43.
94. Patel, P. R. & Grant, D. R., *Can. Inst. Food Sci. Technol. J.*, **15** (1982) 24.
95. Schmidt, R. H., Illingsworth, B. L. & Ahmed, E. M., *J. Food Sci.*, **43** (1978) 613.
96. Thanh, V. H. & Shibasaki, K., *J. Agric. Food Chem.*, **26** (1978) 695.
97. Derbyshire, E., Wright, D. J. & Boulter, D., *Phytochemistry*, **15** (1976) 3.
98. Badley, R. A., Atkinson, D., Houser, H., Oldani, G. & Stubbs, J., *J. Biochim. Biophys. Acta*, **412** (1975) 2.
99. German, B., Damodaran, S. & Kinsella, J. E., *J. Agric. Food Chem.*, **30** (1982) 807.
100. Damodaran, S. & Kinsella, J. E., *J. Biol. Chem.*, **256** (1981) 3394.
101. Wolf, W. J. & Tamura, T., *Cereal Chem.*, **46** (1969) 331.
102. Saio, K. & Watanabe, T., *J. Texture Studies*, **9** (1978) 135.
103. Watanabe, T. & Nakayama, O., *Nippon Nagei Kagaku Kaishi*, **36** (1962) 890.
104. Kleef, F. S. M. van, *Biopolymers*, **25** (1986) 31.

105. Mori, T., Nakamura, T. & Utsumi, S., *J. Food Sci.*, **47** (1982) 26.
106. Mori, T., Nakamura, T. & Utsumi, S., *J. Food Sci.*, **30** (1982) 828.
107. Nakamura, T., Utsumi, S., Kitamura, K., Herada, K. & Mori, T., *J. Agric. Food Chem.*, **32** (1984) 647.
108. Nakamura, T., Utsumi, S. & Mori, T., *J. Agric. Food Chem.*, **32** (1984) 349.
109. Utsumi, S., Nakamura, T. & Mori, T., *J. Agric. Food Chem.*, **31** (1983) 503.
110. Utsumi, S. & Kinsella, J. E., *J. Agric. Food Chem.*, **33** (1985) 297.
111. Kamata, Y., Umeya, J., Kimura, M., Tanii, S. & Yamauchi, F., *Nippon-Shokuhin-Kogyo Gakkaishi*, **37**(3) (1991) 184.
112. Damodaran, S., *J. Agric. Food Chem.*, **36**(2) (1988) 262.
113. Kamata, Y., Takahata, H. & Yamauchi, F., *Nippon-Shokuhin Kogyo Gakkaishi*, **36**(7) (1990) 557.
114. Nakamura, T., Utsumi, S. & Mori, T., *Agric. Biol. Chem.*, **50** (1986) 2429.
115. Kakalis, L. T., Baianu, I. C. & Kumosinski, T. F., *J. Agric. Food Chem.*, **38**(3) (1991) 639.
116. Morrisey, P. A., Mulvihill, D. M. & O'Neill, E., In *Developments in Food Proteins, Vol. 5*, ed. B. J. F. Hudson. Elsevier Applied Science, London, 1987, p. 195.
117. King, N. L., *J. Agric. Food Chem.*, **25** (1977) 166.
118. Peng, I. C. & Nielson, S. S., *J. Food Sci.*, **51** (1986) 588.
119. Peng, L. C., Dayton, W. R., Quess, D. W. & Allen, C. E., *J. Food Sci.*, **47** (1982) 1984.
120. Peng, I. C., Dayton, W. R., Quess, D. W. & Allen, C. E., *J. Food Sci.*, **47** (1982) 1976.
121. Shiga, K., Nakamura, Y. & Taki, Y., *Jpn. J. Zootech. Sci.*, **56** (1985) 897.
122. Haga, S. & Ohashi, T., *Agric. Biol. Chem.*, **48** (1984) 1001.
123. Muguruma, M., Lin, L. C. & Ito, T., *Nippon Shokuhin-Kogyo Gakkaishi*, **36**(9) (1990) 754.
124. Halling, P. J., *CRC Crit. Rev. Food Sci Nutr*, **15** (1981) 155.
125. Stainsby, G., In *Functional Properties of Food Macromolecules*, ed. J. R. Mitchell & D. A. Ledward. Elsevier Applied Science, London, 1986, p. 315.
126. Mitchell, J., In *Developments in Food Proteins—4*, ed. B. J. F. Hudson. Elsevier Applied Science, London, 1986, p. 291.
127. Hawthorne, J. R., *Biochim. Biophys. Acta*, **6** (1950) 28.
128. Cotterill, O. J. & Winter, A. R., *Poultry Sci.*, **34** (1955) 679.
129. Garibaldi, J. A., Donovan, J. W., Davis, J. G. & Cimino, S. L., *J. Food Sci.*, **33** (1968) 514.
130. Rhodes, M. B. & Feeney, R. E., *Poultry Sci.*, **36** (1957) 891.
131. Robinson, D. S. & Monsey, J. B., *J. Sci. Food Agric.*, **23** (1972) 893.
132. Nakai, S. & Kason, C. M., *Biochim. Biophys. Acta*, **351** (1974) 21.
133. Ehrenpreis, S. & Warner, R. C., *Arch. Biochem. Biophys.*, **61** (1956) 38.
134. Kato, A., Imoto, T. & Yagishita, K., *Agric. Biol. Chem.*, **39** (1975) 541.
135. Kato, A., Yoshida, K., Matsudomi, N. & Kobayashi, K., *Agric. Biol. Chem.*, **40** (1976) 2361.
136. Dam, R., *Poultry Sci.*, **50** (1971) 1824.
137. Johnson, T. M. & Zabik, M. E., *J. Food Sci.*, **46** (1981) 1226.

138. Johnson, T. M. & Zabik, M. E., *J. Food Sci.*, **46** (1981) 1231.
139. Johnson, T. M. & Zabik, M. E., *J. Food Sci.*, **46** (1981) 1237.
140. McLean, R. A. & Anderson, V. L., *Technometrics*, **8** (1966) 447.
141. MacDonnell, L. R., Feeney, R. E., Hanson, H. L., Campbell, A. & Sugihara, T. F., *Food Technol.*, **9** (1955) 49.
142. Poole, S., West, S. I. & Fry, J. C., *Food Hydrocolloids*, **1** (1987) 227.
143. Clark, D. C., Mackie, A. R., Smith, L. J. & Wilson, D. R., *Nihon Suisan Gakkai-shi*, **2**(3) (1988) 209.
144. Phillips, L. G., Yang, S. T., Schulman, W. & Kinsella, J. E., *J. Food Sci.*, **54**(3) (1989) 743.
145. Joseph, M. S. B. & Mangino, M. E., *Australian J. Dairy Technol.*, **43**(1) (1990) 9.
146. Feeney, R. E., Ducay, E. D., Silva, R. B. & MacDonnell, L. R., *Poultry Sci.*, **31** (1952) 639.
147. Taylor, C., Functional aspects of chemically modified bovine blood plasma and egg albumen proteins. PhD dissertation, University of Surrey, 1988.
148. Howell, N. K. & Taylor, C., *Int. J. Food Sci. Technol.*, **26**(4) (1991) 385.
149. Murphy, M. C. & Howell, N. K., *J. Sci. Food Agric.*, **55** (1991) 489.
150. Low, T. L. K., The amino acid sequence of porcine and bovine albumins. PhD dissertation, University of Texas, Texas, 1970.
151. Brown, J. R., *Fed. Proc. Am. Soc. Exp. Biol.*, **34** (1975) 591.
152. Cann, J. R., In *Methods of Protein Separation*, Vol. 1, ed. I. N. Catsimpoolas. Plenum Press, New York, 1975, p. 1.
153. Elfagm, A. A. & Wheelock, J. V., *J. Dairy Sci.*, **61** (1978) 28–32.
154. Nakashini, F., Takahashi, K. & Imagawa, T., *Rakuno-Kagaku, no. Kenkyu*, **17** (1968) A28.
155. Utsumi, S., Damodaran, S. & Kinsella, J., *J. Agric. Food Chem.*, **32** (1984) 1406.
156. Damodaran, S. & Kinsella, J. E., *Cereal Chem.*, **63** (1986) 381.
157. Shimada, K. & Cheftal, J. C., *J. Agric. Food Chem.*, **37**(1) (1990) 161.
158. Fields, S. & Song, O. K., *Nature UK*, **340**(6230) (1990) 245.

Chapter 3

INTERACTION OF FOOD PROTEINS WITH STARCH

Wayne E. Marshall & Joseph Chrastil

*US Department of Agriculture,
Southern Regional Research Center,
New Orleans, Lousiana, USA*

INTRODUCTION

Many food products, referred to in this chapter as food systems, are composed of protein and starch as their main constituents on a dry weight basis. Some are natural products, such as cereal grains (wheat, corn*), legumes (beans, peas) and tubers (potatoes, yams) which can be cooked and consumed in a minimally processed form. Others are fabricated products, such as bakery products, pasta, snack foods and breakfast cereals which are processed and consumed in a variety of shapes and sizes. Each of these food systems has its own characteristic texture, developed after cooking, which is, in part, due to the interaction between protein and starch. Since texture, along with flavor, are two major attributes of foods, protein–starch interactions are of considerable importance.

This chapter will be devoted to a description of food protein–starch interactions. Much of the literature, which describes protein–starch interactions, covers these interactions either in cereal grains or products derived from cereal grains. Therefore, our discussions will emphasize cereals, because most of the early research sought to understand food systems, such as bread, in terms of protein–starch interactions.

A protein–starch interaction we will not discuss involves the enzymatic synthesis or hydrolysis of starch. These are highly specific interactions between synthesizing and degrading enzymes (protein) and substrates (simple sugars or starch). Since enzymes are normally not considered

*In England, 'maize'.

typical food proteins, no further mention of this interaction will be made. For a comprehensive discussion of enzymes which synthesize or hydrolyze starch, the reader is directed to a review by Robyt.[1]

In this chapter, we wish to give the reader an understanding of protein–starch interactions, how they contribute to food texture, point out the gaps in current knowledge, and present ideas for future experimentation.

CHARACTERISTICS OF THE PROTEIN–STARCH INTERACTION

Location of Protein and Starch in Cereal Grains

In cereal grains, storage protein comprises about 80% of the total protein. Storage protein can be found either in specialized, spherical membrane bound protein bodies or packed in the cytosol of starchy endosperm cells as part of a continuous protein matrix.[2] The protein bodies are located in the seed endosperm in separate layers called aleurone and starchy endosperm.[2]

Starch (amylopectin+amylose) is constrained to spherical or polyhedral, membrane bound starch granules which are located in the starchy endosperm layers of the grain endosperm.[2] Some cereal cultivars contain 'waxy' starch granules which possess little if any amylose and amylose-associated lipid. The organization of amylopectin, amylose and associated lipid within the granule is only partially understood.[3] Even less understood is the organization of the starch granule membrane, which contains proteins, carbohydrates and lipids. 'Starch' as described in the literature is synonymous with starch granules because starch is usually isolated in intact granules unless extra steps are taken to rupture the granule membrane and extract the amylopectin and amylose inside.

Protein and Starch as Charged Colloids

The earliest studies on the molecular forces which drive food protein–starch interactions considered both proteins and starch as colloids which possess oppositely charged surfaces.[4,5] Proteins derive their surface charge from the amino acid composition of their primary structure. The ratio of basic amino acids (lysine, arginine, histidine) to acidic amino acids (glutamic and aspartic acids) establishes the isoelectric point of the protein and determines the surface charge at a specific pH.

For starch, the situation is more complex. Pure amylose and amylopectin do not possess charged groups and are neutral macromolecules.

Adjusting suspensions of pure amylose or amylopectin to very low or very high pH values causes dehydration and enolization of the glucose hydroxyl groups,[6] thereby imparting a negative charge to these macromolecules. However, in food systems, pH is rarely extreme. In cereal grains, starch granules contain lipid both within and on the surface of the granule membrane.[7] Much of this lipid is phospholipid[7] which carries a net negative charge. As an example, Marsh & Waight[8] determined that a suspension of wheat starch granules had an isoelectric point of 3·7, but the isoelectric point of a potato starch suspension could not be determined in the pH range 2–10. The origin of the granule surface charge was interpreted as arising from surface phospholipid and protein in wheat and surface phospholipid only in potato.[8] Due to the influence of phospholipid and to some degree protein, the starch granule surface will be negatively charged and behave as a negatively charged colloid under pH conditions found in most food systems.

Molecular Forces Involved in Protein–Starch Interactions
Yoshino & Matsumoto[4] conducted titration experiments with wheat proteins and wheat flour and found that wheat proteins had an equivalence point between pH 6 and 7, so the macromolecules behaved as positively charged colloids under acidic conditions. Colloid titration of wheat starch revealed little if any positive charge in the acid pH region. Although the potential existed for interaction between differently charged colloids, no protein–starch interaction was described.

Takeuchi[5] was one of the first to describe definitively the molecular forces involved in a food protein–starch system. Takeuchi used an electrolytic conductance method: potato starch was used as a titrant with a solution of α-casein and the titration conducted at pH 4·0. An equivalent volume of titrant was observed as a break in the titration curve (Fig. 1) and the value of the equivalent volume agreed with the equivalent volume determined by independent electrolytic conductance measurements on both protein and starch. From these observations, Takeuchi[5] concluded that the protein–starch interaction was a charge–charge phenomenon represented by positively charged protein and negatively charged starch.

Dahle[9] conducted a systematic study of food protein–starch interactions using a wheat protein–wheat starch system. Dahle[9] determined protein–starch binding by measuring the starch remaining in solution after precipitation of the protein–starch complex. He investigated the effect of pH and heat on the ability of protein to interact with starch as depicted in

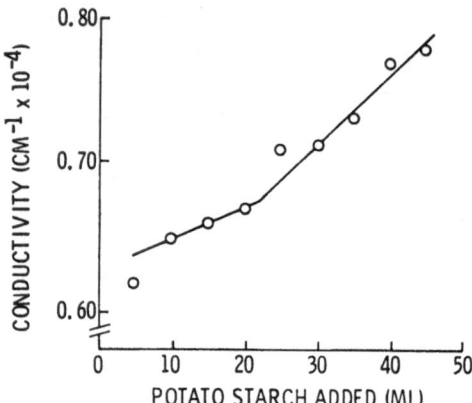

FIG. 1. Relationship between the electrolytic conductance of a mixed solution of α-casein and potato starch and the titration volume of a potato starch suspension. The point at which the slope changes (22 ml) is the equivalence point. The equivalence point defines the initiation of interaction between α-casein and starch. (From Takeuchi.[5])

Tables 1 and 2. Maximum protein–starch interaction occurred at pH 6·5. Binding was slightly less than maximum on the acidic side of pH 6·5 but much less than maximum on the basic side. The pH results can be explained by considering protein and starch to be oppositely charged colloids at acidic pH.[5] At more alkaline pH, the protein becomes less

TABLE 1

EFFECT OF pH ON ABILITY OF WHEAT PROTEIN TO INTERACT WITH STARCH[a]

pH	Absorbance (650 nm) of supernatant following centrifugation of wheat starch–wheat protein mixture	Degree of wheat starch–wheat protein interaction (%)
3·6	0·29	47
4·5	0·27	51
5·2	0·24	56
6·5	0·17	69
7·0	0·26	53
8·3	0·48	13
6·0 (no protein)	0·55	0

[a]Modified from Dahle.[9]

TABLE 2

EFFECT OF HEAT ON ABILITY OF WHEAT PROTEIN TO INTERACT WITH STARCH[a]

Treatment	pH	Absorbance (650 nm) of supernatant following centrifugation of wheat starch–wheat protein mixture	Degree of wheat starch– wheat protein interaction (%)
No protein	5·5	0·51	0
Heat denatured	5·5	0·52	0
Native	5·5	0·15	71
Heat denatured	4·0	0·53	0
Native	4·0	0·17	67

[a]Modified from Dahle.[9]

positively charged and less likely to complex with the negatively charged starch. These data also help explain the results of Hlynka & Chanin[10] who determined loaf volume of breads baked from doughs at various pH values. Optimum loaf volume occurred at pH 5·7 and diminished at lower pH. The smallest loaf volume was at pH 7·0. Dahle[9] found that heating the protein before exposure to starch had an inhibitory effect on the formation of a protein–starch complex (Table 2). The effect was independent of pH (Table 2). Denaturation and subsequent aggregation of the wheat protein effectively prevented starch interaction. These results may provide an explanation for the early observations of Geddes[11,12] who demonstrated loss of loaf volume in baked bread using flour which had been preheated at different moisture levels prior to dough formation. These examples point out the practical importance of Dahle's work and stress the significance of protein–starch interactions in breadmaking.

Dahle et al.[13] extended his earlier work[9] to determine whether sulfhydryl groups or disulfide bonds in the wheat proteins played a role in starch binding. Blocking sulfhydryl groups with N-ethylmaleimide had a minor effect on binding but disruption of disulfide bonds with dithiothreitol almost eliminated binding (Table 3). Apparently, retention of the native structure of wheat proteins is critical to formation of protein–starch complexes and these interactions do not require the involvement of sulfhydryl groups.

Moore & Carter,[14] using various food proteins and hydrolysed starch, indicated that several, predominantly non-covalent factors are required

TABLE 3

EFFECT OF *N*-ETHYLMALEIMIDE (NEM) AND DITHIOTHREITOL (DTT) ON BINDING
ACTION OF DILUTE ACETIC EXTRACT (NEUTRALIZED) OF WHEAT FLOUR WATER SOLUBLES[a]

Agent	Percent protein bound[b]			
	Wheat starch	Regular corn starch	Waxy maize corn starch	High amylose corn starch
None, control	28·5	30·0	30·5	32·0
NEM	20·5	25·0	22·0	26·5
DTT	0	8·0	0	4·5

[a]From Dahle *et al.*[13]
[b]Supernatant protein measured following centrifugation of protein–starch system.

to account for protein–starch interactions. They examined heated and
non-heated, protein–hydrolysed starch systems and developed two al-
ternate models for this interaction (Fig. 2). Both models (Figs 2a and 2b)
require protein aggregation, but one describes linear aggregation with

(a)

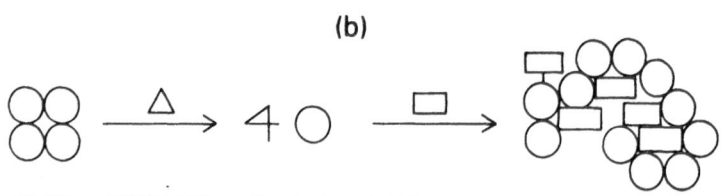

(b)

FIG. 2. (a) Model for linear aggregation of protein under thermal treatment.
Minimal carbohydrate (starch) attachment is depicted through covalent bonding.
(b) Schematic model for interaction of protein with carbohydrate (starch). The
arrows depict how a non-covalent aggregate consisting of protein subunits can
rearrange to form a large aggregate with the entrapment of carbohydrate (starch).
O = protein; □ = carbohydrate; △ = heat treatment. (From Moore &
Carter.[14])

minimal covalent attachment of carbohydrate and the other depicts considerably more aggregation with physical entrapment of carbohydrate within the protein aggregate. Using these models as a guide, Moore & Carter[14] believed that molecular forces, such as charge–charge and hydrogen bond interactions, would maintain the carbohydrate within the protein matrix, but physical entrapment, through protein aggregation, would provide the framework within which protein and starch could interact. Furthermore, disruption of protein aggregation by reduction of disulfide bonds could disrupt the entrapment mechanism and reduce or eliminate protein–starch interaction as observed by Dahle et al.[13] Moore & Carter[14] also found that starch hydrolysates exerted a protective effect against excessive protein aggregation, which would allow extensive precipitation of the protein. Imeson et al.[15] obtained similar results with animal proteins and anionic polysaccharides, where stable polysaccharide–protein complexes were formed following heat denaturation which inhibited extensive protein–protein aggregation.

THE INFLUENCE OF HEAT AND MOISTURE ON PROTEIN–STARCH INTERACTIONS

Heat and Excess Moisture

Protein–starch interactions appear to be of two types, depending on temperature. At low temperatures, such as room temperature, interactions between protein and starch are charge–charge (ionic) phenomena.[5,9] In unheated doughs, for example, the degree of interaction depends largely on the isoelectric point of the protein. Therefore, the interaction is pH dependent.[13]

The situation appears more complex when protein–starch systems are heated. Heating is usually necessary to develop the final texture of a food system. Under conditions where moisture is not limiting, such as the cooking of rice or oats, high temperatures denature protein. As a result of denaturation, proteins can undergo extensive cross-linking, particularly through disulfide bond formation, thus forming a continuous protein network.

For starch, high temperatures cause gelatinization. During the gelatinization process, water disrupts starch crystallinity, the granules swell, amylose diffuses out of the granules leaving mostly amylopectin, and the granules eventually collapse and are held in a matrix of amylose as part of a gel network.[16] When protein and starch are in contact with each other as in cereal grains or products made from cereals, stable complexes can develop by formation of a protein–starch matrix, where hydrogen bonding and covalent bonding, in addition to charge–charge interactions may be found.[14] These complexes, formed by the application of heat, likely contribute to the formation of texture in the food system and, therefore, should be studied in greater detail.

Heat and Low Moisture

Many food products with low water content but a high percentage of protein and starch are fabricated foods, such as pasta, snack foods and breakfast cereals.[17] Fabricated foods typically are exposed to high heat conditions during processing.[17] A process which is becoming very popular in the food industry is extrusion. Extruders combine high heat and low moisture with high shear and pressure to produce a variety of textures. The water content of the material to be extruded is usually between 15 and 40%. At these moisture levels, proteins denature and starch gelatinizes at higher temperatures than in high moisture systems.[18,19] Therefore, the cooking temperature becomes critical, depending on the exact moisture content of the food system.

The fate of protein under extrusion conditions is poorly understood. There is evidence to suggest that the proteins should have the correct charge, size, and shape to align themselves properly during their time of residence in the extruder, so that after the extruded mass has cooled, the protein molecules being in the correct alignment, stabilizing interactions can take place and stabilize the final product.[20]

The extrusion of starch has recently been reviewed by Lai & Kokini.[21] They point out that high shear forces in the extruder play a major role in determining the properties of the extruded product. In the production of extruded breakfast cereals and snack foods, product texture is largely determined by shear forces tearing apart starch granules[22] and mechanically disrupting the molecular forces which help maintain starch crystallinity (Fig. 3).[23] Since moisture is usually limited during extrusion, starch can exist in both melted and gelatinized states and the amylopectin/amylose polymers can be reduced to dextrinized chain fragments by a combination of high heat, high shear and low moisture (Fig. 3).[24] The final texture of extruded starch products is due to a combination of heat, moisture, shear and pressure, all working in concert with each other.

Thus far no satisfactory model has been developed to predict the final texture of extruded protein or starch products under a given set of experimental conditions. Therefore, not surprisingly, information on protein–

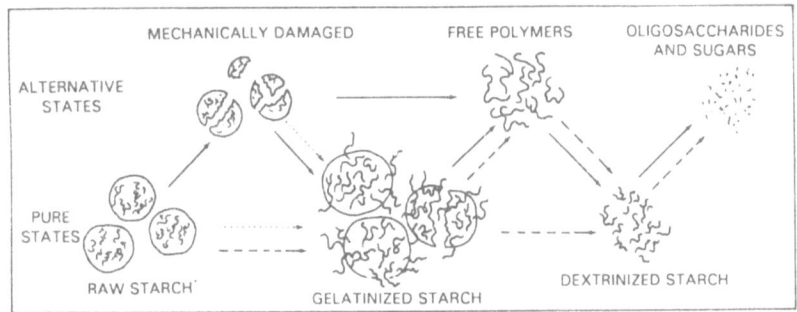

FIG. 3. Proposed model of starch degradation during extrusion. → = shear forces; ····> = heat; ·····> = moisture. (From Gomez & Aguilera.[24])

starch interactions during extrusion processing and their effect on product texture is scarce. Hauck[25] and Meuser et al.[26] have added protein to starch material and noted improvements in texture and expansion volume of the extruded product. However, much remains to be investigated. A description of protein–starch interactions during extrusion is needed and the molecular basis of texture in extruded protein–starch systems requires attention. Considering the strong current interest in extrusion processing, this would appear to be an important and fertile field of research.

ROLE OF THE STARCH GRANULE IN PROTEIN–STARCH INTERACTIONS

Importance of the Starch Granule to Protein–Starch Interactions in Bread Baking

Harris[27] was one of the first investigators to notice the importance of starch in bread baking when he observed that starches of different wheat varieties performed differently in bake tests. These observations were expanded by Sandstedt[28] who used rice, corn and potato starch in addition to other wheat starches to show that when mixed with gluten, rice and corn starch made bread of inadequate volume and texture while wheat and potato starch produced bread of adequate volume and texture. Apparently, the starches differ in physicochemical properties to the extent that when combined with gluten they help produce a range of loaf volumes and loaf textures.

The two most obvious differences in starches of various plant sources is granule size and gelatinization temperature. Unfortunately, there appears to be no consensus in the literature regarding the relationship between granule size and bread baking performance. Kulp[29] observed that small wheat starch granules produced doughs which had varying degrees of protein–starch binding but produced breads of inferior baking qualities compared with regular starch. In partial agreement with Kulp's[29] observations, Soulaka & Morrison[30] determined that there was an optimum percentage of small wheat starch granules that could be tolerated before loaf volume decreased. Lelievre et al.[31] discovered a complex relationship between granule size and bread quality. Granule size was an important quality factor but gluten concentration was also essential. These two elements acted together in a complex manner to define bread-making performance. A balance of granule size and gluten content appeared to be necessary for optimum bread quality.

In contrast to the above observations, D'Appolonia et al.[32] reported the same loaf volumes of breads baked with large or small starch granules. Hoseney et al.[33] examined both granule size and gelatinization temperature using potato starch and a variety of cereal starches in a breadmaking system in which changes in water absorption and loaf volume were monitored (Table 4). No definitive correlations could be made between baking properties, starch granule size or gelatinization temperature. The authors concluded that neither granule size nor gelatinization temperature could account for bread baking characteristics of the starch and that some unidentified starch property was involved. In reference to this conclusion, Bushuk[34] commented that wheat starch granules may have unique surfaces which interact optimally with gluten. Since Hoseney et al.[33] also found barley, rye and small-granule wheat starch to yield water absorption and loaf volume values similar to the prime wheat starch control (Table 4), these granules may have similar, gluten-binding, surface properties although they are from disparate sources. These surface properties, at least for wheat starch, could reside in the 'adhering matter' which surrounds the granules.[34] Adhering matter is a complex of protein, carbohydrate and lipid[35] which is bound to the granule surface in an unknown manner.

Wehrli & Pomeranz[36] examined protein–starch interactions in unbaked bread dough and in the baked product by an autoradiographic technique. They used tritium-labelled glycolipid and determined the location of the label with respect to gluten and starch granules. In dough, protein and starch appear to interact through the glycolipid but in baked

TABLE 4

BAKING DATA AND STARCH GRANULE CHARACTERISTICS FOR RECONSTITUTED FLOURS
CONTAINING GLUTEN AND WATER-SOLUBLES FROM WHEAT AND STARCH FROM VARIOUS
SOURCES[a]

Source of starch	Water absorption (%)	Loaf volume (cm³)	50% gelat. (°C)	Starch granule size
Wheat (control)	61	80	55	Small/large
Corn	75	48	68	Medium
Milo	68	54	67	Medium
Oat	70	58	55	Medium
Barley	61	78	60	Medium
Rye	61	77	51	Large
Rice	73	68	71	Small
Potato	73	60	55	Large
Small-granule Wheat	68	77	56	Small

[a]Modified from Hoseney et al.[33]

bread, a complex between gelatinized starch and glycolipid was observed. Starch–protein interactions may be more involved in the early stages of breadmaking than in maintaining the texture of the finished product.

Definitive studies have yet to be done to determine which components of the starch granule surface are responsible for specific bread baking characteristics such as loaf volume, water absorption and crumb texture.

Importance of the Starch Granule in Protein–Starch Interactions to the Development of Quality Factors and Functional Properties in Wheat

In wheat, starch granules are embedded in a protein matrix within the endosperm tissue. At the protein–starch interface or starch granule surface is water soluble material consisting of protein and a carbohydrate portion which yields only glucose upon hydrolysis.[37] This protein–starch complex appears to act as adhesive material which anchors the granules to the protein matrix[37] and has been implicated as the basis for grain hardness in wheat.[38] Furthermore, the amount of water soluble material extracted from the starch granule surface has been shown to be a function of wheat hardness. These were some of the first studies to relate starch–protein interactions to a functional property in a cereal grain.

Although Simmonds et al.[38] associated a protein–starch complex with grain hardness, they did not identify any specific protein(s) responsible for differences in hardness. Greenwell & Schofield[39] identified a 15 kDa protein present in soft wheat varieties which could barely be detected in hard wheat. They inferred that this protein played an important role in conferring endosperm softness in wheat. Since the protein is associated with the surface of the starch granule, its presence may modulate the adhesiveness of the protein–starch complex identified by Barlow et al.,[37] and thereby help control wheat softness. Evidence for the direct involvement of the 15 kDa protein in the control of wheat softness was reported by Anjum & Walker[40] in a recent review. The authors alluded to results obtained by using a reconstituted tablet technique in which protein, starch, water and the water-soluble fraction from wheat can be combined, compressed into tablet form and tensile strength measured on the tablet. When the 15 kDa protein was removed from wheat starch granules, the reconstituted tablets exhibited high tensile strengths which could be correlated with increased grain hardness. Anjum & Walker[40] noted that these results were a direct indication that the 15 kDa protein imparts a 'softness' character to the wheat kernel.

The proteins associated with the surface of the wheat starch granule exist as both water and salt extractable proteins. Lowy et al.[41] determined that only 8% of the total protein associated with the starch granule was salt extractable, and identified the major salt extractable protein as a basic 30 kDa protein with an isoelectric point in excess of 10. The function of this protein was not determined but Lowy et al.[41] speculated that it might bind to and alter the surface charge of the starch granule, thus modifying the dispersion characteristics of the starch.

Seguchi[42–44] studied the relationship between the surface of the wheat starch granule and its oil-binding properties. He reported that treatment of granules with proteases or amylases diminished oil binding (Table 5).[42] Therefore, a protein–starch complex at the starch granule surface was involved in this important functional property. Bleaching (chlorination) of the wheat flour also improved oil-binding performance[42] by its effect on the protein–starch complex. Bleaching was also found to improve the oil-binding capacity of rice, potato and corn starches (Table 5).[43] Seguchi[44] also observed that heat treatment of non-chlorinated wheat starch improved oil-binding performance which was lost upon exposure of the starch to a protease or an amylase. Seguchi[44] speculated that the starch granule surface was changed from a hydrophilic to a lipophilic surface by modification of the protein portion of the protein–starch complex. Perhaps the various treatments unfolded the protein to expose more hydrophobic (lipophilic) amino acid side-chains on the protein surface.

TABLE 5

EFFECTS OF PROTEOLYTIC AND SACCHAROLYTIC TREATMENTS ON OIL BINDING CAPACITY OF VARIOUS STARCHES[a]

Treatment	Relative oil-binding capacity of chlorinated starch (%)			
	Wheat	Potato	Corn	Rice
None	100 (1·4)[b]	100 (1·0)	100 (9·0)	100 (1·8)
α-Amylase	14 (0·2)	20 (0·2)	17 (1·5)	17 (0·3)
Pepsin	0 (0·0)	29 (0·3)	83 (7·5)	100 (1·8)

[a]Modified from Seguchi.[43]
[b]Milliliters of oil/g of starch.

More recently, Seguchi[45] used a protein dye-binding procedure to observe a protein film on the surface of wheat starch granules. He correlated the disappearance of the film with swelling of the granules and gelatinization of the starch. The film appeared to control the permeation of water into the granule, thus controlling swelling and eventual gelatinization.

Eliasson & Tjerneld[46] examined the adsorption of wheat proteins on starch granules from several different food crops. They wanted to determine the extent to which pH, ionic strength, temperature and size of the starch granule surface affected adsorption. Kulp & Lorenz[47] had suggested that the surface of the wheat starch granule might be important for protein–starch interactions during dough formation and baking performance. Eliasson & Tjerneld[46] found that increasing the pH from 3·1 to 7·6 increased adsorption. Adsorption reached a maximum at a NaCl concentration of 0·0025 M, then declined at higher levels, due primarily to protein precipitation. Hruskova & Hampl[48,49] also found that NaCl had a complex effect on wheat protein–wheat starch interactions, possibly due to alteration of surface charge on the starch granule. Heating the starch granules also increased adsorption.[46] Wheat proteins were adsorbed to a much greater extent on potato starch than on wheat or corn starch (Fig. 4).[46] Wheat proteins may form multiple layers on potato starch as opposed to single layers on other starches, possibly due to a stronger negative surface charge on the potato starch.

D'Egidio et al.[50] noted that protein–starch interactions were important in determining semolina quality of durum wheat. They noticed a direct relationship between the interaction of starch and gliadins isolated from

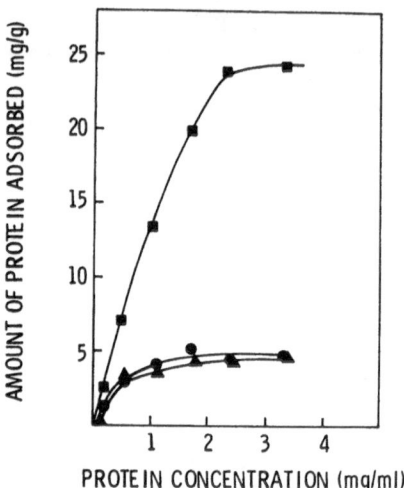

FIG. 4. Adsorption of non-heated wheat protein on granules of different starches. ● = wheat starch; ▲ = maize starch; ■ = potato starch. (From Eliasson & Tjerneld.[46])

different semolina samples and the quality of pasta produced from these samples.

The interaction of starch and protein at the surface of the wheat starch granule is undoubtedly important to the function of the granule in imparting specific functional properties to the food system. Important wheat properties, such as baking performance and hardness, appear to depend on protein–starch interactions at the granule surface. The literature explored in this section yields information about the relationship between these interactions and their effect on the functional properties of food systems. Further information is needed to develop a predictive model which will allow the food processor to alter the protein–starch complex on the granule in a well-defined manner to obtain the proper functionality desired.

Importance of the Starch Granule and Protein–Starch Interactions in Other Cereals

Wheat starch is not unique among the cereals in having protein associated with the granule surface in which the protein–starch complex may influence the functional properties of the particular cereal grain. Protein has been found in connection with sorghum[51] and rice[52] starch granules. In both cases, a correlation was made between the presence of protein and the functional properties of the grain.

Grain Sorghum

Chandrashekar & Kirleis[51] determined that the presence of protein surrounding the starch granules in sorghum influenced water uptake and starch gelatinization. Starch granules in part of the sorghum endosperm are surrounded by protein bodies. This protein could act as a barrier to swelling of the starch granule in the presence of water, thus inhibiting both water absorption and gelatinization. The barrier concept was strengthened by the observation that 2-mercaptoethanol, a disulfide bond reducing agent, added to sorghum flour, increased water uptake and gelatinization. Apparently, the starch granules are covered by a protein network which is loosened after the reduction of disulfide bonds.

Rice

Chrastil[53] investigated the interaction between purified preparations of the major rice storage protein, oryzenin, and purified starch from medium- and long-grain varieties of rice at two storage temperatures (4 and 40°C). Oryzenin was found to be reversibly bound to starch. However, the manner in which this binding occurred was not explored. The molar binding ratio (O/S; *O*ryzenin/*S*tarch) was a linear function of oryzenin molecular weight (M_w) where M_w was dependent on the number of disulfide bonds present in the protein. Both the M_w of oryzenin and/or O/S were related to swelling, water uptake and cooking time of rice grains and dough leavening of rice flour doughs (Chrastil, unpublished observations). O/S decreased linearly as M_w increased. Chrastil[53] was able to directly relate the interaction between oryzenin and starch, from the change in absorbance pattern of the protein, to stickiness of both long- and medium-grain rice kernels (Fig. 5). The different kernel samples were cooked and stickiness of cooked rice grains were determined by the rice cluster distribution method.[54] Chrastil[53] also observed that storage of rice at high temperatures (40°C) decreased O/S and stickiness and increased kernel swelling, cooking time and dough leavening. Changes in rice functional properties were related to increased M_w and increased hydrophobicity of oryzenin. Changes in starch structure during storage were very small.

Hamaker *et al.*[52] investigated the influence of a 60 kDa starch granule-associated protein on stickiness in rice. The 60 kDa protein has been shown by Villareal & Juliano[55] and Goldner & Boyer[56] to be present in

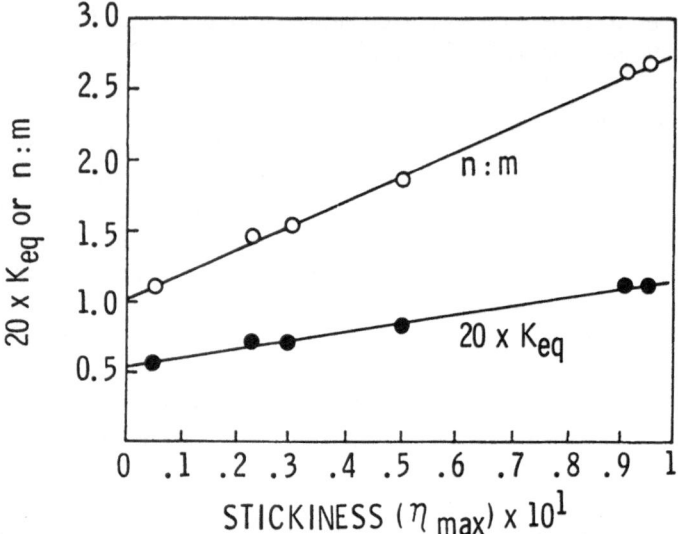

FIG. 5. Relationship between oryzenin:starch (n:m) binding and stickiness of cooked medium- and long-grain rice. K_{eq} can be calculated from the binding ratio (n:m) and the change in absorbance of oryzenin due to the binding of starch. To accommodate the y-axis, K_{eq} was multiplied by a factor of 20. The three largest and smallest values of n:m (K_{eq}) correspond to medium- and long-grain rice, respectively, stored at three different temperatures. (From Chrastil.[53])

non-waxy but virtually absent in waxy rice varieties. Hamaker & Griffin[57] and Hamaker et al.[52] observed a significant negative correlation between rice kernel stickiness after cooking and amount of the 60 kDa protein (Table 6). Also, this specific protein may be implicated in the increase of cooked rice stickiness following treatment of the kernels with a reducing agent.[57] Hamaker et al.[52] speculated that the 60 kDa protein forms a polymeric matrix held together by disulfide bonds which confers rigidity to the starch granule and inhibits maximum swelling. The lack of significant swelling is related to a diminished stickiness seen in the cooked product.

Additionally, flour from kernels exposed to reducing agent had lower paste viscosity. In isolated rice starch, reducing agents had no effect on the paste viscosity. These observations indicate that protein is required for changes to occur in certain viscoelastic properties of rice. The involvement of protein–starch interactions in these viscoelastic properties is not clear at this time. In addition, Hamaker & Ball[58] have evidence for the

TABLE 6
CORRELATION COEFFICIENTS FOR SELECTED PARAMETERS
POTENTIALLY INVOLVED IN COOKED RICE TEXTURE[a]

Parameter	Stickiness	60 kDa protein
60 kDa protein	−0·848[b]	—
Amylose	−0·868[b]	0·954[b]
Total protein	−0·357	0·363
L/W ratio[c]	−0·231	0·145

[a]From Hamaker et al.[52]
[b]Denotes significance at $P < 0.01$.
[c]Length-width ratio of milled rice kernels.

existence of a strong interaction between rice endosperm proteins and the amylopectin moiety of starch. Whether these interactions directly affect rice quality remains to be determined.

The effect of proteins on starch gelatinization in whole grain milled rice was investigated by Marshall et al.[59] Evidence for protein–starch interactions as they influence starch gelatinization could be important in determining the quality of cooked rice. Marshall et al.[59] used a proteolytic enzyme to treat rice kernels before subjecting them to simulated cooking in a differential scanning calorimeter. Pronounced changes in starch gelatinization were observed, especially a large decrease in the peak gelatinization temperature, when 53–69% of the total protein was removed from the kernels before simulated cooking. However, most of the change in starch gelatinization could be attributed to disruption of kernel structure that occurred upon enzyme treatment. Removal of rice protein did appear to produce a minor but measurable, decrease in the gelatinization temperature. Takahashi et al.[60] found that the addition of soy protein to various tuber, cereal and legume starches produced a small increase in gelatinization temperature but had no effect on the extent to which the starches were gelatinized. These results and the observations of Marshall et al.[59] are consistent in that addition/removal of protein in a protein–starch system induces an increase/decrease in starch gelatinization temperature.

The study of protein–starch interactions in rice and their effect on rice quality is an important and developing area of rice research. Results obtained thus far indicate that protein–starch interactions do occur, but more information is needed to quantify their involvement in determining rice quality.

PROTEIN–STARCH INTERACTIONS IN STARCH GELS

As noted earlier, starch granules swell in the presence of moisture and heat. The starch within the granules gelatinizes to form a gel structure composed of swollen starch granules dispersed in a continuous polysaccharide matrix.[61] To study the influence of protein on starch gel texture, Lindahl & Eliasson[62] and Ojima et al.[63] examined changes in viscoelastic properties of various starch gels in the presence of wheat and soybean proteins, respectively. Interactions between wheat protein and starch gels were observed for wheat, rye and maize but not for potato, barley and triticale.[62] Soy protein interacted with potato and corn starch.[63] Both Lindahl & Eliasson[62] and Ojima[63] felt these results were due to the surface properties of the different starch granules. In the case of wheat and rye starch, the wheat proteins could bind easily to the granule surface and promote a strong network formation of starch granules embedded in a wheat protein matrix. Where the viscoelastic properties of the starch gel was not changed (potato, barley, triticale) in the presence of protein, the granule surface was such that little if any protein–starch binding occurred.[62] In contrast, Ojima et al.[63] observed a decrease in the gel strength of potato starch in the presence of soy protein, while corn starch gels showed opposite effects. Additionally, a soy protein–corn starch gel exhibited remarkable stability when cooked to 120°C and may make this gel useful as part of a food system cooked to high temperatures. Lindahl & Eliasson[62] pointed out the need to characterize the surface properties of different starch granules from different species, with special attention given to the presence of specific groups, particularly charged groups and lipids.

Hermansson[64] developed protein–starch complexes by heating mixtures of different starches with casein or caseinates. Depending on the ratio of starch to protein, reaction time, reaction temperature, and the ions in the reaction mixture, the physical properties (viscosity, gumminess, stickiness) of the starch were modified. The addition of protein improved the emulsion stability and fat holding properties of the starch and the complex could then be used in food applications not normally reserved for starch. Complex formation occurred by interaction of the protein with the starch granule components, amylose and amylopectin which leached out during swelling and heating of the granule. Hermansson[64] concluded that the starch granule surface apparently played a minor role in the interaction and that the forces involved in complex formation and retention were electrostatic forces, hydrophobic bonds, hydrogen bonds, and van der Waal's forces.

CONCLUSION

This chapter has described protein–starch interactions with special emphasis on understanding these interactions in complex food systems. Additional research on this topic remains to be accomplished. As Pomeranz & Chung[65] pointed out in their review of protein–starch–lipid interactions in cereal systems, these interactions are important in the processing of cereals but they have not been studied extensively. Difficulties have included the complexity of cereal systems, limited knowledge of how these components might interact, and the scarcity of methods to quantify the interactions.

In our opinion, future work in elucidating protein–starch interactions is needed in two general areas:

First, the molecular events which occur during development of a food system need to be described in greater detail. Protein–starch interactions would represent a part of the total system, but interactions among protein, starch, lipid and water must be considered, especially the effect of heat on these interactions. All food systems have characteristic functional properties and an enhanced understanding of food component interactions is required to explain the molecular basis for food functionality.

Second, analytical techniques must be further developed to describe the molecular events which occur during protein–starch interactions in food systems. These techniques should strive to describe these molecular events *in situ* rather than after isolation and purification of individual components. If an *in situ* description is attempted, instrumentation will be required which can monitor protein–starch interactions in food systems where optical clarity will be a serious problem. Spectra from such analytical methods as thermal analytical techniques, nuclear magnetic resonance spectroscopy and Fourier-transform infrared and Raman spectroscopies are largely unaffected by the optical properties of the sample. Also, most of these methods are non-destructive to the system under study. Therefore, they may be useful in determining *in situ* changes which occur in food systems, especially before, during and after heat treatment.

REFERENCES

1. Robyt, J. F., Enzymes in the hydrolysis and synthesis of starch. In *Starch: Chemistry and Technology*, 2nd Edition, ed. R. L. Whistler, J. N. BeMiller & E. F. Pashall. Academic Press, New York, 1984, pp. 87–123.

2. Lorenz, K. J. & Kulp, K., *Handbook of Cereal Science and Technology.* Marcel Dekker, New York, 1991.
3. Manners, D. J., Some aspects of the structure of starch. *Cereal Foods World,* **30** (1985) 461–467.
4. Yoshino, D. & Matsumoto, H., Colloid titration of wheat proteins, dough, and flour. *Cereal Chem.,* **43** (1966) 187–195.
5. Takeuchi, I., Interaction between protein and starch. *Cereal Chem.,* **46** (1969) 570–579.
6. Hodge, J. E. & Osman, E. M., Carbohydrates. In *Principles of Food Science. Part I. Food Chemistry,* ed. O. R. Fennema. Marcel Dekker, New York, 1976, pp. 41–138.
7. Morrison, W. R., Lipids in cereal starches: A review. *J. Cereal Sci.,* **8** (1988) 1–15.
8. Marsh, R. A. & Waight, S. G., The effect of pH on the zeta potential of wheat and potato starch. *Starch/Stärke,* **34** (1982) 149–152.
9. Dahle, L. K., Wheat protein–starch interaction. I. Some starch binding effects of wheat–flour proteins. *Cereal Chem.,* **48** (1971) 706–714.
10. Hlynka, I. & Chanin, W. G., Effect of pH on bromated and unbromated doughs. *Cereal Chem.,* **34** (1957) 371–378.
11. Geddes, W. F., Chemical and physico-chemical changes induced in wheat and wheat products by elevated temperatures—I. *Can. J. Res.,* **1** (1929) 528–557.
12. Geddes, W. F., Chemical and physico-chemical changes induced in wheat and wheat products by elevated temperatures—II. *Can. J. Res.,* **2** (1930) 65–90.
13. Dahle, L. K., Montgomery, E. P. & Brusco, V. W., Wheat protein–starch interaction. II. Comparative abilities of wheat and soy proteins to bind starch. *Cereal Chem.,* **52** (1975) 212–225.
14. Moore, W. E. & Carter, J. L., Protein–carbohydrate interactions at elevated temperatures. *J. Text. Stud.,* **5** (1974) 77–88.
15. Imeson, A. P., Ledward, D. A. & Mitchell, J. R., On the nature of the interaction between some anionic polysaccharides and proteins. *J. Sci. Food Agric.,* **28** (1977) 661–668.
16. Remsen, C. H. & Clark, J. P., A viscosity model for a cooking dough. *J. Food Process. Eng.,* **2** (1978) 39–64.
17. Matz, S. A., *The Chemistry and Technology of Cereals as Food and Feed,* 2nd Edition. Van Nostrand Reinhold, New York, 1991.
18. Kitabatake, N., Tahara, M. & Doi, E., Denaturation temperature of soy protein under low moisture conditions. *Agric. Biol. Chem.,* **53** (1989) 1201–1202.
19. Donovan, J. W., Phase transitions of starch–water system. *Biopolymers,* **18** (1979) 263–275.
20. Ledward, D.A. & Mitchell, J. R., Protein extrusion—more questions than answers? In *Food Structure—Its Creation and Evaluation,* ed. J. M. V. Blanshard & J. R. Mitchell. Butterworths, London, 1988, pp. 219–229.
21. Lai, L. S. & Kokini, J. L., Physicochemical changes and rheological properties of starch during extrusion (A review). *Biotechnol. Prog.,* **7** (1991) 251–266.

22. Burros, B. C., Young, L. A. & Carroad, P. A., Kinetics of corn meal gelatinization at high temperature and low moisture. *J. Food Sci.*, **52** (1987) 1372–1376.
23. Wen, L. F., Rodis, P. & Wasserman, B. P., Starch fragmentation and protein insolubilization during twin-screw extrusion of corn meal. *Cereal Chem.*, **67** (1990) 268–275.
24. Gomez, M. H. & Aguilera, J. M., A physicochemical model of corn starch. *J. Food Sci.*, **49** (1984) 40–43.
25. Hauck, B. W., Process variables and their control for the production of expanded products by extrusion cooking. Paper presented at 66th American Association of Cereal Chemists Meeting (Chicago, Illinois, 15–19 October 1981).
26. Meuser, F., Van Lengerich, B., Pfaller, W. & Harmuth-Hoene, A. E., The influence of HTST-extrusion cooking on the protein nutritional value of cereal based products. In *Extrusion Technology for the Food Industry, Part II. Aspects of Technology*, ed. P. Colonna. Elsevier Applied Science, London, 1987, pp. 35–53.
27. Harris, R. H., The baking qualities of gluten and starch prepared from different wheat varieties. *Baker's Dig.*, **16** (1942) 217–222, 230.
28. Sandsted, R. M., The function of starch in the baking of bread. *Baker's Dig.*, **35** (1961) 36–42, 44.
29. Kulp, K., Characteristics of small granule starch of flour and wheat. *Cereal Chem.*, **50** (1973) 666–679.
30. Soulaka, A. B. & Morrison, W. R., The bread baking quality of six wheat starches differing in composition and physical properties. *J. Sci. Food Agric.*, **36** (1985) 719–727.
31. Lelievre, J., Lorenz, K., Meredith, P. & Baruch, D. W., Effects of starch particle size and protein concentration on breadmaking performance. *Starch/Stärke*, **39** (1987) 347–352.
32. D'Appolonia, B. L. & Gilles, K. A., Effect of various starches in baking. *Cereal Chem.*, **48** (1971) 625–636.
33. Hoseney, R. C., Finney, K. F., Pomeranz, Y. & Shogren, M. D. Functional (breadmaking) and biochemical properties of wheat flour components. VIII. Starch. *Cereal Chem.*, **48** (1971) 191–201.
34. Bushuk, W., Protein-lipid and protein-carbohydrate interactions in flour–water mixtures. In *Chemistry and Physics of Baking*, ed. J. M. V. Blanshard, P. G. Frazier & T. Galliard. Royal Society of Chemistry, London, 1986, pp. 147–154.
35. Kulp, K., Ke, T.-L. & Chiou, H.-Y., Composition of adhering matter to wheat starch. Paper presented at 64th American Association of Cereal Chemists Meeting, Washington, D.C., 28 October–1 November, 1979.
36. Wehrli, H. P. & Pomeranz, Y., A note on autoradiography of tritium-labeled galactolipids in dough and bread. *Cereal Chem.*, **47** (1970) 221–224.
37. Barlow, K. K., Buttrose, M. S., Simmonds, D. H. & Vesk, M., The nature of the starch–protein interface in wheat endosperm. *Cereal Chem.*, **50**, (1973) 443–454.
38. Simmonds, D. H., Barlow, K. K. & Wrigley, C. W., The biochemical basis of grain hardness in wheat. *Cereal Chem.*, **50** (1973) 553–562.

39. Greenwell, P. & Schofield, J. D., A starch granule protein associated with endosperm softness in wheat. *Cereal Chem.*, **63** (1986) 379–380.
40. Anjum, F. M. & Walker, C. E., Review on the significance of starch and protein to wheat kernel hardness. *J. Sci. Food Agric.*, **56** (1991) 1–13.
41. Lowy, G. D. A., Sargeant, J. G. & Schofield, J. D., Wheat starch granule protein: the isolation and characterisation of a salt-extractable protein from starch granules. *J. Sci. Food Agric.*, **32** (1981) 371–377.
42. Seguchi, M., Oil-binding capacity of prime starch from chlorinated wheat flour. *Cereal Chem.*, **61** (1984) 241–244.
43. Seguchi, M., Comparison of oil-binding ability of different chlorinated starches. *Cereal Chem.*, **61** (1984) 244–247.
44. Seguchi, M., Oil-binding ability of heat-treated wheat starch. *Cereal Chem.*, **61** (1984) 248–250.
45. Seguchi, M., Dye binding to the surface of wheat starch granules. *Cereal Chem.*, **63** (1986) 518–520.
46. Eliasson, A.-C. & Tjerneld, E., Adsorption of wheat proteins on wheat starch granules. *Cereal Chem.*, **67** (1990) 366–372.
47. Kulp, K. & Lorenz, K. J., Starch functionality in white pan breads—new developments. *Baker's Dig.*, **55** (1981) 24–25, 27–28, 36.
48. Hruskova, M. & Hampl, J., Effect of NaCl on technological properties of dough. V. Protein/starch interaction. *Mlynsko-Pekarensky Prumysl*, **25**, (1979) 313–317.
49. Hruskova, M. & Hampl, J., Effect of sodium chloride on interactions of cereal proteins and starch. *Sbornik Vysoke Skoly Chemicko-Technologicke v Praze*, **54** (1982) 183–199.
50. D'Egidio, M. G., de Stefanis, E., Fortin, S., Nardi, S. & Sgrulletta, D., Interaction between starch and a specific protein fraction extracted from semolina of *T. durum*. *Can. J. Plant Sci.*, **64** (1984) 785–796.
51. Chandrashekar, A. & Kirleis, A. W., Influence of protein on starch gelatinization in sorghum. *Cereal Chem.*, **65** (1988) 457–462.
52. Hamaker, B. R., Griffin, V. K. & Moldenhauer, K. A. K., Potential influence of a starch granule-associated protein on cooked rice stickiness. *J. Food Sci.*, **56** (1991) 1327–1329.
53. Chrastil, J., Protein-starch interactions in rice grains. Influence of storage on oryzenin and starch. *J. Agric. Food Chem.*, **38** (1990) 1804–1809.
54. Chrastil, J., Chemical and physicochemical changes of rice during storage at different temperatures. *J. Cereal Sci.*, **11** (1990) 71–85.
55. Villareal, C. P. & Juliano, B. O., Waxy gene factor and residual protein of rice starch granules. *Starch/Stärke*, **38** (1986) 118–119.
56. Goldner, W., R. & Boyer, C. D., Starch granule-bound proteins and polypeptides: The influence of the waxy mutations. *Starch/Stärke*, **41** (1989) 250–254.
57. Hamaker, B. R. & Griffin, V. K., Changing the viscoelastic properties of cooked rice through protein disruption. *Cereal Chem.*, **67** (1990) 261–264.
58. Hamaker, B. R. & Ball, P. H., Starch-protein interaction in rice endosperm. Paper presented at 75th American Association of Cereal Chemists Meeting, Dallas, Texas, 14–18 October 1990.

59. Marshall, W. E., Normand, F. L. & Goynes, W. R., Effects of lipid and protein removal on starch gelatinization in whole grain milled rice. *Cereal Chem.*, **67** (1990) 458–463.
60. Takahashi, S., Kobayashi, R., Watanabe, T. & Kainuma, K., Effects of addition of soybean protein on gelatinization and retrogradation of starch. *Nippon Shokuhin Kogyo Gakkaishi*, **30** (1983) 276–282.
61. Eliasson, A.-C. & Bohlin, L., Rheological properties of concentrated wheat starch gels. *Starch/Stärke*, **34** (1982) 267–271.
62. Lindahl, L. & Eliasson, A.-C., Effects of wheat proteins on the viscoelastic properties of starch gels. *J. Sci. Food Agric.*, **37** (1986) 1125–1132.
63. Ojima, T., Yamaura, I. & Kumagaya, T., Effect of food components on physical properties of starch. IV. Effects of soy protein and oil addition on starch gel strength. *Denpun Kagaku*, **33** (1986) 183–190.
64. Hermansson, A.-M. I., Protein/starch complex. U.S., Patent 4,159,982, 3 July 1979.
65. Pomeranz, Y. & Chung, O. K., Starch–lipid–protein interactions in cereal systems. In *Utilization of Protein Resources*, ed. D. W. Stanley, E. D. Murray & D. H. Lees. Food and Nutrition Press, Westport, 1981, pp. 129–157.

Chapter 4

THE MAILLARD REACTION

JENNIFER M. AMES

Department of Food Science and Technology, University of Reading, Whiteknights, Reading RG6 2AP, UK

INTRODUCTION

The Maillard reaction is a type of non-enzymic browning which involves the reaction of carbonyl compounds, especially reducing sugars, with compounds which possess a free amino group, such as amino acids, amines and proteins. In most foods, the ε-amino groups of the lysine residues of proteins are the most important source of free amino groups, and the ease with which they take part in the reaction explains why the Maillard reac tlon is the most important route to nutritional damage of food proteins.[1,2] The Maillard reaction in fact comprises a complex network of intertwining reactions and takes place during food processing, especially when heat treatment is involved, and also on storage. Apart from resulting in nutritional damage, the Maillard reaction is also primarily responsible for the development of aroma and colour, which may be desirable or undesirable, in heated foods. It also results in the formation of potentially toxic compounds and in the development of components with antioxidant properties.[3] In addition, it occurs *in vivo*. The Maillard reaction and its ramifications are so important that four symposia have been devoted to it over the last 12 years.[4-7]

The primary aims of this chapter are to give an up-to-date account of the chemistry of the Maillard reaction, and to survey the effects of the reaction on the properties of the proteins involved. Other consequences of the reaction and means of control are briefly reviewed.

CHEMISTRY OF THE REACTION

The chemistry of the Maillard reaction has been reviewed on many occasions over the last 40 years.[8-16] The earliest article was by Hodge,[8] who comprehensively reviewed the work carried out up to the early 1950s, while the most recent reviews are those by Ledl[15] and Ledl & Schleicher.[16] Hodge,[8] in 1953, produced a scheme for the Maillard reaction, which is still valid nearly 40 years later, and a modified version is shown in Fig. 1. The chemical reactions which occur can be broadly divided into three main stages: (a) the early stage, consisting of the formation and degradation of the N-substituted glycosylamine to the rearrangement product or

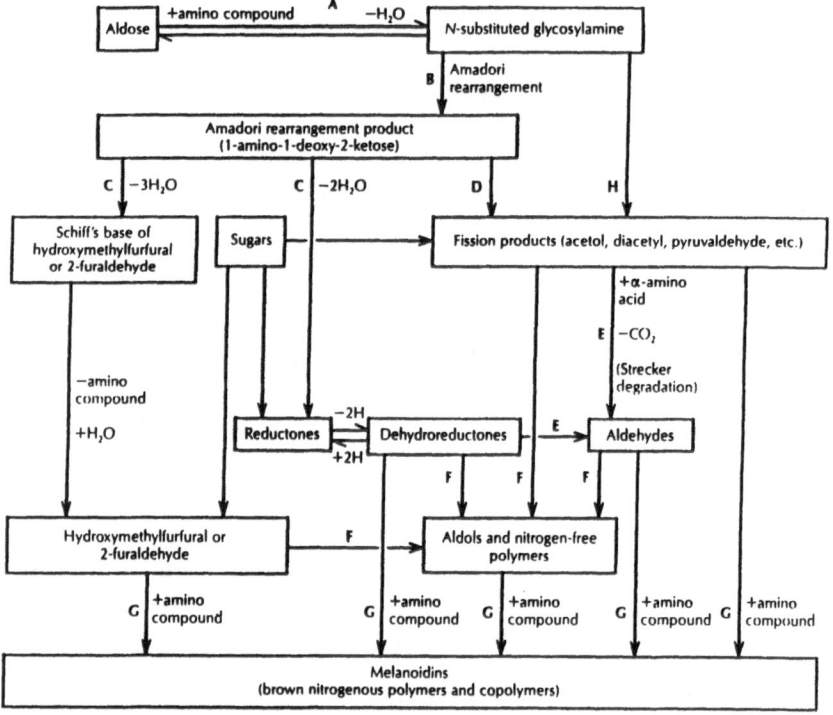

FIG. 1. An outline of the Maillard reaction.[8,17,18] Adapted from References 8 and 18, with permission. Reprinted with permission from J. M. Ames, *Trends in Food Science and Technology*, **1** (1990) 150–4. © Elsevier Science Publishers, Cambridge.

fission products; (b) the advanced stage, comprising degradation of the rearrangement product, and subsequent secondary reactions; and (c) the final stage, typified by the production of brown polymers and co-polymers (the melanoidins).[8,14]

Although the ε-amino group of lysine residues in polypeptide chains is the most important source of free amino groups in foods, investigations of the chemistry of the Maillard reaction have frequently used amino acids or amines, in order to simplify model systems.

Early Stage

The first step in the Maillard reaction involves the addition of a carbonyl group from, e.g. glucose 1, to an unprotonated amino group, to give an addition compound 2, which forms a Schiff base 3, with the elimination of water. Formation of the N-substituted glycosylamine 4, follows as a result of cyclisation[8] (Fig. 2). Normally, only very low levels of glycosylamines can be detected in sugar-amine systems, since they readily undergo further reactions.[16] This initial phase of the Maillard reaction is reversible and well-understood, and is step A in Fig. 1.

Degradation of N-Substituted Glycosylamines to Rearrangement Products
When the sugar is an aldose, the N-substituted aldosylamine formed undergoes an acid-catalysed rearrangement to yield the l-amino-l-deoxy-2-ketose 7, or Amadori Rearrangement Product (ARP) (Fig. 1, step B and Fig. 2).[8,19,20] ARPs may undergo further reactions with reducing sugars and amino compounds to give dideoxyketosamines.[21,22] When ketose sugars, e.g. fructose, react, rearrangement of the ketosylamine yields the 2-amino-2-deoxy-1-aldose or Heyns Rearrangement Product (HRP). HRPs readily undergo further reactions to give, e.g. ARPs.[23] This stage of the Maillard reaction is very well established and documented, e.g.;[8,12] however, it is not clear why it is reversible.[18] The irreversible step of the Amadori rearrangement involves loss of a proton from the cationic form of the Schiff base 5 (formed from the glycosylamine) to give the enol form of the ARP 6, which tautomerises to the more stable keto form 7. Reactions involving protons are usually very rapid and reversible (Nursten, H. E., pers. comm.).

Degradation of the N-Substituted Glycosylamine to Fission Products via Free Radicals
Namiki et al.[14] have shown, by the use of electron spin resonance (ESR) spectroscopy, that free radicals can form from the glycosylamine without

formation of the ARP, and that intermediates of the reaction include two-carbon fission products of the sugar, e.g. the highly reactive eneaminol, glycolaldehyde alkylamine $8^{14,24}$ (Fig. 1, step H). The free radicals were identified as N,N'-disubstituted pyrazine cation radicals 9,[25-27] and, along with the fission compounds produced, lead to the formation of brown

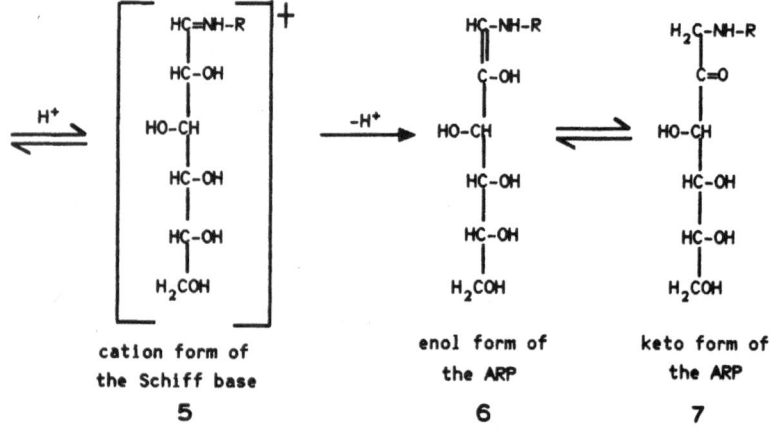

FIG. 2. Formation of a 1-amino-1-deoxy-2-ketose (Amadori Rearrangement Product, ARP) from a hexose and an amino compound.[8,19] Reprinted with permission from J. E. Hodge, *J. Agric. Food. Chem.*, **1** (1953) 928–43. © American Chemical Society.

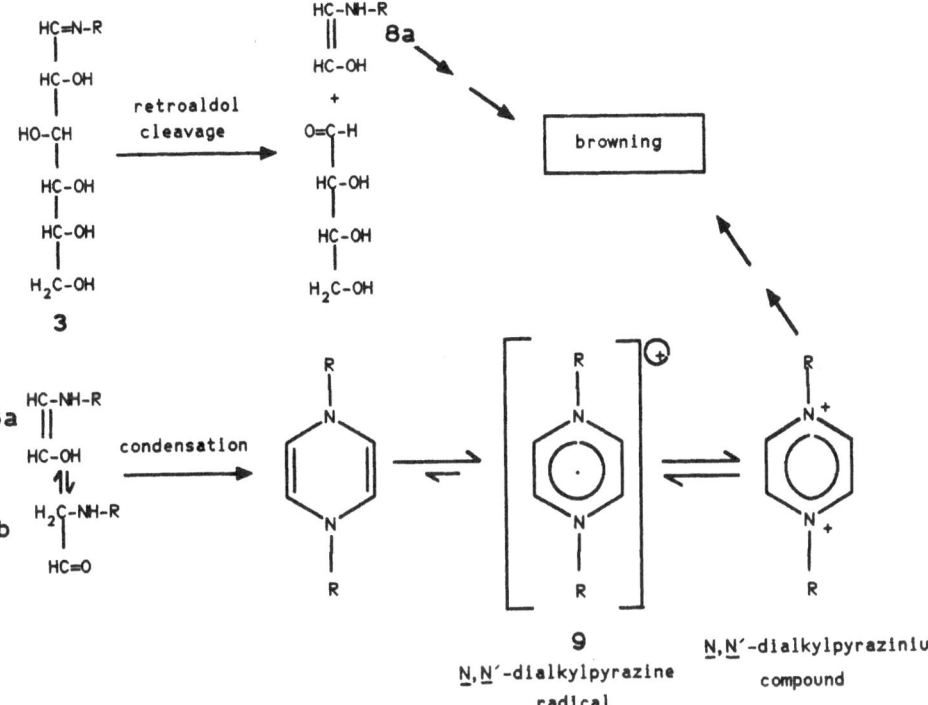

FIG. 3. A possible pathway to the formation of sugar fragmentation products, N,N'-dialkylpyrazine radicals and browning.[14,26,27] Adapted with permission from M. Namiki & T. Hayashi, in *The Maillard Reaction in Foods and Nutrition* (M. R. Waller & M. S. Feather, ed.) (1983). © American Chemical Society.

products (Fig. 3). Three-carbon fission compounds are produced from ARPs just after two-carbon fragments. (See the section on Fission and Strecker Degradation.) The two- and three-carbon fission products, such as glyoxal and methylglyoxal, have very high browning potentials[28] and may well account for the development of the brown colour observed early on in the Maillard reaction, particularly at neutral and alkaline pH.[14] Their development also parallels the increase in the N/C ratio of melanoidins.[28] (See the section on The Final Stage of The Chemistry of The Reaction.)

It is known that reactions between dehydroascorbic acid and amino acids yield free radicals, one of which is blue.[29] Recent work[30] has shown that several blue pigments are produced at an early stage of the Maillard reaction between xylose and glycine, and it has been suggested that they could be related to the previously identified blue free radical.[18]

Advanced Stage

Four possible routes have been established for the degradation of the ARP. The first three involve the formation of deoxyosones, i.e. 1-deoxyosones **12**, 1-amino-1,4-dideoxyosones **13**, and 3-deoxyosones **14**, respectively, while the fourth pathway involves fission and the Strecker degradation,[15,16] (Fig. 4). It is possible that other routes may also exist; for example, studies on the degradation of ARPs during electron impact mass spectrometry have suggested an alternative mechanism of degrada-

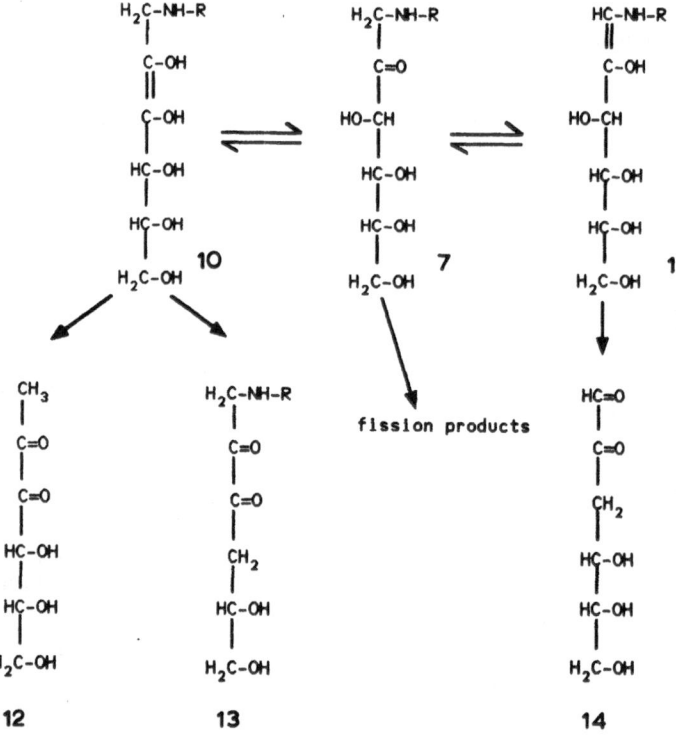

FIG. 4. Degradation of the Amadori Rearrangement Product (ARP) to deoxyosones and fission products.[16] Reprinted with permission from F. Ledl & E. Schleicher, *Angew. Chem. Int. Ed. Engl.*, **29** (1990) 565–94. © VCH Publishers.

tion, e.g.[31] The routes followed in any one system and the balance between them are determined by the reaction conditions. An additional route is via the 4-deoxyosone, but the formation of this intermediate does not involve production of the ARP.

3-Deoxyosone Route

The formation of 3-deoxyosones from the ARP via 1,2-enolisation is well-established (Fig. 4) and results in the formation of 5-hydroxymethylfurfural (HMF) **15** (from hexoses)[19,32] or furfural (from pentoses), with loss of three molecules of water.[19] This pathway is favoured by low pH and is represented by step C in Fig. 1. The first stage involves the formation of the 1,2-eneaminol **11**, which gives the 3-deoxyosone **14** by an irreversible step.[19] 3-Deoxyosones, furfural and HMF have all been isolated and identified in Maillard systems.[14] However, the degradation products of 3-deoxyosones are not limited to furfural and HMF, and some of the other compounds formed from 3-deoxyhexosones are given in Figs 5a and 5b.

In solution, the 3-deoxyosone is likely to exist as an equilibrium mixture of its straight chain form and four cyclic forms.[16] One of these cyclic forms, **14a**, may undergo isomerisation of the keto group to give metasaccharinic acid lactone **16**,[34] while the formation of a furanone derivative **17**, from the same cyclic form, would appear to be possible, but has not been proved.[16] Another possible transformation is cyclisation of 3-deoxyhexosone, with elimination of water, to give a pyranone **18**.[15] This compound can condense with **15** to give a yellow bicyclic product **19**.[34] Alternatively, **18** can react with 2 molecules of furfural to yield a tricyclic structure **20**.[34]

In the presence of large amounts of amines, degradation of 3-deoxyosones to furfural and HMF may be limited, in favour of nitrogen-containing compounds. Pyrrolealdehydes **21**[35,36] and pyridiniumbetaines **22**[37] have been identified (but only in very low yields) in HMF-amine mixtures, indicating that a precursor of HMF, rather than HMF itself, is important for their formation.[16] The pyridiniumbetaine is formed on autoclaving N^α-acetyllysine with lactose at 130°C and pH 7 and is stable in hot solution.[37] It may therefore lead to some protein cross-linking in foods. These pyrrolealdehydes and pyridiniumbetaines are also formed when reducing disaccharides, e.g. maltose and lactose, react with primary amines.[16] An example of such a pyrrolealdehyde was identified in protein linked to the peptide chain via the ε-amino group of a lysine residue.[38] It appears that the pyrrolealdehydes identified can undergo further reactions, e.g. dimerisation, to yield bispyrroles **23** and **24**.[39] In the presence

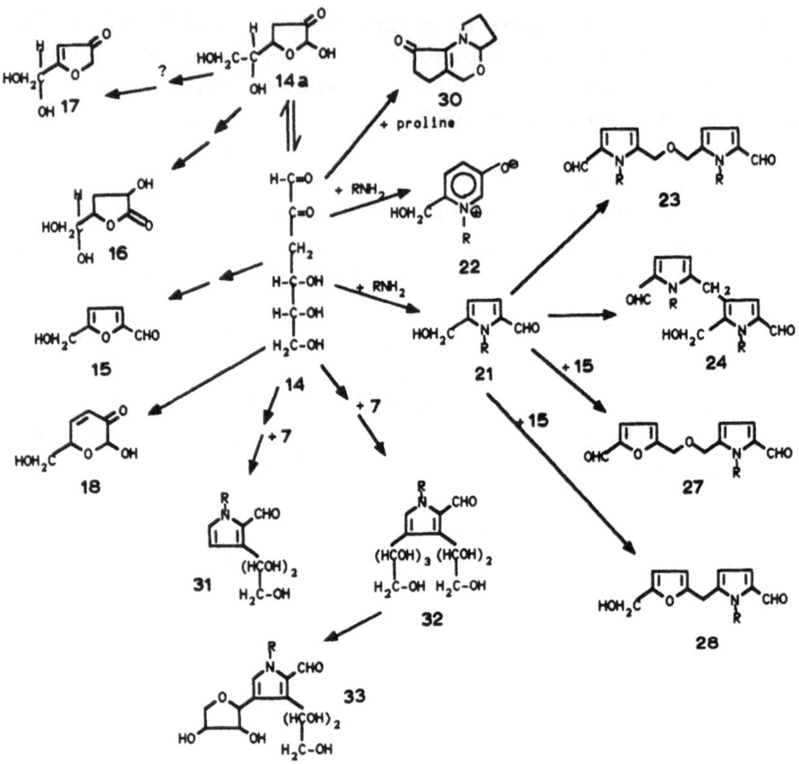

FIG. 5a. Some compounds formed from 3-deoxyhexosones 14.[15,16,19,32,33,35-37,39,42,45]

of amino acids, HMF yields lactones 25,[40] and lactams 26.[41] In the presence of pyrrolealdehydes, HMF forms other bicyclic heterocycles, e.g. 27 and 28.[42] The bispyrroles 23 and 24 may undergo further reactions with free amino groups.[43] The reactions of pyrrolealdehydes of structure 23 could result in the development of inter- and intra-molecular cross-links in proteins *in vivo*.[15]

An interesting compound, formed from HMF and the Strecker aldehyde of lysine, is 29,[42,44] which is produced as a result of aldol condensation and cyclisation. Both amino groups of lysine are required for the formation of 29, which is therefore unlikely to be involved in protein cross-linking.

Very recently,[15] a lactone and formamide have been identified in 3-deoxyosone-amino compound model systems. Although their mech-

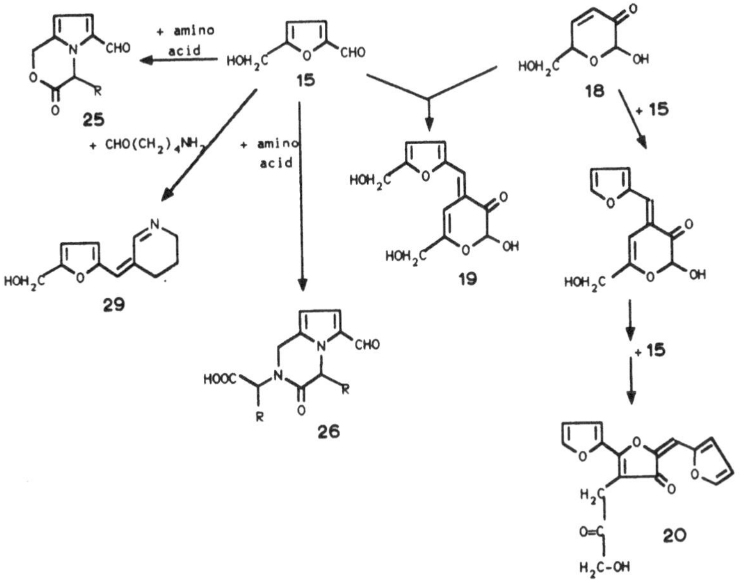

FIG. 5b. Some compounds formed from 5-hydroxymethylfurfural (HMF) 15 and 2-hydroxy-5-hydroxymethylpyran-4-en-3-one 18.[19,32,34,40–42,44]

anisms of formation are unclear, it has been suggested that a disproportionation step is involved.

Secondary amines give different reaction products from those obtained from primary amines, and frequently, carbocyclic derivatives are among the reaction products.[16] For example, secondary amines give either very low yields or no HMF on reaction with hexoses; instead, when the imino acid, proline, was heated with 3-deoxyhexosone, maltoxazine 30 was the main reaction product.[45] Also, an orange carbocyclic compound is obtained from pentoses (via the 3-deoxypentosone) and 4-hydroxy-5-methyl-3(2H)-furanone in the presence of secondary amines.[46] Yields were increased when the pentose was replaced by cyclopent-1-en-3,4-dione, indicating that the sugar was the precursor for this key compound.[16]

3-Deoxyosones can also react with the ARP to give further pyrrolealdehydes, and structures 31 and 32 are formed by competing reactions.[47,48] Compound 32 may undergo intramolecular condensation to yield the ether 33.[48] These compounds may themselves take part in yet further reactions. The presence of hydroxyl groups suggests that they may take part in cross-linking reactions in proteins.[16]

1-Deoxyosone Route

1-Deoxyosones 12 are formed from the ARP via 2,3-enolisation (Fig. 4), and lead to the production of reductones after the loss of two molecules of water.[19] Although 1-deoxyosones have not been identified from any Maillard system,[49] the identification of the expected quinoxaline on heating the glucose ARP with o-phenylenediamine proves that it is formed as an intermediate.[50] This pathway (Fig. 1, step D) is favoured by a higher pH over that of 1,2-enolisation. The formation of the 2,3-enediol 12a from the ARP is often described as irreversible, e.g.;[19] however, the fact that this step comprises only simple tautomerism makes this unlikely (Nursten, H. E., pers. comm.).

It is probable that 1-deoxyosones exist as a mixture of three cyclic forms in solution 12b–12d, possibly together with the open-chain structure 12a,[16] as shown in Fig. 6, which also gives the structures of some other 1-deoxyhexosone degradation products. The hemiketal form of the 1-deoxyhexosone 12b can undergo enolisation and loss of water to give a furanone derivative 34,[51] which can further yield 4-hydroxy-5-methyl-3(2H)-furonone 35, (typically formed from pentoses via the 1-deoxyosone route) plus formaldehyde. When 6-deoxyhexoses, e.g. rhamnose, degrade via the 1-deoxyosone route, 2,5-dimethyl-4-hydroxy-3(2H)-furonone formed.[52]

The 1-deoxyosone route can also lead to pyranone derivatives, via the cyclic hemiacetal form 12d of the 1-deoxyosone,[16] to give structures, such as 36[33,53] and 37.[36] Compound 36, an unstable β-pyranone, may rearrange to give a lactic acid ester,[33] which, on further heating, hydrolyses to produce lactic acid and β-hydroxypropionic acid. It has been proposed very recently that 12 may oxidise to an α,β-triketone, which may lead to a further ester 38. This compound may undergo hydrolysis to yield acids 39 and 40. Compounds 38–40 have all been identified in a glucose-glycine model system.[33] Another pyranone, which may be assumed to be formed from 1-deoxyosones, is the β-diketone 41.[16] It is a precursor of the highly reactive compound, acetylformoin 42,[36,52] which can react with secondary amines to give aminohexosereductones, which possess a cyclopentene ring.[53] Acetylformoin has also been shown to be a precursor of pyr-rolinones 43,[54] and, since these components are formed on reaction of 42 with primary amines, they possess the ability to cross-link peptide chains via reaction with the ε-amino group of lysine residues.[15] Fission of the β-

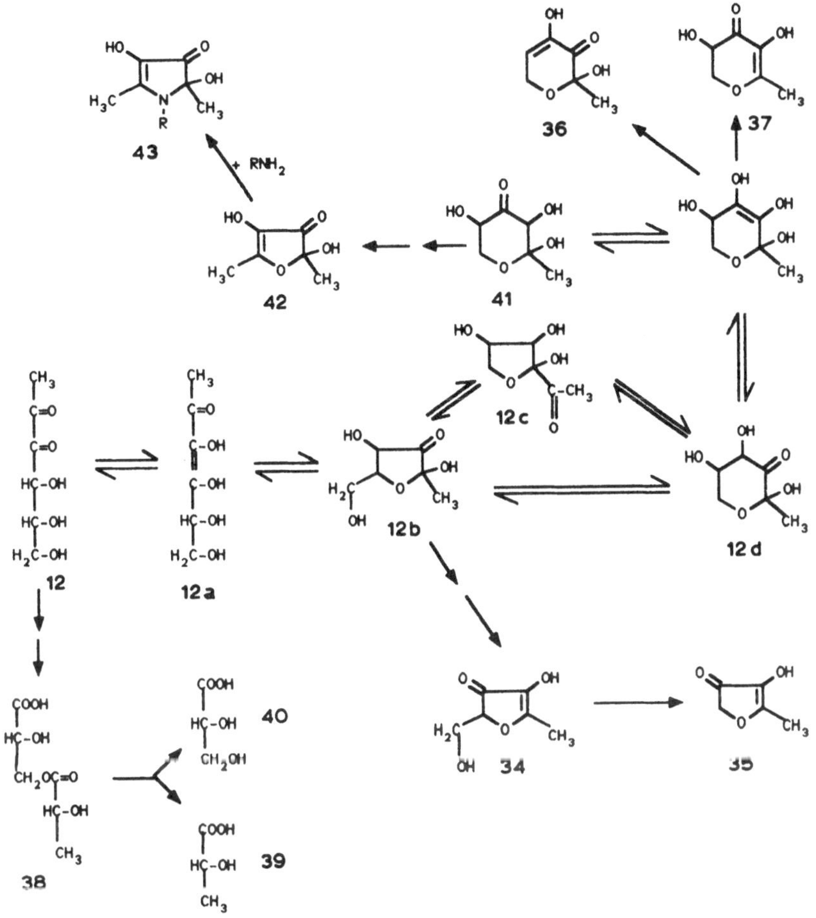

FIG. 6. Some compounds formed from 1-deoxyhexosones 12.[16,33,36,51-54]

diketone **41** may explain the formation of C_4 aminoreductones[55,56] and other fragmentation products.[57]

When disaccharides give rise to 1-deoxyosones, they degrade in a somewhat different manner from those derived from monosaccharides (Fig. 7). The γ-pyranone **37** can only form when there is a free hydroxyl group on C_4, and so is a monosaccharide-specific product.[16] The β-pyranone **44** (which is analogous to **36**) is a key intermediate in disaccharide degradation and is present in, e.g. heated milk[58] and red

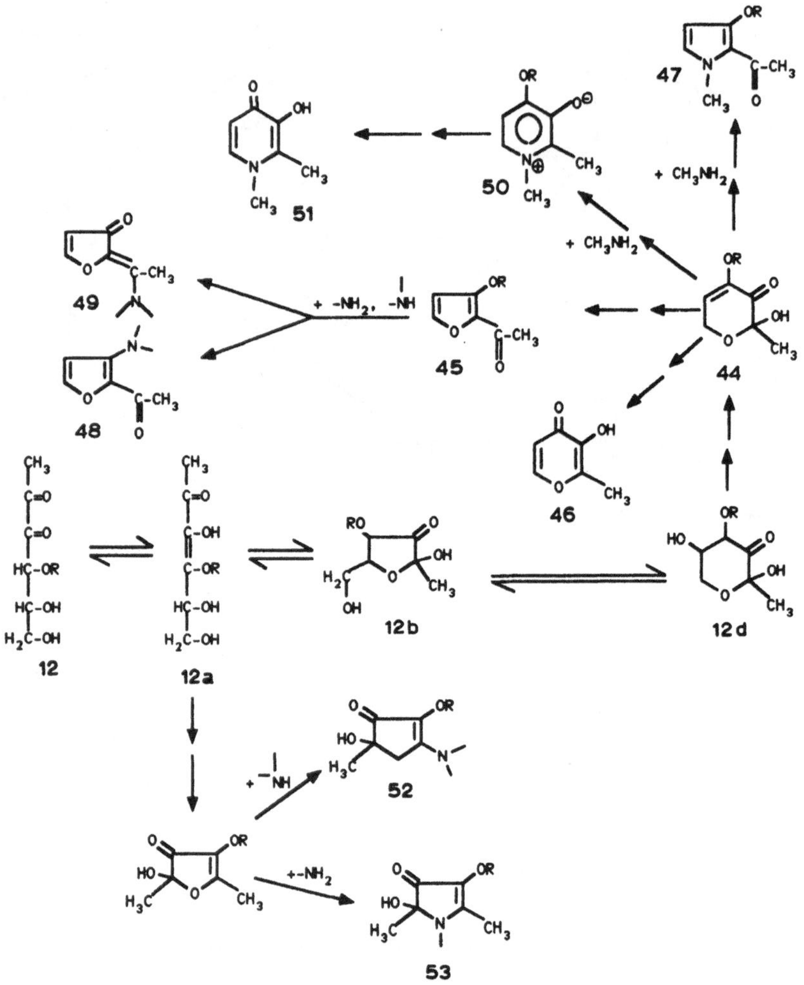

FIG. 7. Some compounds formed from 1-deoxyosones derived from disaccharides. (R = β-gal or α-glu).[16,49,53,58–64]

ginseng.[59] When lactose is the reducing sugar, β-galactosylisomaltol **45** is the predominant product,[60] but, when a monosaccharide is used, the corresponding compound, maltol **46**, is formed in insignificant amounts.[16] When maltose is the disaccharide involved, maltol is formed mainly,[61] implying a significant degree of splitting of the glycosidic bond. The pyrrole derivative of galactosylisomaltol **47** is also formed in small amounts in lactose systems.[58] Heating galactosylisomaltol with primary and secondary amines yields furan derivatives bearing amino groups, e.g. **48** and **49**.[62] When disaccharides are reacted with primary aliphatic amines, the main product to be extracted with organic solvents is the pyridone **51**,[63] which is probably formed from the pyridiniumbetaine **50**, on cleavage of the glycosidic bond.[64] Compound **51** forms on heating disaccharides with proteins, and the ε-amino group of lysine residues becomes incorporated into the heterocycle,[49,64] and could thus lead to cross-linking. The pyrrole isomer is formed in small amounts.

Some compounds produced from disaccharides are analogous to those formed from monosaccharides, e.g. derivatives of **52** and **53**, which possess a glycosyl residue.[53]

1-Amino-1,4-Dideoxyosone Route

It has been established only very recently that ARPs can degrade to 1-amino-1,4-dideoxyosones **13**,[50] and that they are formed in larger amounts from disaccharides.[65] As shown in Fig. 8, these compounds themselves degrade to furan derivatives **54** and **55**.[16] Compound **54** is able to undergo a number of condensation and rearrangement reactions, and

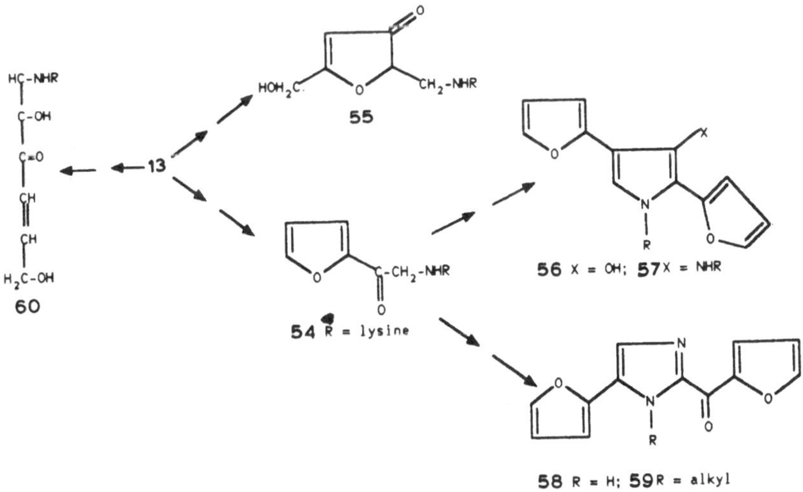

FIG. 8. Some compounds formed from 1-amino-1,4-dideoxyhexosones **13**.[16,65-68]

some of the reaction products are tricyclic structures.[65-67] One such compound is an aminopyrrole 57, and it is possible that the formation of such compounds may lead to cross-links in proteins through involvement of two ε-amino groups of peptide-bound lysine.[16] Compound 54 may also react with ammonia with the formation of imidazole derivatives 58 and 59.[66,67] The isolation of 58 from protein-sugar mixtures has been shown to be due to its formation as a result of the acid hydrolysis of proteins.[16,66,67] Recently, an open-chain form of an aminoreductone, 60, was isolated from a Maillard reaction mixture[68] and was assumed to derive from a 1-amino-1,4-dideoxyosone intermediate.

4-Deoxyosone Route

The formation of 4-deoxyosones 61 during the Maillard reaction has not been proved, but can be assumed, since certain reaction products have been identified, i.e. 62–64, which are most likely to be produced via 4-deoxyosones[16] (Fig. 9). In the presence of large amounts of primary amines, formation of the pyrrole 63,[36] and the pyridiniumbetaine 64,[37] occurs in preference to that of the furan 62. The keto group of pyrroles of structure 63 is unable to react with amino groups to form Schiff bases, and so this reaction will not lead to cross-linking in proteins.[43] 4-Deoxyosones are not formed via the ARP, but rather via the Lobry de Bruyn-Alberda van Ekenstein rearrangement, and subsequent dehydration.[16] Therefore their presence in Maillard systems indicates that some reactions are occurring without the involvement of the ARP.[69]

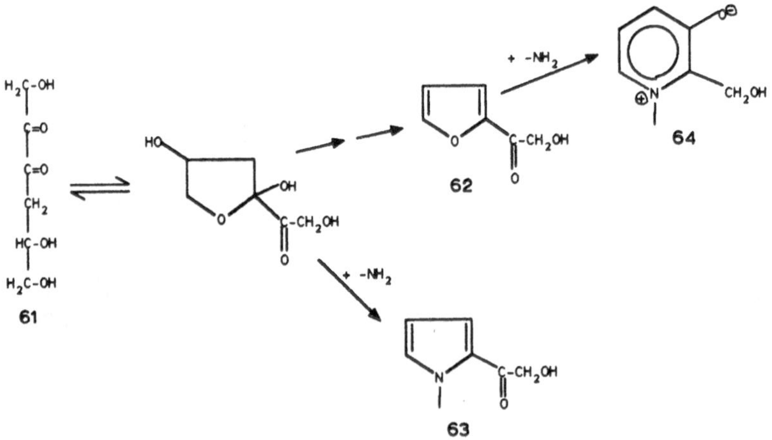

FIG. 9. Some compounds formed from 4-deoxyhexosones 61.[16,36,37]

Fission and Strecker Degradation

Retroaldol-type reactions and oxidative fission of the carbohydrate chain are important routes to the formation of a range of dicarbonyl and hydroxycarbonyl compounds, including butanedione, glyoxal, acetoin and glyceraldehyde. It has been shown that ARPs may undergo oxidative cleavage to give carboxymethylamines[70] and C_3 imines, e.g. methylglyoxal dialkylimine.[27,71] Fission is favoured by higher pH, and represents step D in Fig. 1. Many of the products are highly reactive and can greatly increase the overall rate of the Maillard reaction.

The α-dicarbonyls **65**, produced by retroaldolisation and fission reactions, can then take part in the Strecker degradation of amino acids (Fig. 1, step E), with the formation of Strecker aldehydes **66** (with one less carbon atom than the initial amino acids), aminoenols **67** and carbon dioxide (Fig. 10). Most of the carbon dioxide formed in the Maillard reaction comes from the carboxyl groups of the amino acids.[12] Pyridines, pyrazines and imidazoles can be formed from **67** as a result of the Strecker degradation.

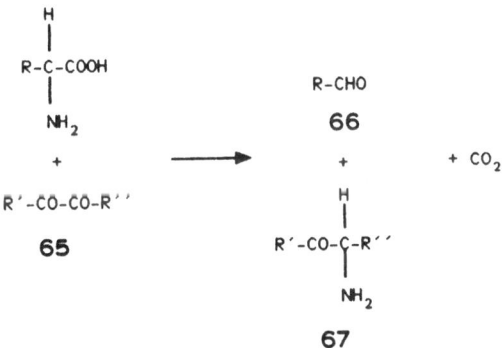

FIG. 10. The Strecker degradation of amino acids.

Deoxyosones can also take part in the Strecker degradation. For example, 3-deoxyosones give the furan **68**, and, in the presence of ammonia, the pyrrole **69** and the pyridine **70**[72] (Fig. 11a). It would seem likely that the 1-deoxyosone would lead to 2-acetylfuran, its pyrrole derivative and 2-methyl-3-hydroxypyridine, but it has been shown that the methyl group is not derived from C_1 of the hexose.[74] It has therefore been proposed that the pyrrole **71** and the pyridine **72** may arise from the Schiff base, formed by dehydration of the addition compound arising from the

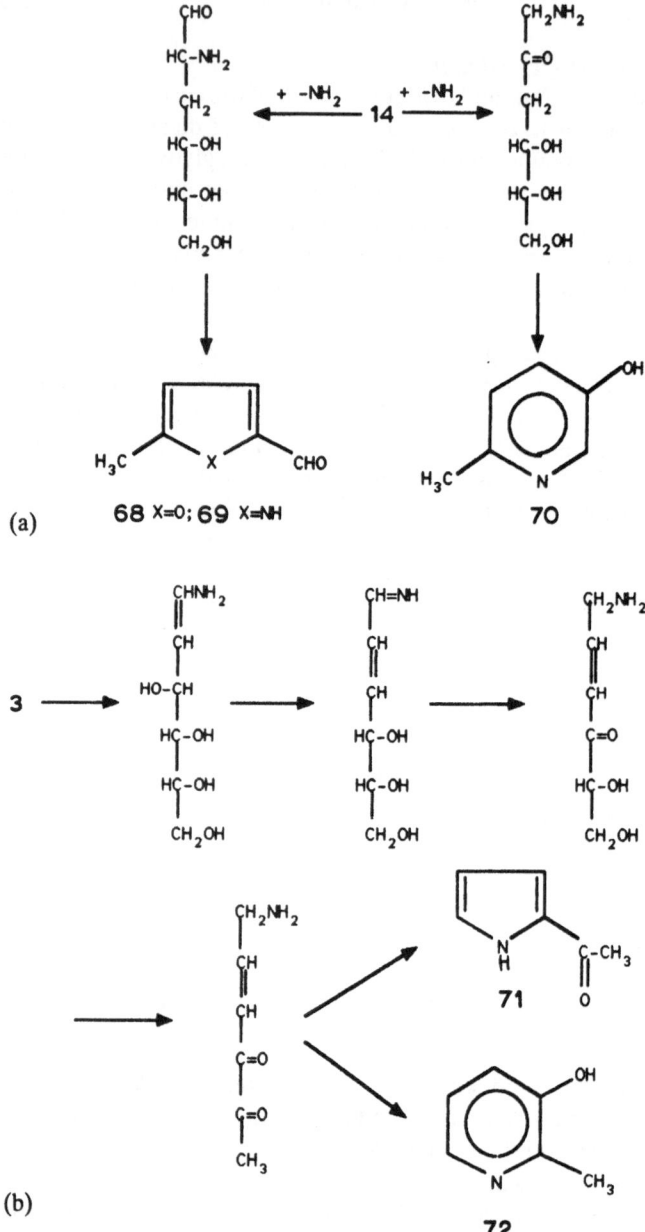

(a)

68 X=O; 69 X=NH   70

(b)

71

72

FIG. 11. Formation of selected pyrroles and pyridines from deoxyhexosones.[72–74] (a) Strecker degradation of the 3-deoxyhexosone **14**. (b) Dehydration of the Schiff base **3**.

sugar and the amino compound, as shown in Fig. 11b.[74] In protein-containing systems, the keto group of 2-acetylpyrrole 71 could form cross-links by reacting with free amino groups.[43]

The possible participation of amines (in place of amino acids) in the Strecker degradation is of interest, since the ε-amino groups of lysine residues in peptide chains constitute an important source of primary amines in foods. However, so far, it has not been proved that such reactions form part of the Maillard reaction.[16]

Final Stage

The final stage of the Maillard reaction is characterised by the formation of unsaturated, brown nitrogenous polymers and copolymers, otherwise known as melanoidins, and is shown by step G in Fig. 1.[8] Nitrogen-free polymers are also formed in this stage of the reaction, as shown by step F in Fig. 1. Melanoidins possess high molecular weights, but little is known of their chemical structures or the final steps leading to their formation.[3,14,16,75] It has been suggested that melanoidins possess a backbone of repeating units,[76,77] possibly furan rings, and some data for polymers prepared from glucose and glycine supports this.[78] Other studies suggest that furan rings are not present,[42,77,79,80] although repeating units possessing conjugated double bonds and tertiary nitrogen may be involved.[77]

Nuclear magnetic resonance spectrometry (NMR) has proved to be the technique yielding the most structural information for melanoidins so far, and has been used by several groups, e.g.[42,77,81–83] [13]C Cross polarisation-magic angle spinning (CP-MAS) NMR of the non-dialysable melanoidins formed from glucose and glycine at 95°C indicated the presence of saturated or aliphatic carbons, and smaller amounts of unsaturated or aromatic carbons and carbonyl or carboxyl groups.[83,84] A separate study involving the reaction of aqueous xylose–glycine systems at 22, 68 and 100°C demonstrated that the proportion of non-dialysable material formed (membrane cut-off 12 000 Da) increased with reaction temperature. Methyl groups were present in the materials formed at all temperatures, but temperature-dependent differences in chemical properties were apparent.[85] A temperature of 22°C gave a water-insoluble melanoidin with less aromaticity and unsaturation, different types of aliphatic carbons, fewer carbonyl groups and a higher nitrogen:carbon ratio than melanoidins prepared at the higher temperatures. In fact, the elemental composition of the material formed at 22°C closely corresponded to the loss of 3 mol of water, implying that all the nitrogen was retained.[3,85] A lower nitrogen:carbon ratio at higher temperatures presumably corresponds to loss of some nitrogen as volatiles. In another study, elemental analysis data obtained for melanoidins prepared from glucose or fructose and glycine by refluxing in aqueous buffer at pH 3·5 were similar to those obtained for the xylose–glycine polymer at 100°C.[80]

A study of the properties of the melanoidins formed at 68°C with time of reaction (up to 70 days) showed that there was a gradual increase in carbon content paralleled by a decrease in nitrogen content with reaction time, and that maximum yield corresponded to the disappearance of xylose at about 45 days.[81] ^{13}C CP-MAS NMR indicated that structural differences also occurred with time of reaction, and that the amount of unsaturated carbon increased with time up to 42 days, and then levelled off, and was accompanied by a decrease in total carbon double-bonded to oxygen. ^{15}N CP-MAS NMR showed that longer reaction times resulted in more pyrrole and/or indole nitrogen and less amine nitrogen. Changes in melanoidins formed in glucose–glycine systems with molecular weights in excess of 16 000 Da have been monitored using radiochemical tracing techniques,[82] and confirm that changes in the melanoidins (mainly rearrangements and dehydrations) do indeed occur with time of reaction. It has been shown that not all the C_1 of the amino acid is lost as carbon dioxide, since significant amounts of the C_1 of glycine were incorporated into the melanoidin formed from glucose and glycine as carboxyl (or carbonyl) groups.[80] The C_2 of glycine was assimilated as a substituted methyl group. Subsequent experiments on the polymer prepared from glucose and alanine showed that C_1 and C_2 of the amino acid were incorporated as a carboxyl carbon and a substituted amino carbon, respectively.[82] Elemental analysis data for the polymer prepared from glucose and methionine showed that both nitrogen and sulphur were incorporated, suggesting that the intact amino acid participates in the reactions leading to melanoidin formation.[80]

C_1 of the sugar was assimilated into the glucose–alanine melanoidin as a substituted methyl group.[82] In contrast, a group in Japan[83,84] showed that, in their glucose–glycine polymer, C_1 of the sugar was incorporated into a variety of moieties: $C^1H_3CH_2-$, C^1H_3CO-, $C^1H(H$ or $OH)=C=$, $HOOC-CH_2-NHC^1H=$, $HOOC-CH_2-N=C^1H-$, $=N-C^1O-$ and HC^1O, indicating that the sugar molecule undergoes considerable modification during browning.

The dialysable material formed in several sugar–amino acid systems has been shown to comprise mainly unreacted starting materials and ARPs (e.g.[80,85]). ^{13}C NMR has shown that the polymer prepared from glucose and glycine yields spectra very similar to those of the analogous

ARP, unsaturated and aromatic carbons being almost absent.[42] Although
the polymer cannot comprise linked ARPs, since elemental analysis data
indicates the loss of 3 mol of water during the reaction,[80] dehydrated
sugar-derived intermediates are probably involved,[42,77,80] and a possible
repeating unit of melanoidin is shown in Fig. 12. The water-soluble fractions of aqueous xylose–glycine mixtures prepared at 22°C for 150 days
and 68°C for 3 days contained a 12 carbon eneaminol as a major reaction
product, which has been shown to be an intermediate in melanoidin
synthesis.[81] Hayase *et al.*[84] speculated, from their ^{15}N CP-MAS data, that
a conjugated eneamine linkage accounted for most of the amino acid nitrogen in their glucose–glycine melanoidin prepared at 95°C. (Reductones such as eneaminols are considered to play important roles in the
metal chelating and antioxidative activities of melanoidins.[14]) In contrast,
the water-soluble fraction of the mixtures heated at 100°C for 38 h was
much more complex, and ^{13}C CP-MAS NMR showed evidence of
azomethine, ester, anhydride, acid and/or amide groups, in addition to
unsaturated and aromatic carbons.[85]

Other techniques applied to the analysis of melanoidins include Curie-
point pyrolysis GC-MS[86] and ESR spectroscopy,[87] but so far they have
yielded very little structural information.

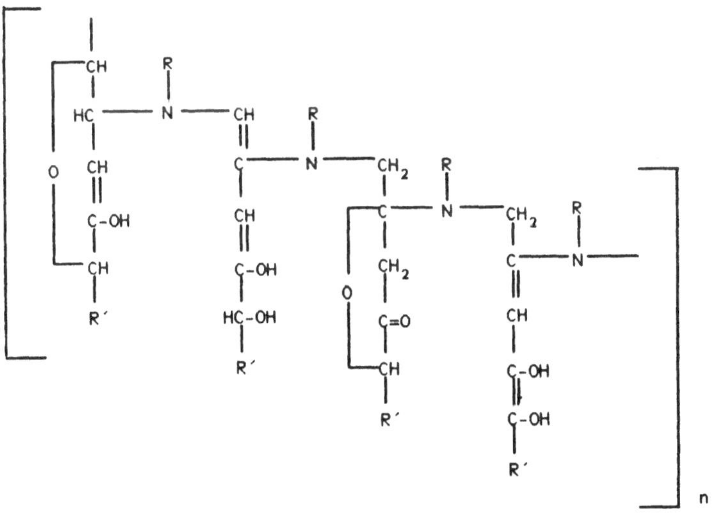

FIG. 12. A possible repeating unit of melanoidin.[77] Reprinted with permission
from H. Kato & H. Tsuchida, in *Maillard Reactions in Food* (C. Eriksson, ed.);
Progress in Food and Nutrition Science, **5** (1–6) (1981) © Pergamon Press, plc.

Little work has been carried out on the high molecular weight coloured compounds formed in protein-containing systems and the studies which have been performed were done about 20 years ago.[88-90] Studies focused on casein– or insulin–glucose systems stored at 55°C and 75% relative humidity (rh). The results obtained were complex but showed that most of the colour formed was protein bound and that the pigments formed were probably heterogeneous. They probably represent a distribution of highly degraded carbohydrate structures bonded to a variety of sites on the protein,[88] and protein bound pigments of 7800 and 1500 Da were isolated from the insulin system.[90] Analysis of the stored insulin–glucose system indicated cross-linking between the peptide chains due to condensation of the sugar residues (some of which had already reacted with the free amino groups of the amino acid residues) giving aggregates of up to 31 residues in the higher molecular weight pigment.[90]

EFFECT ON PROTEINS

Reactivity of Amino Acid Residues
Most of the work aimed at understanding the early stage of the Maillard reaction in heated and stored food-related protein/sugar systems, has been carried out using milk or milk proteins, (e.g.[1,91-95]). Certainly, the Maillard reaction is of special importance in milk, since milk is the only naturally occurring protein food which also possesses a high content of reducing sugars[96] and, in addition, it also possesses a high lysine content. Regardless of the protein involved, it is the ε-amino groups of the lysine residues which play the most important role in the reaction,[97] although the α-amino groups of the terminal amino acid residues also react. Protein-bound ARPs are formed, by the steps outlined in the sections on The Early Stage of The Reaction, and Degradation of the N-Substituted Glycosylamines to Rearrangement Products, and in Fig. 2. Subjecting a range of plant and animal proteins to the Maillard reaction showed that the percentage of lysine destroyed increased with its percentage content in the protein.[98] A subsequent study showed that the amino acid profile of a protein may be maintained by processing the hydrolysed material, which results in a more uniform destruction of the amino acids.[99] In heated or stored milk proteins, typical early stage reaction products are lactuloselysine and fructoselysine,[97,100] and, in UHT milk stored for 6 months to 3 years at 30–37°C, lactuloselysine formation involved 10–30% of the lysine residues.[101] The fructoselysine content was about 10% of that

of lactuloselysine. The reactions concerned, in both protein-containing model systems and food, lead to losses in available lysine, but no discoloration of the system and no reduction in protein solubility.[97,102] It is worth noting that whereas the Maillard reaction is usually considered to be a zero-order reaction, loss of lysine during the reaction generally obeys first-order kinetics, at least up to 75% loss.[103]

Reactions which occur after lysine ARP formation involve other amino acid residues, especially arginine and histidine, in addition to lysine.[93,102] There is some evidence that cysteine may also be involved[102,104] (see the section on Nutritional Effects). An important study on whole milk powder by Nielsen et al.[105] has shown that only small losses of tryptophan occur when foods are processed. The losses suffered by whole milk powder on heating at 100°C for 6 h or on storage at 50°C for 9 weeks or at 60°C for 4 weeks were considered to be minor. This group showed that earlier studies, demonstrating significant losses of tryptophan on food processing, can be attributed to inappropriate methods of analysis.[106] Increasing the time of heating or storage results in degradation and further reaction of ARPs, and, in the final stage of the Maillard reaction, the content of lysine ARPs is very low,[107,108] with a general reduction in amino acid availability taking place.[109] Cross-linking of the peptide chains (see the section on Cross-linking) leads to reduced availability of all the amino acids and reduced protein digestibility,[108,110–112] probably due to the formation of enzyme-resistant structures.[1,96] (See the section on Nutritional Effects.)

Effect on the Physicochemical Properties of Proteins

The Maillard reaction results in changes in the physicochemical properties of the protein, e.g. thermal stability and solubility.

It has been known for many years that sugars[113] and low molecular weight aldehydes, including formaldehyde,[113,114] may maintain or increase the heat stability of milk and other proteins. More recently it has been shown that sugars and polyols can increase the thermal stability and solubility of a number of proteins (including blood serum proteins, ovalbumin and lysozyme) in the early stages of the Maillard reaction.[115–117] The contrasting results obtained from the classical studies carried out by Lea & Hannan,[118,119] and Mohammad et al.,[120] about 40 years ago on milk proteins, showing that proteins lose much of their solubility on taking part in the Maillard reaction,[118–120] may possibly be attributed to those systems having undergone more extensive reaction.

Shalabi & Fox[121] showed that low molecular weight carbonyl compounds, including pentose sugars, increase the heat stability of milk, especially in the presence of urea. It was suggested that interactions between the ε-amino groups of the lysine residues account for the increased stability which was enhanced by urea. However, further studies by the same group showed that several dicarbonyls, including butanedione and cyclohexa-1,2-dione, react preferentially with arginine residues and give a marked increase in heat stability of lactose-free milk, especially in the presence of urea.[122] This led to the belief that arginine residues play an important role in heat stability, and the suggestion that urea may act by converting lysine residues to homocitrulline, a compound which is structurally very similar to arginine, and which may go on to react with dicarbonyls in a similar manner to arginine.[122] The effect of butanedione on the heat stability of concentrated milks appears to be more complex.[123] (Interactions between amino acid residues and dicarbonyl compounds are discussed in more detail in the section on Protein Polymerisation and Cross-Link Formation.)

Studies carried out in the last 10–15 years on changes to protein brought about as a result of the Maillard reaction, have frequently involved the study of model systems comprising lysozyme or ovalbumin and glucose, stored at 37–50°C and 65–75% rh. Such studies have a greater bearing on the Maillard reaction *in vivo* (see the section on The Maillard Reaction *in vivo*) than on food systems, but they do have relevance to foods stored under adverse conditions.

Egg white proteins, stored both with and without various reducing sugars at 50°C and 65% rh, showed a minimum solubility value after 8 h, followed by an increase to a maximum value after about 12 h[124] (see Fig. 13). For the system with glucose, the solubility increased to a level above the original value and was maintained at this high level up to 40 h. This was followed by a gradual decrease in solubility to a level of about 62% after 68 h. For the system containing protein only, the original level of solubility was regained after about 12 h, followed by a very rapid decrease to 20% after 40 h, and a slow decrease to 13% after 68 h. Similar patterns of heat stability behaviour were observed for each reducing sugar tested. Studies on ovalbumin, in place of egg white protein, gave comparable results.[125] The authors concluded that the sugar appeared to maintain or increase the solubility and heat stability of the protein in the early stages of the reaction.[124,125] This is in line with the studies mentioned above, showing that the addition of sugars and carbonyl compounds to proteins before heating maintains or increases solubility and other functional

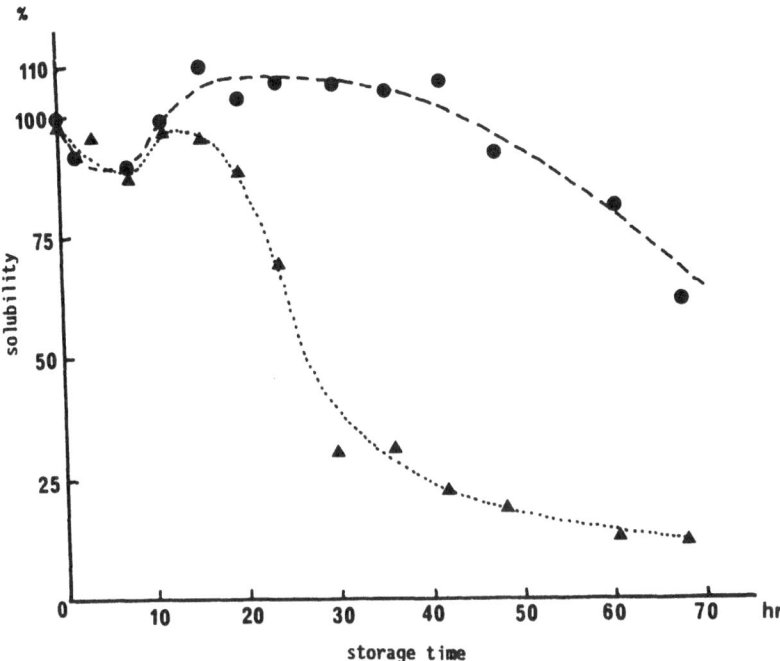

FIG. 13. Solubility of freeze-dried egg white and freeze-dried egg white–glucose solids in citrate buffer at pH 6·0 containing 2M sodium chloride.[124] Solubility is expressed as a percentage of the soluble protein present in unstored egg white. ▲, egg white solid; ●, egg white–glucose solid. Reprinted with permission from Y. Kato et al., Agric. Biol. Chem., 42 (1978) 2233–7. © Japan Society for Bioscience, Biotechnology, and Agrochemistry.

properties.[115,116,122] Eventually, the ovalbumin–glucose complex became insoluble, due to cross-link formation, and sensitive to heat as a result of loss of bound water.[124,125] An earlier study[116] showed that the increased heat stability of a BSA–sugar complex over that of the protein alone was due to protection against precipitation of the denatured molecules, rather than to changes in the internal protein configuration.

Damage to Protein Structure

An ovalbumin–glucose mixture, stored at 50°C and 65% rh, showed losses of lysine, arginine, serine and threonine, compared with the unstored system.[125] Losses were greatest for arginine (68% after 8 days and 83% after 19 days). Circular dichroism spectra showed that these losses of amino acids correlated with a decrease in the α-helix form of the protein, and with an increase in unordered secondary protein structure.[125] Both the content of the α-helix form, and also the amount of available lysine, decreased with time of storage and degree of protein aggregation,[126] and cupric and ferric ions promoted these changes.[127] Analysis of the monomer form of the ovalbumin–glucose complex formed by heating an aqueous solution of the protein and sugar at 50°C for 90 min, showed that the free lysine ε-amino group content corresponded to about 7·6 residues, compared with 20 residues for ovalbumin alone. Of the bound lysine, about 6·4 residues (i.e. about half the bound lysine) was accounted for by fructoselysine.[128] Circular dichroism spectra showed that the ovalbumin–glucose monomer was only partially denatured in spite of 62% of the ε-amino groups being blocked, and that this modified protein–sugar monomer also appeared to be more stable than that of ovalbumin alone, possibly due to stabilisation by intrachain non-covalent interactions between the protein and the carbohydrate moieties bound in the forms of the Schiff base and the ARP.[128]

The pattern of lysozyme glycosylation is dependent on the water activity of the system.[129] On reaction of lysozyme with glucose in aqueous solution, the major glycosylation sites were α-lys-1 (4%), ε-lys-1 (14%), lys-13 (11%), lys-33 (20%), lys-96 (19%) and lys-116 (33%), with the remaining lysine residues described as relatively inactive.[130] No arginine residues were glycosylated. In contrast, storage of lysozyme with glucose at 50°C and relative humidities of 4, 30 and 90% resulted in glycosylation of lys-1 and several arginine residues, rather than further lysine residues. This glycosylation pattern in restricted water environments was independent of the precise relative humidity. The preferential glycosylation of certain lysine residues was also illustrated by storage of RNAase with glucose in phosphate buffer at 37°C.[131] The most reactive sites were lys-1,7 and 41 (all of which are in or near the active site of the enzyme), and lys-37 (which is adjacent to an acidic amino acid). Lys-1 was the most reactive site for Schiff base formation but lys-41 was the major site for formation of the ARP. As far as lys-1 was concerned, it was shown that the Amadori rearrangement proceeded about four times as fast at the ε-amino group as at the α-amino group.[131]

After storing aqueous lysozyme–glucose systems at 50°C and 75% rh for 30 days, impairment of the lysine, arginine and tryptophan residues was observed in whole lysozyme, and also in the lysozyme monomers isolated from the system.[132,133] Incubation of the lysozyme monomers for a further 10 days, in the absence of glucose, resulted in polymerisation, with additional impairment of the lysine and arginine (but not tryptophan) residues, indicating that the sugar was only required for the first steps of the reaction. This latter observation is supported by an earlier study on an *in vitro* RNAase A-glucose model system, in which it was shown that the glycosylated protein continued to polymerise even after removal of free glucose.[134]

Protein Polymerisation and Cross-Link Formation

It has been known for 40 years that cross-linking of the peptide chains occurs in the advanced stage of the Maillard reaction.[119,120] In 1949 it was suggested that lysine residues were required for the first step of the reaction and that this was followed by interactions involving arginine residues.[120] Most of the information obtained recently relating to the effects on the protein comes from *in vitro* physiological studies.

Studies with unmodified and chemically modified lysozymes, stored with and without glucose at 50°C and 75% rh, demonstrated the roles played by lysine and arginine in protein polymerisation.[132,135] Acetylation of the protein, resulting in blocking of the free amino groups of lysine, prevented both browning and protein polymerisation, indicating that reaction of the free amino groups is an essential first step. This was confirmed by a study on acetylated lysozyme, which was incubated with glucose in the presence and absence of free lysine,[136] and which showed that the protein polymerised only in the presence of free lysine. It was also shown that protein-bound lysine is involved in the production of appreciable amounts of glucose adducts, e.g. Schiff's bases and ARPs on incubation,[135] and suggested that arginine and, also possibly tryptophan, residues could subsequently be attacked by compounds formed from these glucose adducts.[135,136]

When the arginine residues of lysozyme were masked with cyclohexane-1,2-dione, browning and protein polymerisation occurred in a similar manner to that in the protein–glucose system,[135] suggesting that the dione was released from the arginine residues, and reacted in place of glucose, resulting in the development of colour and protein polymerisation. It therefore appeared that dicarbonyls were key components in the secondary stage of the reaction between proteins and reducing sugars.[135] This hypothesis was backed up by the observation that lysozyme monomers, isolated from a glucose–lysozyme system, demonstrated the ability to cross-link untreated lysozymes, even after removal of glucose,[132] and it was suggested that glucose-induced polymerisation of lysozyme could proceed as follows. Certain bifunctional agents, probably α-dicarbonyl compounds, which are generated from reactions between the ε-amino groups of the lysine residues with glucose, attach to arginine, lysine and tryptophan residues via one of their two functional groups. Then the proteins, with the bifunctional agents attached, may polymerise by binding of the second functional site with the remaining lysine and arginine (but not tryptophan) residues of the protein molecules. This was supported by a subsequent study which showed that α-dicarbonyls and hydroxycarbonyls, e.g. butanedione, methylglyoxal and dihydroxyacetone, had the ability to polymerise both lysozyme and acetylated lysozyme and to impair arginine residues.[133,136] The tryptophan residues seem to react only with the first functional group of the bifunctional agents, i.e. they do not appear to play a role in polymerisation of the protein–sugar complexes, but further work is required to confirm this.[132,136] Decreases in arginine, lysine and tryptophan residues were 3·3, 2·0 and 0·9, and 6·5, 2·7 and 1·9 mol/mol lysozyme after storage of the lysozyme–glucose system for 10 and 30 days, respectively.[135] Since there was a steady decrease in free arginine throughout storage, whereas the rate of decrease of free lysine and tryptophan slowed after the first 10 days, it appears that arginine residues are more involved than lysine residues in reactions of the second functional group of the protein-bound bifunctional agents.[135,136]

The key carbonyl compound was identified as 3-deoxyglucosone (3DG),[133,137] and this component was shown to bring about similar polymerisation behaviour to glucose in protein–sugar systems based on lysozyme, ovalbumin, BSA and insulin,[133,137] but the fluorescence intensity and the rate of protein polymerisation were greater with 3DG.[138] It was later suggested that the ε-amino groups of lysine act as the generators of 3DG in protein–glucose mixtures and that the arginine residues subsequently attack these protein–bound 3DG moieties,[139] and it has been shown that aminoguanidine inhibits the cross-linking effects of 3DG in glycated proteins.[138] Different behaviour was apparent in protein–fructose systems for which greater rates of polymerisation were observed. Fructose gives rise to the HRP which is not a precursor for 3DG, but 3DG can form directly from fructose at pH 5·5 and 100°C.[140] There is evidence that this route to 3DG also operates in in vitro physiological systems.[139] Storage of fructose with succinylated lysozyme gave some impairment of the arginine residues and some protein polymerisation, but no such changes to the protein were observed for the corresponding glucose system, showing that 3DG was generated from fructose without involvement of the amino groups. Interestingly, the degree of polymerisation of the succinylated lysozyme was much less than that for the intact lysozyme on incubation with fructose, although impairment of the arginine residues was about the same in both systems, suggesting that intramolecular, rather than intermolecular cross-links formed in the succinylated protein system.[139] Storage of aqueous solutions of collagen or lysozyme with fructose at 37°C resulted in a greater rate of protein polymerisation than for systems based on glucose,[141] and the results indicated that binding of fructose to the protein, by-passing the HRP, may be responsible for the acceleration of the advanced stage of the Maillard reaction *in vivo*.[141] The next step in understanding cross-link formation in protein–sugar systems is to elucidate the chemical structure of the 3DG mediated cross-link.[139] 3DG also appears to be required for the formation of protein-bound pyrroles, formed as a result of non-enzymic glycosylation, *in vivo*.[39] Reaction of glucose with a lysine residue leads to the ARP, which then breaks down to yield the 3DG which can further react with the lysine residue to give the protein-bound 5-hydroxymethyl-l-alkylpyrrole-2-carboxaldehyde.[39] The presence of carbonyl and hydroxyl groups on this structure will allow it to take part in cross-linking reactions. The nature of one cross-link formed *in vivo* has been established in a separate study (see section on The Maillard Reaction *in vivo*). The only other study reporting the nature of a cross-link in a Maillard system is that of Moller,[142] who produced evidence suggesting the presence of a lysine–carbohydrate–aspartic acid cross-link in the caseins of UHT milk stored at 37°C for 3 years.

Although work reported in the last 6 years has done much to improve our understanding of the chemical nature of the changes occurring in proteins undergoing the Maillard reaction in *in vitro* physiological systems, few studies are directly related to food, and, from the information available at present, it is not clear how the changes taking place under food processing conditions compare with those occurring at lower temperatures for longer times. Some compounds formed during the advanced stage of the Maillard reaction may also be involved in the formation of cross-links in proteins (see section on The Advanced Stage of The Chemistry of The Reaction). In addition, disulphide bonds have been implicated in cross-linking, e.g. one study has shown that heating β-lactoglobulin with lactose at 95°C for 1–4 h results in the progressive development of new bands in the acidic region by isoelectrofocusing, accompanied by a gradual reduction in the β-lactoglobulin bands.[94] These changes did not occur on the storage of the protein in the absence of lactose. The new bands appear to be due to the presence of macromolecular reaction products produced by cross-linking of the peptide chains by the formation of disulphide bridges.[104] These disulphide bridges appear to protect the lysine from attack by the lactose so that almost 40% remains chemically available after 4 h heating at 95°C. The possible role played by disulphide bridges and other bonds in the cross-linking of peptide chains in an egg white–glucose system has also been discussed.[124]

It is important to bear in mind that some cross-linking of poly-peptide chains on heat treatment or storage occurs independently of sugars.[97,100,143] For example, reactions of the ε-amino groups of lysine residues with the amide groups of asparagine and glutamine residues, and the reduction and re-oxidation of disulphide bonds.[144,145]

RELATIVE REACTIVITIES OF SUGARS TOWARDS PROTEINS

Although some reports have been published on the relative reactivity of sugars towards amino acids, e.g.[146,147] very little work has been reported on the relative reactivity of reducing sugars towards proteins. However, an early study by Lewis & Lea[148] showed that when casein was stored with reducing sugars at 37°C and 70% rh, the order of reactivity (as assessed by % drop in amino nitrogen/day) was xylose > arabinose > glucose > lactose > maltose > fructose, fructose being only one-tenth as re-active as glucose. The order arabinose > glucose > lactose was con-firmed by a later study.[149] In contrast, fructose and aldopentoses led to more extensive polymerisation of lysozyme than aldohexoses on stor-age of aqueous solutions at 37°C.[139] Amongst the aldohexoses exam-ined, galactose and mannose gave greater degrees of polymerisation than glucose. Kato et al.[150,151] investigated the reactivity of a series of aldohexoses towards ovalbumin at 50°C and 65% rh. Glucose and man-nose showed very similar behaviour, and the protein from the systems based on these sugars was almost completely soluble, even after 10 days storage. In contrast, no more than 40% of the protein from the galactose system was soluble after the same time.[150,151] The development of colour in each system increased with time of storage and with increasing sugar ratio, most colour developing in the system based on galactose. Protein insolubilisation reached a maximum at a protein:sugar ratio of 1:0·5 (cor-responding to an amino group:sugar molar ratio of 1:6), and then levelled out. The results suggest that protein polymerisation and the development of colour are not necessarily connected, and that they may occur independently via the Maillard reaction. This was backed up by a later study, in which it was shown that if free glucose was removed from the ovalbumin–glucose system at an early stage of the reaction, colour development was strongly suppressed, whereas protein polymerisation still took place.[152] The fact that the reaction with galactose proceeds at a greater rate than with either glucose or mannose was attributed to the configuration of the C_4 hydroxyl group in galactose.[150,151] This was confirmed by the study of a system based on talose (an aldohexose with the C_4 hydroxyl group in the same configuration as in galactose). Colour development and protein polymerisation behaviours were very similar for the galactose and talose systems, and differed from those based on glucose and mannose.[150,151]

It has been shown that, on reacting various monosaccharides with haemoglobin in aqueous solution, the rate of Schiff base formation depends on the proportion of the sugar which exists in the carbonyl form (rather than in the ring form) in aqueous solution,[153] and a strong correlation between the rate of reaction of 15 sugars and the proportion of the sugar existing in the carbonyl form was observed. Of the aldohexoses tested, glucose showed the lowest rate of reaction. The proportion of sugar existing in the carbonyl form is higher for galactose and talose than for glucose and mannose[154] and therefore it would be expected that the initial binding of the sugar to the free amino group would be faster for galactose and talose. However, in studies on ovalbumin–aldohexose systems, it has been shown that there is no significant difference in amino group decrease in ovalbumin stored with glucose, mannose or galactose.[150] Therefore it appears that in these systems of lower moisture content, the sugar-dependent differences in reactivity cannot be explained by the carbonyl:hemiacetal ratio of the sugar in aqueous solution, but are likely to depend on rates of formation of ARP degradation products and subsequent reactions. It has been suggested that ovalbumin–glucose and –mannose complexes convert to intermediates with an unstable cis-chair; chair configuration, while those from galactose and talose form the energetically favourable trans-chair;chair configuration, and that this may account for differences in reactivity of the two groups of sugars.[151]

In a study in which the rates of reaction of glucose and lactose with ovalbumin were compared,[152] it was shown that the free amino group contents of the systems based on both sugars were very similar after 26 days incubation, although colour development and protein polymerisation were greater in the glucose system. It was suggested that the overall increased rate of reaction of the glucose system could be due to the rapid degradation of fructosyllysine to degradation products with a high reactivity towards other components in the system, resulting in colour development and protein polymerisation. This study also illustrates the importance of the C_4 hydroxyl group of the sugar moiety reacting with the protein. With lactose, the glucopyranoside reacts with the protein, and so certain of the rearrangement steps in the glucose and lactose systems are the same in both systems. However, the C_4 hydroxyl group is protected by the galactopyranoside moiety in the lactose system, resulting in reduced browning and protein polymerisation.[152] When a series of disaccharides was reacted with ovalbumin under the same conditions,[155] systems based on isomaltose and melibiose, which both have sugar moieties linked at the C_6 hydroxyl group of the glucopyranose structure, protein polymerisation was the same as that for systems based on glucose. When disaccharides, in which the second sugar moiety was bound to the C_4 hydroxyl group of the glucopyranose were reacted, protein polymerisation was the same as for systems based on lactose.[155] The importance of the C_4 hydroxyl group in determining the rate of degradation of the ARP was confirmed by reacting 4-O-methyl-D-glucose with ovalbumin: the system showed similar behaviour to that based on lactose.[152]

NUTRITIONAL EFFECTS

The nutritional consequences of the Maillard reaction have been extensively reviewed (e.g.[1,97,102,156]). Reductions in nutritional value may be caused by involvement of amino acids, especially the essential amino acid, lysine, involvement of certain vitamins, and complexing of trace metals.[102]

Amino Acids
The Maillard reaction can lead to a serious reduction in the nutritional value of protein, mainly due to the ease with which the ε-amino groups of the lysine residues in peptide chains can react with reducing sugars, leading to the formation of derivatives, including ARPs, from which either lysine cannot be regenerated, or which are only poor sources of lysine.[102] The formation of ARPs is particularly important during the processing of milk and formulae intended for consumption by babies. Although these materials are rich in lysine, they also contain a high level of lactose, and a lack of temperature control during processing results in significant blockage of the lysine residues, and therefore reduced PER.[102] Processing

of cow's milk by freeze-drying, UHT heating by steam injection and spray-drying, result in little blocking of lysine residues. In contrast, HTST sterilisation of milk and spray-drying of infant formulae (which possess a higher lactose to protein ratio) give 5–10% blocking, while roller drying of milk can give blocking of 45–75%.[95,102] The processing of milks in which the lactose has either been hydrolysed or replaced by glucose leads to 37% and 15–55% blocking of the lysine residues, respectively.[95] Storage of processed milk also leads to blocking of the lysine residues. Spray-dried milk containing 2·5% moisture and stored at 60°C resulted in 13 and 36% blocking after 1 week and 8 weeks, respectively,[157] while UHT milk, processed by indirect heating at 140°C for 3·2 sec, and stored at 4–37°C for 6 months to 3 years showed that 10–30% of the lysine residues of the caseins had become blocked, due to the formation of both lactuloselysine and fructoselysine.[101,142] In this last study, the greatest losses appeared to occur on storage at 30°C, possibly due to a storage temperature of 37°C favouring further reactions involving the ARPs.[142]

Losses of other amino acids may also occur as a result of the Maillard reaction, as discussed in the section on The Reactivity of Amino Acid Residues, and a reduction in protein quality is the most nutritionally significant consequence of the Maillard reaction, particularly in infant foods.[102] Although lysine is an essential amino acid, in many food and feed proteins it is the sulphur amino acids which are limiting and which lead to reductions in PER.[158] A recent study showed that baby food sterilised in jars, suffered a reduction in PER of 34%.[102] When the diet of rats fed this processed food was supplemented with either lysine or tryptophan, there was no effect on the PER, while supplementation with methionine resulted in an increase in the PER to 95% of the original value. Since the rat can convert methionine to cysteine, the increase in PER on addition of methionine to the diet could be due to its acting as a source of either methionine or cysteine. However, methionine is more stable than cysteine to the processing conditions used and Hurrell[102] suggests that the methionine is more likely to be acting as a source of cysteine, and that losses of this amino acid in the baby food could occur via reactions with premelanoidins. The formation of cross-links in peptide chains (see the section on Protein Polymerisation and Cross-Link Formation) may further reduce protein digestibility,[159] due to blocking of the cleavage of peptide bonds by trypsin and carboxypeptidase B.[160] (The effect of various Maillard reaction products on the activities of several enzymes involved in the digestion of proteins and carbohydrates has recently been reviewed).[156]

Vitamins

Ascorbic acid can take the place of reducing sugars in the Maillard reaction and thus lead to a reduction in the amount of Vitamin C available to the organism. Such losses are only likely to be of importance when the foods concerned are used as a primary source of ascorbic acid in the diet, e.g. orange juice for consumption by babies and young children, and when processing and/or storage conditions favour the loss of this nutrient. Vitamins B_1, B_6, B_{12} and pantothenic acid can all react with Maillard reaction products leading to reductions in availability.[161]

Trace Metals

Certain products of the Maillard reaction, e.g. premelanoidins, are known to possess the ability to complex metals, e.g. zinc and iron, which are essential trace dietary components. This effect is particularly important for individuals being fed intravenously, and it has been shown that dramatic increases occur in urinary zinc, copper and iron excretion when intravenous feeds contain Maillard reaction products. Orally administered Maillard reaction products do not increase excretion of these metals to the same extent.[102]

FLAVOUR

The Maillard reaction is the main route to aroma development in heated foods and the types of compounds formed have been extensively reviewed, e.g.[12,13,162,163] Mostly, the Maillard reaction leads to desirable aromas, e.g. in bread, meat and coffee, but occasionally unwanted notes are produced, such as burnt odours in over-processed products, or cooked odours in foods in which it is desirable to maintain the original fresh aroma, e.g. in fruit juices. Due largely to the use of gas chromatograpy–mass spectrometry as a routine analytical technique, it has become possible to separate and identify the many, often hundreds of, components responsible for the aroma of a single heated food or model system. Aroma components formed by the Maillard reaction can best be divided into three groups:[12] (1) 'Simple' sugar dehydration/fragmentation products: furans 73, pyrones 74, cyclopentenes 75, carbonyls 76, acids 77; (2) 'simple' amino acid degradation products: aldehydes 76, sulphur compounds, e.g. 78; and (3) volatiles produced by further interactions: pyrroles 79, pyridines 80, imidazoles 81, pyrazines 82, oxazoles 83, thiazoles 84, compounds from aldol condensations. Some of the parent compounds

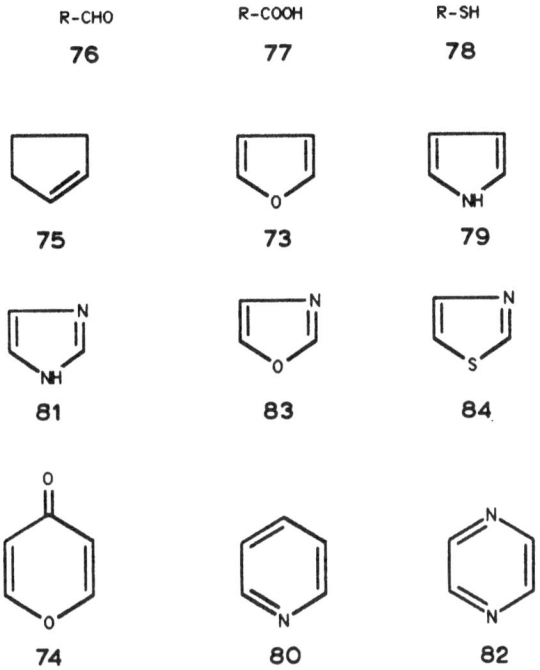

FIG. 14. Some aroma components formed by the Maillard reaction.[12] *Note:* The parent compounds are shown.

are shown in Fig. 14, and routes to the formation of some of their derivatives have been summarised.[12,162]

A great deal of work on the formation of aroma components as a result of the Maillard reaction has been carried out on sugar-amino acid model systems, and it is impossible to review fully the wealth of information available in this chapter. Instead, the reader is directed to the proceedings of symposia and reviews which deal with this subject in detail, e.g.[4–7,13,162–164] Far fewer studies have been carried out on protein–sugar systems, mainly because it is generally considered that it is the free amino acids, rather than the intact proteins which act as aroma precursors in heated foods.[165] Nevertheless, one study on freeze-dried mixtures of lactose and casein stored at 80°C and 75% rh has led to the identity of 80 aroma components comprising furans, lactones, pyrazines, pyrroles and a variety of miscellaneous compounds.[166,167] In a separate study, the identification of an additional 13 aroma components on the incorporation of starch and gluten into a glucose–glutamic acid model system heated at 150°C, led to the suggestion that gluten may be important in the development of crusty aroma during the cooking of bakery products.[168] Sugar-amino acid model systems which would be expected to be closest to sugar–protein systems would be those containing lysine or models for peptide-bound lysine, i.e. amines, as the amino compound (since this amino acid is damaged to the greatest extent as a result of the Maillard reaction). Relatively few studies have used lysine, no doubt due to the fact that more interesting aromas are obtained from, e.g. cysteine.[164] However, recent studies on sugar–lysine (or arginine) systems have revealed the formation of various bicyclic and fused ring derivatives including 2,3-dihydro-1*H*-pyrrolizines, 2-(2-furyl)pentahydroazepines and cyclopent-(b)-azocin-9-ones.[164,169,170] Unfortunately, no studies have been carried out so far looking at the formation of these components in model systems containing proteins, and the reaction mechanisms proposed for some of them[169] suggest that they would not in fact be formed from peptide-bound lysine. The ε-amino groups of lysine residues in peptide chains can react with, e.g. HMF, to give a pyrrolealdehyde[39] (see the section on Degradation of ARPs via The 3-Deoxyosone Route). However, since this compound is protein-bound, it possesses no aroma properties.

In addition to aroma components, the Maillard reaction can also result in the formation of certain taste compounds, including some very bitter components, e.g. the tricyclic compounds shown in Fig. 15,[171] which are formed from proline and sucrose. These components possess taste thresholds of 5–10 ppm and are likely to contribute to the bitterness of roasted coffee and malt.

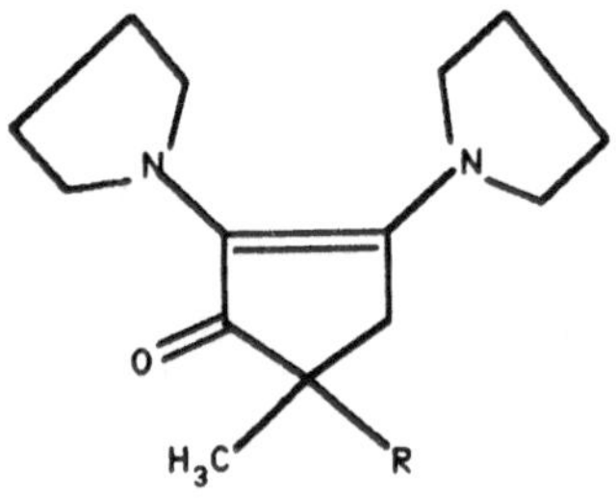

85 R=OH; **86** R=H

FIG. 15. Some bitter compounds formed by proline–glucose or pyrrolidine-glucose interactions.[171]

COLOUR

The development of colour is the most obvious effect of the Maillard reaction, although no colour is produced in the early stages. In spite of this, surprisingly little is known about the structures of the coloured compounds formed,[3] especially in comparison with the wealth of information available relating to the nature of the aroma components. The formation of colour may be either desirable as in bread, or undesirable as in concentrated milks. The colour compounds may be divided into the melanoidins, which are brown and possess molecular weights of several thousand daltons (see the section on The Final Stage of The Chemistry of The Reaction) and the low molecular weight structures, which typically comprise two to four linked rings.[75]

Very little work has involved the investigation of colour development in protein–sugar systems. However, an interesting study on the loss of lysine and the development of colour in various glucose–protein systems has shown that lysine losses do not seem to correlate with colour development,[172] possibly due to the involvement of other amino acid residues in colour development, as well as their availability for reaction. No attempt was made to characterise the colour compounds formed in this study. Many of the data available on the low molecular weight colour components have been obtained by Ledl et al.[16] who have studied a series of sugar–amino acid (or amine) model systems. Such colour compounds have not so far been isolated and identified from foods or even from model systems containing proteins as the amino compound, since the reaction product mixtures obtained from model systems based on amino acids are extremely complex and present a number of analytical challenges.[173] They may be formed from 3-deoxyosones 14 and/or 1-deoxyosones 12 as intermediates.[16] A selection of low molecular weight colour compounds formed during the Maillard reaction is shown in Fig. 16. The yellow compound 87 forms by condensation of 15 with the pyranone 18, both of which are produced from 14,[35] whilst 42, which derives from 12, can react with carbonyl compounds to yield, e.g. the yellow compound 88.[174] The formation of the yellow compound 89 on heating pentoses with primary amines or amino acids (including lysine)[175,176] (Ames, J. M., unpublished data) involves condensation of furfural with 35, showing that some of the low molecular weight colour components require both 1,2- and 2,3-enolisation for their production. Analogous compounds are formed on reaction of 35 with pyrrole derivatives. The methyl group of 89 may condense with carbonyls to give tricyclic struc-

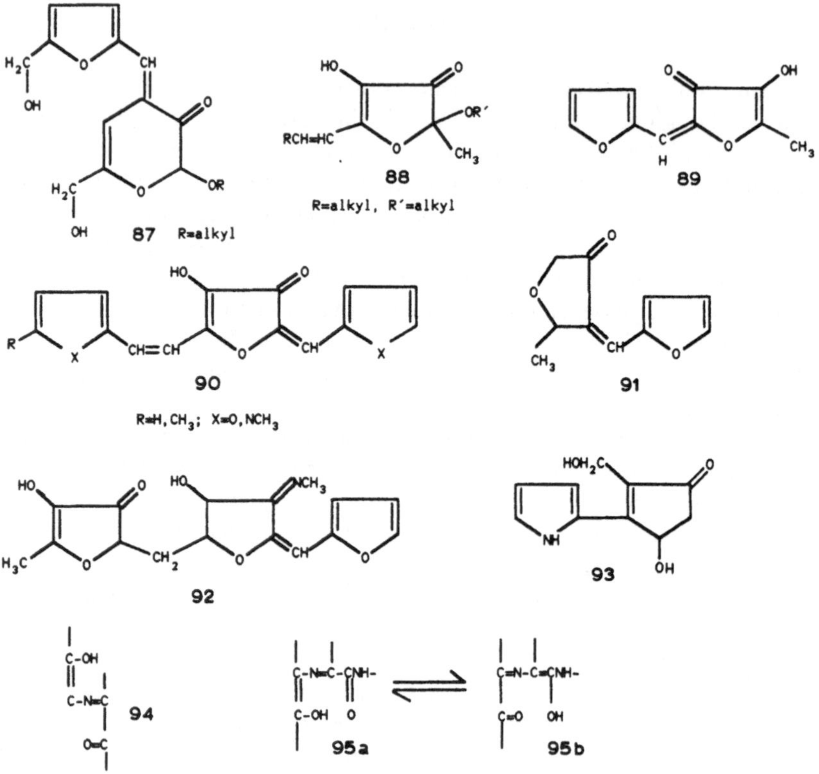

FIG. 16. Some coloured compounds formed by the Maillard reaction.[35,174-182]
(Ames, J. M., unpublished data.)

tures, e.g. 90[3] which is deep orange.[177] Compound 35 is formed in yields
of up to 5% under favourable conditions, and is considered to play a key
role in colour formation as it easily condenses with a range of carbonyl
compounds to yield a variety of colour components.[16] Other colour
compounds isolated from xylose–glycine or xylose–lysine model systems
include 91,[176] 92[178] and 93.[179] The last compound is a reminder that
cyclopentane derivatives are important products of the Maillard reaction,
although their mechanisms of formation are unclear.[18] The cyclopentane
ring of 93 is probably formed from a one-carbon unit and a four-carbon
unit, each derived from separate xylose molecules.

Coloured compounds may also be formed during the Maillard reaction by the condensation of α-dicarbonyl compounds with aminoenols **67**, formed from the Strecker degradation of amino acids, or peptides to give compounds of types **94**[180] and **95**[181], respectively.

ANTIOXIDATIVE EFFECTS

It is well known that the Maillard reaction can lead to the formation of components which possess antioxidative activity towards lipids in a variety of food systems, including milk products, cereals, biscuits and meat,[14,182-184] as well as in sugar–amino acid and sugar–peptide model systems.[184] Melanoidins are among the Maillard reaction products known to possess antioxidative activities,[185] and it has been suggested that they may also possess anticarcinogenic activity, possibly by scavenging free oxygen radicals.[185]

TOXICOLOGICAL EFFECTS

The presence of Maillard reaction products in the diet can lead to a number of toxicological effects.[156] It has become clear over the last 10 years that some steps of the Maillard reaction comprise a section of a route to the formation of certain mutagenic and carcinogenic compounds in heated protein-rich foods, especially fish and meat.[156,186-189] For example, a mixture of creatinine, fructose and glycine, containing 14% water, leads to the formation of 2-amino-3-methylimidazo [4,5-f]quinoline (IQ) **96** and 2-amino-3,8-dimethylimidazo[4,5-f]quinoxaline (MeIQx) **97** (Fig. 17).[187] Both these compounds are mutagenic and carcinogenic, in fact, IQ compounds are the most potent mutagens analysed so far by the Ames test.[188] They have both been identified in beef extract and MeIQx has also been detected in broiled beef, broiled chicken and fried beef, while IQ has been identified in broiled, sun-dried sardine.[186,189] The formation of IQ and IQx compounds has been discussed by Jagerstad *et al.*[190] It appears that yields of these components are mainly dependent on the processing conditions, and that by reducing the temperature from, e.g. 130 to 100°C, their yields will be decreased from ppb to below the limits of detection.[188] Further, protein-bound amino acids lead to no detectable mutagenicity.[188]

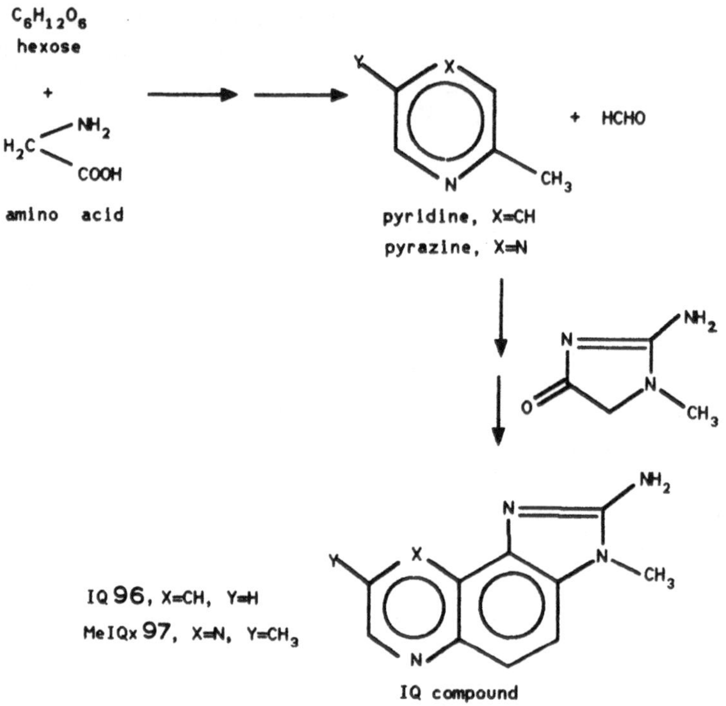

FIG. 17. Proposed mechanism of formation of 2-amino-3-methylimidazo[4,5-f]quinoline (IQ) and 2-amino-3,8-dimethylimidazo[4,5-f]quinoxaline (MeIQx).[187] Reprinted with permission from T. Nyhammar *et al.*, in *Amino-Carbonyl Reactions in Food and Biological Systems* (M. Fujimaki, ed.) 1986. © Elsevier Science Publishers, Amsterdam.

THE MAILLARD REACTION *IN VIVO*

The last 5–10 years have seen a great increase in studies of the Maillard reaction *in vivo*. It has been recognized as the cause of glucose-dependent cross-linking of proteins, implicated in both the ageing process and in some of the physiological complications faced by diabetic patients.[16,191] It is known that ageing processes, such as atherosclerosis and the development of cataracts, occur at an earlier age in diabetics than in normal people, and therefore, in some respects, diabetes may be considered to be premature ageing. The development of protein cross-links, brought about by the Maillard reaction, may explain the reduced functionality of long-lived

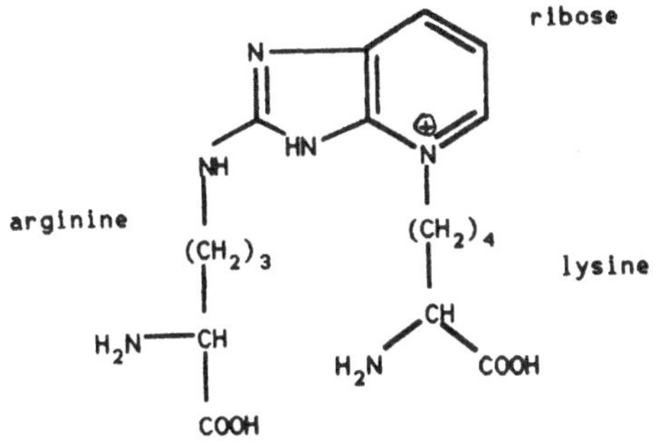

FIG. 18. Structure of the pentose-derived protein cross-link 'pentosidine.[191]

components in the body, e.g. collagen.[191] The nature of these cross-links in humans is only just beginning to be elucidated.[191] The presence of an imidazo[4,5b]pyridinium cross-link has been identified in the insoluble dura mater of elderly individuals after death, involving the sidechains of both lysine and arginine (Fig. 18). The compound was named pentosidine, since it could be synthesized *in vitro* from pentose, lysine and arginine, and it could be detected in a range of tissues.[191,192] Interestingly, its accumulation in human skin increased with age and also in the presence of diabetes.[191]

CONTROL OF THE MAILLARD REACTION

Both the extent and the course of the Maillard reaction in food systems are affected by a number of determinants, the most important being: temperature and duration of heating or storage, composition of the system, a_w, pH and the addition of sulphur dioxide.[17,193] Attempts to control the reaction usually centre on manipulation of one or more of these factors,[3,17,194,195] and may aim either to prevent or minimise the reaction, or to steer it, in order to optimise the types and amounts of various possible reaction products and to optimise food quality.

Prevention of the Maillard Reaction

The Maillard reaction increases with temperature and duration of heating or storage, and therefore reducing these parameters will minimise the extent of reaction. Nevertheless, the reaction can occur at temperatures as low as 2°C.[16] It is well documented that the Maillard reaction may be prevented by removing one of the reactants, i.e. either reducing sugars or amino compounds, or by moving the a_w away from intermediate values, i.e. 0·5–0·8.[196] The kinetics of the Maillard reaction, especially with respect to the influence of a_w, have been extensively studied by e.g. Labuza et al.[103] For example, the importance of careful control of temperature and a_w in order to prevent the Maillard reaction and thus to maintain the nutritional quality of whey powders, has been demonstrated.[197] pH also influences the course of the Maillard reaction by affecting the degradation pathway followed by ARPs, and it has been established that the route to furfurals possesses a lower browning potential than that to reductones.[3] Therefore, the rate of colour development can be reduced by lowering the pH. The Maillard reaction can also be impeded by the addition of sulphur dioxide, the use of which delays colour development.[198]

Steering of the Maillard Reaction

As far as the food manufacturer is concerned, efforts to control the Maillard reaction are of interest in order to improve or maintain food quality, especially aroma and colour,[17] and, in certain cases, nutritional value (see the section on Nutritional Effects). All the processing parameters mentioned above, i.e. temperature, time, a_w and pH, can influence the course of the reaction, and this is well illustrated by the work of Shu & Ho[199] who studied a model meat flavour system, in which the effect of manipulating each processing parameter on the chemical and sensory aroma profiles was demonstrated. Temperature and duration of heating also influence the chemical nature of the melanoidins produced (see the section on The Final Stage of The Chemistry of The Reaction). In addition, the reactant amino acid greatly influences the aroma of the system.[17] Many amino acids produce a caramel aroma, but some lead to highly specific notes, e.g. cysteine gives meaty aromas while proline and hydroxyproline lead to bread and cracker notes.[17]

CONCLUSION

Although a great deal is known about the chemistry of the Maillard reaction, there are still some large gaps in our knowledge. Much is known about the pathways leading to the formation of the volatile aroma components and, details of the pathways leading to the formation of low molecular weight non-volatile reaction products are gradually being elucidated.

Rather less information is available about the reactions leading to the formation of melanoidins, and the chemical nature of these compounds is still obscure. Most studies have looked at sugar–amino acid model systems. While this approach is undoubtedly very valuable, it is the lysine residues of polypeptide chains which form the major source of free amino groups in many foods, and the conformation of the protein will affect the availability of these groups for reaction with reducing sugars. Most of the work on protein-containing systems has been carried out very recently and studies have used physiological conditions. Therefore, studies on sugar–protein systems more relevant to food processing conditions are called for. Lastly, the control of the Maillard reaction is of great interest to the food industry as well as to food chemists, especially in view of the current interest in minimising the use of food additives, and more work could very usefully be carried out in this area.

ACKNOWLEDGEMENTS

The author's thanks are due to the originators of Figures 1, 2, 3, 4, 12, 13, 17 and 18 who have given their permission for material to be reproduced here. Thanks are also due to Professor H. E. Nursten for his comments on the manuscript, and to C. J. Ames for his help with translating some of the German papers.

REFERENCES

1. Hurrell, R. F., Reactions of food proteins during processing and storage and their nutritional consequences. In *Developments in Food Proteins*, Vol. 3, ed. B. J. F. Hudson. Elsevier, London, 1984, pp. 213–244.
2. Hurrell, R. F., Food manufacturing processes and their influence on the nutritional quality of foods. In *Nutritional Impact of Food Processing*, ed. J. C. Somogyi & H. R. Miller. Bibl. Nutr. Dieta. No. 43. Karger, Basel, 1989, pp. 125–139.
3. Nursten, H. E., Maillard browning reactions in dried foods. In *Concentration and Drying of Foods*, ed. D. MacCarthy. Elsevier Applied Science, London, 1986, pp. 53–68.
4. Eriksson, C., ed., *Maillard Reactions in Food. Progress in Food and Nutrition Science*, Vol. 5 (1–6). Pergamon, Oxford, 1981.
5. Waller, G. R. & Feather, M. S., ed. *The Maillard Reaction in Foods and Nutrition*. ACS Symp. Ser. 215, ACS, Washington DC, 1983.

6. Fujimaki, M., Namiki, M. & Kato, H., ed. *Amino-Carbonyl Reactions in Food and Biological Systems*. Developments in Food Science, Vol. 13. Elsevier, Amsterdam, 1986.

7. Finot, P. A., Aeschbacher, H. U., Hurrell, R. F. & Liardon, R., ed. *The Maillard Reaction in Food Processing, Human Nutrition and Physiology*. Birkhäuser, Basel, 1990.

8. Hodge, J. E., Chemistry of browning reactions in model systems. *J. Agric. Food Chem.*, 1 (1953) 928–943.

9. Ellis, G. P., The Maillard reaction. *Adv. Carbohydr. Chem.*, 14 (1959) 63–134.

10. Reynolds, T. H., Chemistry of nonenzymic browning. I. The reaction between aldoses and amines. *Adv. Food Res.*, 12 (1963) 1–52.

11. Reynolds, T. H., Chemistry of nonenzymic browning. II. *Adv. Food Res.*, 14 (1965) 167–283.

12. Nursten, H. E., Recent developments in studies of the Maillard reaction. *Food Chem.*, 6 (1981) 263–277.

13. Danehy, J. P., Maillard reactions: nonenzymatic browning in food systems with special reference to the development of flavor. *Adv. Food Res.*, 30 (1986) 77–138.

14. Namiki, M., Chemistry of Maillard reactions: Recent studies on the browning reaction mechanism and the development of antioxidants and mutagens. *Adv. Food. Res.*, 32 (1988) 115–184.

15. Ledl, F., Chemical pathways of the Maillard reaction. In *The Maillard Reaction in Food Processing, Human Nutrition and Physiology*, ed. P. A. Finot, H. U. Aeschbacher, R. F. Hurrell & R. Liardon. Birkhäuser, Basel, 1990, pp. 19–42.

16. Ledl, F. & Schleicher, E., New aspects of the Maillard reaction in foods and in the human body. *Angew. Chem. Int. Ed. Engl.*, 29 (1990) 565–594.

17. Ames, J. M., Control of the Maillard reaction in food systems. *Trends Food Sci. Technol.*, 1 (1990) 150–154.

18. Nursten, H. E., Key mechanistic problems posed by the Maillard reaction. In *The Maillard Reaction in Food Processing, Human Nutrition and Physiology*, ed. P. A. Finot, H. U. Aeschbacher, R. F. Hurrell & R. Liardon. Birkhäuser, Basel, 1990, pp. 145–153.

19. Hodge, J. E., Origin of flavors in foods: non-enzymatic browning reactions. In *Symp. Foods: Chemistry and Physiology of Flavors*, ed. H. W. Schultz, E. A. Day & L. M. Libbey. AVI, Westport, Conn., 1967, pp. 465–491.

20. Mills, F. D., Baker, B. G. & Hodge, J. E., Amadori compounds as non-specific flavor precursors in processed foods. *J. Agric. Food Chem.*, 17 (1969) 723–727.

21. Anet, E. F. L. J., Chemistry of non-enzymic browning. II. Some crystalline amino acid-deoxy sugars. *Aust. J. Chem.*, 10 (1957) 193–197.

22. Baltes, W., Franke, K., Hörtig, W., Otto, R. & Lessig, U., Investigations on model systems of Maillard reactions. In *Maillard Reactions in Food*, ed. C. Eriksson. *Progress in Food and Nutrition Science*, Vol. 5 (1–6). Pergamon Press, Oxford, 1981, pp. 137–145.

23. Heyns, K., Müller, G. & Paulsen, H., Quantitative studies on the reactions

of hexoses with amino acids. *Justus Liebigs Ann. Chem.*, **703** (1967) 202–214.

24. Hayashi, T., Mase, S. & Namiki, M., Formation of the N,N'-dialkyl-pyrazine cation radical from glyoxal dialkylimine produced on reaction of a sugar with an amine or amino acid. *Agric. Biol. Chem.*, **49** (1985) 3131–3137.

25. Hayashi, T., Ohta, Y. & Namiki, M., Electron spin resonance spectral study on the structure of the novel free radical products formed by the reactions of sugars with amino acids or amines. *J. Agric. Food Chem.*, **25** (1977) 1282–1287.

26. Namiki, M. & Hayashi, T., A new mechanism of the Maillard reaction involving sugar fragmentation and free radical formation. In *The Maillard Reaction in Foods and Nutrition*, ed. G. R. Waller & M. S. Feather. ACS Symp. Ser. 215, ACS, Washington DC, 1983, pp. 21–46.

27. Hayashi, T., Mase, S. & Namiki, M., Formation of three-carbon sugar fragment at an early stage of the browning reaction of sugar with amines or amino acids. *Agric. Biol. Chem.*, **50** (1986) 1959–1964.

28. Hayashi, T. & Namiki, M., Role of sugar fragmentation in an early stage browning of amino-carbonyl reaction of sugar with amino acids. *Agric. Biol. Chem.*, **50** (1986) 1965–1970.

29. Yano, M., Hayashi, T. & Namiki, M., Formation of free radical products by the reaction of dehydroascorbic acid with amino acid. *J. Agric. Food Chem.*, **24** (1976) 815–819.

30. Gomyo, T., Haiyan, L., Miura, M., Hayase, F. & Kato, H., Kinetic aspects of the blue pigment formation in a Maillard reaction between D-xylose and glycine. *Agric. Biol. Chem.*, **53** (1989) 949–957.

31. Yaylayan, V., Jocelyn Pare, J. R., Laing, R. & Sporns, P., Intramolecular nucleophilic substitution reactions of tryptophan and lysine Amadori rearrangement products. In *The Maillard Reaction in Food Processing, Human Nutrition and Physiology*, ed. P. A. Finot, H. U. Aeschbacher, R. F. Hurrell & R. Liardon. Birkhäuser, Basel, 1990, pp. 115–120.

32. Anet, E. F. L. J., Formation of furan compounds from sugars. *Chem. Ind.* (1962) 262.

33. Beck, J., Ledl, F., Sengl, M. & Severin, T., Formation of acids, lactones and esters through the Maillard reaction. *Z. Lebensm. Unters. Forsch.*, **190** (1990) 212–216.

34. Ledl, F., Hiebl, J. & Severin, T., Formation of coloured β-pyranones from hexoses and pentoses. *Z. Lebensm. Unters. Forsch.*, **177** (1983) 353–355.

35. Kato, H., Chemical studies on amino-carbonyl reaction III. Formation of substituted pyrrole-2-aldehydes by reaction of aldoses with alkylamines. *Agric. Biol. Chem.* **31** (1967) 1086–1090.

36. Jurch, G. R. & Tatum, J. H., Degradation of D-glucose with acetic acid and methylamine. *Carbohydr. Res.*, **15** (1970) 233–239.

37. Pachmayr, O., Ledl, F. & Severin, T., Formation of 1-alkyl-3-oxopyridiniumbetaines from sugars. XXI. Investigations relating to the Maillard reaction. *Z. Lebensm. Unters. Forsch.*, **182** (1986) 294–297.

38. Hayase, F., Nagaraj, R. H., Miyata, S., Njoroge, F. G. & Monnier, V. M.,

Aging of proteins: immunological detection of a glucose-derived pyrrole formed during Maillard reaction *in vivo. J. Biol. Chem.*, **264** (1989) 3758–3764.

39. Olsson, K., Pernemalm, P. Å., Popoff, T. & Theander, O., Formation of aromatic compounds from carbohydrates. V. Reaction of D-glucose and methylamine in slightly acidic aqueous solution. *Acta Chem. Scand. Ser. B,* **31** (1977) 469–474.

40. Shigematsu, H., Kurata, T., Kato, H. & Fujimaki, M., Formation of 2-(5-hydroxymethyl-2-formylpyrrol-1-yl)alkyl acid lactones on roasting alkyl-α-amino acid with D-glucose. *Agric. Biol. Chem.*, **35** (1971) 2097–2105.

41. Olsson, K., Pernemalm, P. Å. & Theander, O., Formation of aromatic compounds from carbohydrates. VII. Reaction of D-glucose and glycine in slightly acidic aqueous solution. *Acta Chem. Scand. Ser. B,* **32** (1978) 249–256.

42. Olsson, K., Pernemalm, P. Å. & Theander, O., Reaction products and mechanism in some simple model systems. In *Maillard Reactions in Food,* ed. C. Eriksson. *Progress in Food and Nutrition Science,* Vol. 5 (1–6). Pergamon Press, Oxford. 1981, 47–55.

43. Njoroge, F. G., Sayre, L. M. & Monnier, V. M., Detection of D-glucose-derived pyrrole compounds during Maillard reaction under physiological conditions. *Carbohydr. Res.*, **167** (1987) 211–220.

44. Miller, R., Olsson, K. & Pernemalm, P. Å., Formation of aromatic compounds from carbohydrates. IX. Reaction of D-glucose and L-lysine in slightly acidic, aqueous solution. *Acta Chem. Scand. Ser. B,* **38** (1984) 689–694.

45. Tressl, R., Helak, B. & Rewicki, D., Malzoxazin, eine tricyclische Verbindung aus Gerstenmalz. *Helv. Chim. Acta.,* **65** (1982) 483–489.

46. Ledl, F., Krönig, U., Severin, T. & Lotter, H., Investigations relating to Maillard-reaction XVIII. Isolation of N-containing coloured products. *Z. Lebensm. Unters. Forsch.,* **177** (1983) 267–270.

47. Njoroge, F. G., Fernandes, A. A. & Monnier, V. M., 3-(D-erythro-Trihydroxypropyl)-1-neopentylpyrrolecarboxaldehyde, a novel nonenzymatic browning product of glucose. *J. Carbohydr. Chem.,* **6** (1987) 553–568.

48. Farmar, J. G., Ulrich, P. C. & Cerami, A., Novel pyrroles from sulfite-inhibited Maillard reactions: Insight into the mechanism of inhibition. *J. Org. Chem.,* **53** (1988) 2346–2349.

49. Ledl, F., Fritsch, G., Hiebl, J., Pachmayr, O. & Severin, T., Degradation of Maillard products. In *Amino-Carbonyl Reactions in Food and Biological Systems,* ed. M. Fujimaki, M., Namiki &. H. Kato. Developments in Food Science, Vol. 13. Elsevier, Amsterdam, 1986, pp. 173–182.

50. Beck, J., Ledl, F. & Severin, T., Formation of 1-deoxy-D-*erythro*-2,3-hexodiulose from Amadori compounds. *Carbohydr. Res.,* **117** (1988) 240–243.

51. Hiebl, J., Ledl, F. & Severin, T., Isolation of 4-hydroxy-2-(hydroxymethyl)-5-methyl-3(2H)-furanone from sugar amino acid reaction mixtures. *J. Agric. Food Chem.,* **35** (1987) 990–993.

52. Hodge, J. E., Fisher, B. E. & Nelson, E. C., Dicarbonyls, reductones and

heterocyclics produced by reactions of reducing sugars with secondary amine salts. *Proc. Am. Soc. Brew. Chem.* (1963) 84–92.

53. Ledl, F., Formation of aminoreductones from disaccharides. *Z. Lebensm. Unters. Forsch.*, **179** (1984) 381–384.

54. Ledl, F. & Fritsch, G., Formation of pyrrolinone reductones by heating hexoses with amino acids. *Z. Lebensm. Unters. Forsch.*, **178** (1984) 41–44.

55. Mills, F. D., Baker, B. G. & Hodge, J. E., Thermal degradation of 1-deoxy-1-piperidino-D-fructose. *Carbohydr. Res.*, **15** (1970) 205–213.

56. Ledl, F. & Severin, T., Formation of aminoreductones from glucose and primary amines. XIV. Investigations relating to Maillard reaction. *Z. Lebensm. Unters. Forsch.*, **169** (1979) 173–175.

57. Simon, H., Heubach, G., Bitterlich, W. & Gleinig, H., Reaction of bromodiacetyl and of alicyclic 1-chloro-2,3-diones with primary and secondary amines to reductones and some properties of the products. *Chem. Ber.*, **98** (1965) 3692–3702.

58. Ledl, F., Ellrich, G. & Klostermeyer, H., Proof and identification of a new Maillard compound in heated milk. *Z. Lebensm. Unters. Forsch.*, **182** (1986) 19–24.

59. Matsuura, H., Hirao, Y., Yoshida, Y., Kumihiro, K., Fuwa, T., Kasai, R. & Tanaka, O., Study of red ginseng: new glucosides and a note on the occurrence of maltol. *Chem. Pharm. Bull.*, **32** (1984) 4674–4677.

60. Hodge, J. E. & Nelson, E. C., Preparation and properties of galactosylisomaltol and isomaltol. *Cereal Chem.*, **38** (1961) 207–221.

61. Patton, S., The formation of maltol in certain carbohydrate–glycine systems. *J. Biol. Chem.*, **184** (1950) 131–134.

62. Peer. H. G., van den Ouweland, G. A. M. & de Groot, C. N., The reaction of aldopentoses and secondary amine salts, a convenient method of preparing 4-hydroxy-5-methyl-2,3-dihydrofuran-3-one. *Recl. Trav. Chim. Pays-Bas*, **87** (1968) 1011–1020.

63. Severin, T. & Loidl, A., Formation of pyridone derivatives from maltose and lactose. XII. Investigations of the Maillard reaction. *Z. Lebensm. Unters. Forsch.*, **161** (1976) 119–124.

64. Ledl, F., Osiander, H., Pachmayr, O. & Severin, T., Formation of maltosine, a product of the Maillard reaction with a pyridone structure. *Z. Lebensm. Unters. Forsch.*, **188** (1989) 207–211.

65. Huber, B. & Ledl, F., Formation of 1-amino-1,4-dideoxy-2,3-hexodiuloses and 2-aminoacetylfurans in the Maillard reaction. *Carbohydr. Res.*, **204** (1990) 215–220.

66. Huber, B., Ledl, F., Severin, T., Stangl, A. & Pfleiderer, G., Formation of 2-(2-furoyl)-4(5)-(2-furyl)-1*H*-imidazole in the Maillard reaction. *Carbohydr. Res.*, **182** (1988) 301–306.

67. Njoroge, F. G., Fernandes, A. A. & Monnier, V. M., Mechanism of formation of the putative advanced glycosylation end product and protein cross-link 2-(2-furoyl)-4,5-(2-furanyl)-1*H*-imidazole. *J. Biol. Chem.*, **263** (1988) 10646–10652.

68. Estendorfer, S., Ledl, F. & Severin, T., Formation of an aminoreductone from glucose. *Angew. Chem. Int. Ed. Engl.*, **29** (1990) 536–537.

69. Beck, J., Ledl, F., Sengl, M. & Severin, T., Formation of glucosyl deoxyosones from Amadori compounds of maltose. *Z. Lebensm. Unters. Forsch.*, **188** (1989) 118–121.
70. Ahmed, M. U., Thorpe, S. R. & Baynes, J. W., Identification of N^ε-carboxymethyllysine as a degradation product of fructoselysine in glycated protein. *J. Biol. Chem.*, **261** (1986) 4889–4894.
71. Hayashi, T. & Namiki, M., Role of sugar fragmentation in the Maillard reaction. In *Amino-Carbonyl Reactions in Food and Biological Systems*, ed. M. Fujimaki, M. Namiki & H. Kato. Developments in Food Science, Vol. 13. Elsevier, Amsterdam, 1986, pp. 29–38.
72. Shaw, P. E. & Berry, R. E., Hexose-amino acid degradation studies involving formation of pyrroles, furans and other low molecular weight products. *J. Agric. Food Chem.*, **25** (1977) 641–644.
73. Nyhammar, T., Olsson, K. & Pernemalm, P. Å., In *The Maillard Reaction in Foods and Nutrition*, ed. G. R. Waller & M. S. Feather. ACS Symp. Ser. 215, ACS, Washington D.C.,1983, pp. 71–82.
74. Nyhammar, T., Olsson, K. & Pernemalm, P. Å., On the formation of 2-acylpyrroles and 3-pyridinols in the Maillard reaction through Strecker degradation. *Acta Chem. Scand. Ser. B*, **37** (1983) 879–889.
75. Ames, J. M. & Nursten, H. E., Recent advances in the chemistry of coloured compounds formed during the Maillard reaction. In *Trends in Food Science*, ed. W. S. Lien & C. W. Foo. Singapore Institute of Food Science and Technology, 1989, pp. 8–14.
76. Motai, H., Viscosity of melanoidins formed by oxidative browning. Validity of the equation for a relationship between color intensity and molecular weight of melanoidin. *Agric. Biol. Chem.*, **40** (1976) 1–7.
77. Kato, H. & Tsuchida, H., Estimation of melanoidin structure by pyrolysis and oxidation. In *Maillard Reactions in Food*, ed. C. Eriksson. *Progress in Food and Nutrition Science*, Vol. 5 (1–6). Pergamon Press, Oxford, 1981, pp. 147–156.
78. Barbetti, P. & Chiappini, I., Fractionation and spectroscopic characterization of melanoidic pigments from a glucose–glycine non-enzymic browning system. *Ann. Chim. (Rome)*, **66** (1976) 293–304.
79. Feather, M. S., Some aspects of the chemistry of non-enzymatic browning (the Maillard reaction). In *Chemical Changes in Food During Processing*, ed. T. Richardson & J. W. Finley. AVI, Westport, 1985, pp. 289–303.
80. Feather, M. S. & Nelson, D., Maillard polymers derived from D-glucose, D-fructose, 5-(hydroxymethyl)-2-furaldehyde, and glycine and methionine. *J. Agric. Food Chem.*, **32** (1984) 1428–1432.
81. Benzing-Purdie, L. M. & Ratcliffe, C. I., A Study of the Maillard reaction by ^{13}C and ^{15}N CP-MAS NMR: Influence of time, temperature and reactants on major products. In *Amino-Carbonyl Reactions in Food and Biological Systems*, ed. M. Fujimaki, M. Namiki & H. Kato. Developments in Food Science, Vol. 13. Elsevier, Amsterdam, 1986, pp. 193–205.
82. Feather, M. S. & Huang, R. -D., Some studies on a Maillard polymer derived from L-alanine and D-glucose. In *Amino-Carbonyl Reactions in Food and Biological Systems*, ed. M. Fujimaki, M. Namiki & H. Kato. Developments in Food Science, Vol 13. Elsevier, Amsterdam, 1986, pp. 183–192.

83. Kato, H., Kim, S. B. & Hayase, F., Estimation of the partial chemical structures of melanoidins by oxidative degradation and ^{13}C CP-MAS NMR. In *Amino-Carbonyl Reactions in Food and Biological Systems*, ed. M. Fujimaki, M. Namiki & H. Kato. Developments in Food Science, Vol 13. Elsevier, Amsterdam, 1986, pp. 215–223.

84. Hayase, F., Kim, S. B. & Kato, H., Analysis of the chemical structures of melanoidins by ^{13}C NMR, ^{13}C and ^{15}N CP-MAS NMR spectrometry. *Agric. Biol. Chem.*, **50** (1986) 1951–1957.

85. Benzing-Purdie, L. M., Ripmeester, J. A. & Ratcliffe, C. I., Effects of temperature on Maillard reaction products. *J. Agric. Food Chem.*, **33** (1985) 31–33.

86. Baltes, W., Application of pyrolytic methods in food chemistry. *J. Anal. Appl. Pyrol.*, **8** (1985) 533–545.

87. Wu, C. H., Russell, G. F. & Powrie, W. D., Paramagnetic behaviour of model system melanoidins. In *Amino-Carbonyl Reactions in Food and Biological Systems*, ed. M. Fujimaki, M. Namiki & H. Kato. Developments in Food Science, Vol 13. Elsevier, Amsterdam, 1986, pp. 135–144.

88. Clark, A. V. & Tannenbaum, S. R., Isolation and characterisation of pigments from protein-carbonyl browning systems. Isolation, purification and properties. *J. Agric. Food Chem.*, **18** (1970) 891–894.

89. Clark, A. V. & Tannenbaum, S. R., Studies on limit-peptide pigments from glucose-casein browning systems using radioactive glucose. *J. Agric. Food Chem.*, **21** (1973) 40–43.

90. Clark, A. V. & Tannenbaum, S. R., Isolation and characterization of pigments from protein–carbonyl browning systems. Models for two insulin-glucose pigments. *J. Agric. Food Chem.*, **22** (1974) 1089–1093.

91. Henry, K. M., Kon, S. K., Lea, C. H. & White, J. C. D., Deterioration on storage of dried skim milk. *J. Dairy Res.*, **15** (1948) 293–356.

92. Lea, C. H. & Hannan, R. S., Studies of the reaction between proteins and reducing sugars in the 'dry' state. II. Further observations on the formation of the casein-glucose complex. *Biochim. Biophys. Acta.*, **4** (1950) 518–531.

93. Lea, C. H. & Hannan, R. S., Studies of the reaction between proteins and reducing sugars in the 'dry' state. III. Nature of the protein groups reacting. *Biochim. Biophys. Acta.*, **5** (1950) 433–454.

94. Ludwig, E., Die Verfolgung der Maillard-Reaktion zwischen β-Lactoglobulin und Lactose mittels Isoelektrofocussierung in Polyacrylamidgelen. *Nahrung*, **18** (1974) 615–620.

95. Finot, P. A., Deutsch, R. & Bujard, E., The extent of the Maillard reaction during the processing of milk. In *Maillard Reactions in Food*, ed. C. Eriksson. *Progress in Food and Nutrition Science*, Vol. 5 (1–6). Pergamon Press, Oxford, 1981, pp. 345–355.

96. Hurrell, R. F. & Finot, P. A., Food processing and storage as a determinant of protein and amino acid availability. In *Nutritional Adequacy, Nutrient Availability and Needs*, ed. J. Mauron. Experientia Basel Supplementum. Birkhäuser, Basel, 1983, pp. 135–156.

97. Erbersdobler, H. F., Protein reactions during food processing and storage —their relevance to human nutrition. In *Nutritional Impact of Food Processing*, ed. J. C. Somogyi & H. R. Muller. Bibl. Nutr. Dieta. No. 43. Karger, Basel, 1989, pp. 140–155.

98. Frangne, R. & Adrian, J., The Maillard reaction. VI. Reactivity of various purified proteins. *Ann. Nutr. Aliment.*, **26** (1972) 97–106.

99. Adrian, J. & Frangne, R., Le comportement des proteolysats au cours de la reaction de Maillard. *Ind. Aliment. Agric.*, **26** (1976) 23–28.

100. Hurrell, R. F. & Carpenter, K. J., Nutritional significance of cross-link formation during food processing. In *Protein Crosslinking. Nutritional and Medical Consequences*, ed. M. Friedman. *Advances in Experimental Medicine and Biology*, Vol. 86B. Plenum Press, New York, 1977, pp. 225–238.

101. Möller, A. B., Andrews, A. T. & Cheeseman, G. C., Chemical changes in ultra-heat-treated milk during storage. II. Lactuloselysine and fructoselysine formation by the Maillard reaction. *J. Dairy Res.*, **44** (1977) 267–275.

102. Hurrell, R. F., Influence of the Maillard reaction on the nutritional value of foods. In *The Maillard Reaction in Food Processing, Human Nutrition and Physiology*, ed. P. A. Finot, H. U. Aeschbacher, R. F. Hurrell & R. Liardon. Birkhäuser, Basel, 1990, pp. 245–258.

103. Labuza, T. P. & Saltmarch, M., The nonenzymatic browning reaction as affected by water in foods. In *Water Activity: Influences on Food Quality*, ed. L. B. Rockland & G. F. Stewart. Academic Press, New York, 1981, pp. 605–650.

104. Ludwig, E., Untersuchungen zur Maillard-Reaktion zwischen β-Lactoglobulin und Lactose. 3. Mitt. Der Einfluß intermolekularer Disulfidbrücken auf die Blockierung von Lysin. *Nahrung*, **23** (1979) 707–714.

105. Nielsen, H. K., De Wecke, D., Finot, P. A., Liardon, R. & Hurrell, R. F., Stability of tryptophan during food processing and storage. 1. Comparative losses of tryptophan, lysine and methionine in different model systems. *Br. J. Nutr.*, **53** (1985) 281–292.

106. Nielsen, H. K., Klein, A. & Hurrell, R. F., Stability of tryptophan during food processing and storage. 2. A comparison of the methods used for the measurement of tryptophan losses in processed foods. *Br. J. Nutr.*, **53** (1985) 293–300.

107. Möller, A. B., Andrews, A. T. & Cheeseman, G. C., Chemical changes in ultra-heat-treated milk during storage. III. Methods for the estimation of lysine and sugar-lysine derivatives formed by the Maillard reaction. *J. Dairy Res.*, **44** (1977) 277–281.

108. Hurrell, R. F. & Carpenter, K. J., The estimation of available lysine in foodstuffs after Maillard reactions. In *Maillard Reactions in Food*, ed. C. Eriksson. *Progress in Food and Nutrition Science*, Vol. 5 (1–6). Pergamon Press, Oxford, 1981, pp. 159–176.

109. Carpenter, K. J. & Booth, V. H. Damage to lysine in food processing: its measurement and its significance. *Nutr. Abstr. Rev.*, **43** (1973) 423–451.

110. Erbersdobler, H., Amino acid availability. In *Protein Metabolism and Nutrition*, ed. D. J. A. Cole, K. N. Boorman, P. J. Buttery, D. Lewis, R. J. Neale & H. Swan. Butterworths, London, 1976, pp. 139–158.

111. Tanaka, M., Lee, T. -C. & Chichester, C. O., Effect of browning on chemical properties of egg albumin. *Agric. Biol. Chem.*, **39** (1975) 863–866.

112. Kato, H., Matsumura, M. & Hayase, F., Chemical changes in casein heated with and without D-glucose in the powdered state or in an aqueous solution. *Food Chem.*, **7** (1981) 159–168.

113. Holt, C., Muir, D. D. & Sweetsur, A. W. M., The heat stability of milk and concentrated milk containing added aldehydes and sugars. *J. Dairy Res.*, **45** (1978) 47–52.

114. Nelson, V., Effects of formaldehyde and copper salts on the heat stability of evaporated milk. *J. Dairy Sci.*, **37** (1954) 825–829.

115. Tybor, P. T., Dill, C. W. & Landmann, W. A., Effect of decolorization and lactose incorporation on the emulsification capacity of spray-dried blood protein concentrates. *J. Food Sci.*, **38** (1973) 4–6.

116. Morales, M., Dill, C. W. & Landmann, W. A. Effect of Maillard condensation with D-glucose on the heat stability of bovine serum albumin. *J. Food Sci.*, **41** (1976) 234–236.

117. Back, J. F., Oakenfull, D. and Smith, M. B., Increased thermal stability of proteins in the presence of sugars and polyols. *Biochemistry*, **18** (1979) 5191–5196.

118. Lea, C. H., The reaction between milk protein and reducing sugar in the dry state. *J. Dairy Res.*, **15** (1948) 369–376.

119. Lea, C. H., Hannan, R. S. & Rhodes, D. N. Studies of the reaction between proteins and reducing sugars in the dry state. IV. Decomposition of the amino-sugar complex and the reaction of acetylated casein with glucose. *Biochim. Biophys. Acta.*, **7** (1951) 366–377.

120. Mohammad, A., Fraenkel-Conrat, H. & Olcott, H. S., The "browning" reaction of proteins with glucose. *Arch. Biochim. Biophys.*, **24** (1949) 157–178.

121. Shalabi, S. I. & Fox, P. F., Heat stability of milk: synergistic action of urea and carbonyl compounds. *J. Dairy Res.*, **49** (1982) 197–207.

122. Shalabi, S. I. & Fox, P. F., Heat stability of milk: influence of modification of lysine and arginine on the heat stability-pH profile. *J. Dairy Res.*, **49** (1982) 607–617.

123. Shalabi, S. I. & Fox, P. F., Effect of diacetyl on the heat stability of concentrated milks. *J. Food Technol.*, **17** (1982) 753–760.

124. Kato, Y., Watanabe, K. & Sato, Y., Effect of the Maillard reaction on the attributes of egg white proteins. *Agric. Biol. Chem.*, **42** (1978) 2233–2237.

125. Watanabe, K., Sato, Y. & Kato, Y., Chemical and conformational changes of ovalbumin due to the Maillard reaction. *J. Food Process. Preserv.*, **3** (1980) 263–274.

126. Kato, Y., Watanabe, K. & Sato, Y., Effect of Maillard reaction on some physical properties of albumin. *J. Food Sci.*, **46** (1981) 1835–1839.

127. Kato, Y., Watanabe, K. & Sato, Y., Effect of some metals on the Maillard reaction of ovalbumin. *J. Agric. Food Chem.*, **29** (1981) 540–543.

128. Kato, Y., Watanabe, K. & Sato, Y., Conformational stability of ovalbumin reacted with glucose in a Maillard reaction. *Agric. Biol. Chem.*, **47** (1983) 1925–1926.

148   JENNIFER M. AMES

129. Wu, H., Govindarajan, S., Smith, T., Rosen, J. D. & Ho, C.-T., Glucose–lysozyme reactions in a restricted water environment. In *The Maillard Reaction in Food Processing, Human Nutrition and Physiology*, ed. P. A. Finot, H. U. Aeschbacher, R. F. Hurrell & R. Liardon. Birkhäuser, Basel, 1990, pp. 85–90.

130. Hull, C. J., Studies on the glycation and Maillard reactions of protein. Ph.D. Dissertation, University of South Carolina, Columbia, South Carolina, 1985.

131. Baynes, J. W., Ahmed, M. U., Fisher, C. I., Hull, C. J., Lehman, T. A., Watkins, N. G. & Thorpe, S. R., Studies on glycation of proteins and Maillard reactions of glycated proteins under physiological conditions. In *Amino-Carbonyl Reactions in Food and Biological Systems*, ed. M. Fujimaki, M. Namiki & H. Kato. Developments in Food Science, Vol. 13. Elsevier, Amsterdam, 1986, pp. 421–431.

132. Cho, R. K., Okitani, A. & Kato, H., Chemical properties and polymerizing ability of the lysozyme monomer isolated after storage with glucose. *Agric. Biol. Chem.*, **48** (1984) 3081–3089.

133. Cho, R. K., Okitani, A. & Kato, H., Polymerisation of proteins and impairment of their arginine residues due to intermediate compounds in the Maillard reaction. In *Amino-Carbonyl Reactions in Food and Biological Systems*, ed. M. Fujimaki, M. Namiki & H. Kato. Developments in Food Science, Vol. 13. Elsevier, Amsterdam, 1986, pp. 439–448.

134. Eble, A. S., Thorpe, S. R. & Baynes, J. W., Nonenzymatic glucosylation and glucose-dependent cross-linking of protein. *J. Biol. Chem.*, **258** (1983) 9406–9412.

135. Okitani, A., Cho, R. K. & Kato, H., Polymerization of lysozyme and impairment of its amino acid residues caused by reaction with glucose. *Agric. Biol. Chem.*, **48** (1984) 1801–1808.

136. Cho, R. K., Okitani, A. & Kato, H., Polymerization of acetylated lysozyme and impairment of their amino acid residues due to α-dicarbonyl and α-hydroxycarbonyl compounds. *Agric. Biol. Chem.*, **50** (1986) 1373–1380.

137. Kato, H., Cho, R. K., Okitani, A. & Hayase, F., Responsibility of 3-deoxyglucosone for the glucose-induced polymerization of proteins. *Agric. Biol. Chem.*, **51** (1987) 683–689.

138. Igaki, N., Saai, M., Hata, F., Yamada, H., Oimomi, M., Baba, S. & Kato, H., The role of 3-deoxyglucosone in the Maillard reaction. In *The Maillard Reaction in Food Processing, Human Nutrition and Physiology*, ed. P. A. Finot, H. U. Aeschbacher, R. F. Hurrell & R. Liardon. Birkhäuser, Basel, 1990, pp. 103–108.

139. Shin, D. B., Hayase, F. & Kato, H., Polymerization of proteins caused by reaction with sugars and the formation of 3–deoxyglucosone under physiological conditions. *Agric. Biol. Chem.*, **52** (1988) 1451–1458.

140. Kato, H., Yamamoto, M. & Fujimaki, M., Mechanisms of browning degradation of D-fructose in special comparison with D-glucose–glycine reaction. *Agric. Biol. Chem.*, **33** (1969) 939–948.

141. Sakai, M., Igaki, N., Nakamichi, T., Ohara, T., Masuta, S., Maeda, Y., Hata, F., Oimomi, M. & Baba, S., Acceleration of fructose-mediated collagen glycation. In *The Maillard Reaction in Food Processing, Human Nutri-*

tion and Physiology, ed. P. A. Finot, H. U. Aeschbacher, R. F. Hurrell & R. Liardon. Birkhäuser, Basel, 1990, pp. 481–486.

142. Möller, A. B., Chemical changes in ultra heat treated milk during storage. In *Maillard Reactions in Food*, ed. C. Eriksson. *Progress in Food and Nutrition Science*, Vol. 5 (1–6). Pergamon Press, Oxford, 1981, pp. 357–368.

143. Dworschak, E., Nonenzymic browning and its effect on protein nutrition. *CRC Crit. Rev. Food Sci. Nutr.*, **13** (1980) 1–40.

144. Feeney, R. E., Overview on the chemical deteriorative changes of proteins and their consequences. In *Chemical Deterioration of Proteins*, ed. J. R. Whitaker & M. Fujimaki. ACS Symp. Ser. 123, ACS, Washington D.C., 1980, pp. 1–47.

145. Matsumoto, J. J., Chemical deterioration of muscle proteins during frozen storage. In *Chemical Deterioration of Proteins*, ed. J. R. Whitaker & M. Fujimaki. ACS Symp. Ser. 123, ACS, Washington D.C., 1980, pp. 95–124.

146. Spark, A. A., Role of amino acids in non-enzymic browning. *J. Sci. Food Agric.*, **20** (1969) 308–316.

147. Ashoor, S. H. & Zent, J. B., Maillard browning of common amino acids and sugars. *J. Food Sci.*, **49** (1984) 1206–1207.

148. Lewis, V. M. & Lea, C. H., A note on the relative rates of reaction of several reducing sugars and sugar derivatives with casein. *Biochim. Biophys. Acta.*, **4** (1950) 532–534.

149. Rao, N. M. & Rao, M. M. Effect of non-enzymatic browning on the nutritive value of casein-sugar complexes. *J. Food Sci. Tech. (Mysore)*, **9** (1972) 66–68.

150. Kato, Y., Matsuda, T., Kato, N., Watanabe, K. & Nakamura, R., Browning and insolubilisation of ovalbumin by the Maillard reaction with some aldohexoses. *J. Agric. Food Chem.*, **34** (1986) 351–355.

151. Kato, Y., Matsuda, T., Kato, N., Watanabe, K. & Nakamura, R., Maillard reaction of some aldohexoses with ovalbumin. In *Amino-Carbonyl Reactions in Food and Biological Systems*, ed. M. Fujimaki, M. Namiki & H. Kato. Developments in Food Science, Vol. 13. Elsevier, Amsterdam, 1986, pp. 115–124.

152. Kato, Y., Matsuda, T., Kato, N. & Nakamura, R., Browning and protein polymerisation induced by amino-carbonyl reaction of ovalbumin with glucose and lactose. *J. Agric. Food Chem.*, **36** (1988) 806–809.

153. Bunn, H. F. & Higgins, P. J., Reaction of monosaccharides with protein: possible evolutionary significance. *Science*, **213** (1981) 222–224.

154. Hayward, L. D. & Angyal, S. J., A symmetry rule for the circular dichroism of reducing sugars, and the proportion of carbonyl forms in aqueous solutions thereof. *Carbohydr. Res.*, **53** (1977) 13–20.

155. Kato, Y., Matsuda, T., Kato, N. & Nakamura, R., Maillard reaction in sugar-protein systems. In *The Maillard Reaction in Food Processing, Human Nutrition and Physiology*, ed. P. A. Finot, H. U. Aeschbacher, R. F. Hurrell & R. Liardon. Birkhäuser, Basel, 1990, pp. 97–102.

156. O'Brien, J. & Morrissey, P. A., Nutritional and toxicological aspects of the Maillard reaction in foods. *CRC. Crit. Rev. Food Sci. Nutr.*, **28** (1989) 211–248.

157. Finot, P. A., Hurrell, R. F., Deutsch, R. & Klein, A., 1979. Unpublished data. In Finot, P. A., Deutsch, R. & Bujard, E. The extent of the Maillard reaction during the processing of milk. In *Maillard Reactions in Food*, ed. C. Eriksson. *Progress in Food and Nutrition Science*, Vol. 5 (1–6). Pergamon Press, Oxford, 1981, pp. 345–355.

158. Knipfel, J. E., Nitrogen and energy availabilities in foods and feeds subjected to heating. In *Maillard Reactions in Food*, ed. C. Eriksson. *Progress in Food and Nutrition Science*, Vol. 5 (1–6). Pergamon Press, Oxford, 1981, pp. 177–192.

159. Erbersdobler, H. F., The biological significance of carbohydrate–lysine crosslinking during heat treatment of food proteins. In *Protein Crosslinking: Nutritional and Medical Consequences*, ed. M. Freidman. Plenum Press, New York, 1977, pp. 367–378.

160. Hansen, L. P. & Millington, R. J., Blockage of protein enzymatic digestion (carboxypeptidase-B) by heat-induced sugar-lysine reactions. *J. Food Sci.*, **44** (1979) 1173–1177.

161. Ford, J. E., Hurrell, R. F. & Finot, P. A., Storage of milk powders under adverse conditions. 2. Influence on the content of water-soluble vitamins. *Br. J. Nutr.*, **49** (1983) 355–364.

162. Nursten, H. E., Aroma compounds from the Maillard reaction. In *Developments in Food Flavour*, ed. G. G. Birch & M. G. Lindley. Elsevier Applied Science, London, 1986, pp. 173–190.

163. Parliment, T. H., McGorrin, R. J. & Ho, C.-T., ed. *Thermal Generation of Aromas*. ACS Symp. Ser. 409, ACS, Washington D.C., 1989.

164. Tressl, R., Processed flavors—scope and limitations. In *Flavour Science and Technology*, ed. Y. Bessière & A. F. Thomas. Wiley, Chichester, 1990, pp. 87–104.

165. Hurrell, R. F., Maillard reaction in flavour. In *Food Flavours Part A. Introduction*, ed. I. D. Morton & A. J. MacLeod. Developments in Food Science, Vol. 3A. Elsevier, New York, 1982, pp. 399–437.

166. Ferretti, A. & Flanagan, V. P., The lactose-casein (Maillard) browning system: volatile components. *J. Agric. Food Chem.*, **19** (1971) 245–249.

167. Ferretti, A., Flanagan, V. P. & Ruth, J. M., Non-enzymic browning in a lactose-casein model system. *J. Agric. Food Chem.*, **18** (1970) 13–18.

168. Berry, S. K. & Gramshaw, J. W., Influence of starch plus gluten on the non-enzymatic browning reaction of the glucose-glutamic acid system. *J. Agric. Food Chem.*, **36** (1988) 1265–1267.

169. Tressl, R., Helak, B., Kersten, E. & Rewicki, D., Related ring enlargement reactions of proline, azetidinic acid, arginine and lysine with reducing sugars. In *The Maillard Reaction in Food Processing, Human Nutrition and Physiology*, ed. P. A. Finot, H. U. Aeschbacher, R. F. Hurrell & R. Liardon. Birkhäuser, Basel, 1990, pp. 121–132.

170. Apriyantono, A. & Ames, J. M., Volatile compounds produced on heating lysine with xylose. In *Flavour Science and Technology*, ed. Y. Bessière & A. F. Thomas. Wiley, Chichester, 1990, pp. 117–120.

171. Pabst, H. M. E., Ledl, F. & Belitz, H.-D., Bitter compounds obtained by heating proline and sucrose. *Z. Lebensm. Unters. Forsch.*, **178** (1984) 356–360.

172. Schnickels, R. A., Warmbier, H. C. & Labuza, T. P., Effect of protein substitution on nonenzymatic browning in an intermediate moisture food system. *J. Agric. Food Chem.*, **24** (1976) 901–903.
173. Nursten, H. E. and O'Reilly, R., The complexity of the Maillard reaction as shown by a xylose-glycine model system. In *Amino-Carbonyl Reactions in Food and Biological Systems*, ed. M. Fujimaki, M. Namiki and H. Kato. Developments in Food Science, Vol. 13. Elsevier, Amsterdam, 1986, pp. 15–28.
174. Ledl, F. & Severin, T., Formation of coloured compounds from hexoses. XVI. Investigations relating to the Maillard reaction. *Z. Lebensm. Unters. Forsch.*, **175** (1982) 262–265.
175. Severin, T. & Krönig, U., Maillard reaction IV. Structure of a colored product from pentoses. *Chem. Mikrobiol. Technol. Lebensm.*, **1** 1972, 156–157.
176. Nursten, H. E. & O'Reilly, R., Coloured compounds formed by the interaction of glycine with xylose. In *The Maillard Reaction in Foods and Nutrition*, ed. G. R. Waller &. M. S. Feather. ACS Symp. Ser. 215, ACS. Washington D.C., 1983, pp. 103–121.
177. Ledl, F. & Severin, T., Browning reactions of pentoses with amines. Investigation of the Maillard reaction XIII. *Z. Lebensm. Unters. Forsch.*, **167** (1978) 410–413.
178. Nursten, H. E. & O'Reilly, R., Coloured compounds formed by the interaction of glycine and xylose. *Food Chem.*, **20** (1986) 45–60.
179. Banks, S. B., Ames, J. M. & Nursten, H. E., Isolation and characterisation of 4-hydroxy-2-hydroxymethyl-3-(2'-pyrrolyl)-2-cyclopenten-1-one from a xylose/lysine reaction mixture. *Chem. Ind.*, (1988) 433–434.
180. Kurata, T., Fujimaki, M. & Sakurai, Y., Red pigment produced by the reaction of dehydro-L-ascorbic acid with α-amino acid. *Agric. Biol. Chem.*, **37** (1973) 1471–1477.
181. Sakurai, H. & Ishii, K., Structural analysis of ninhydrin-positive substance produced by the reaction of dehydroascorbic acid with glycylleucine. *Bull. Coll. Agric. Vet. Med. Nihon Univ.*, **45** (1988) 50–59.
182. Lingnert, H. & Hall, G., Formation of antioxidative Maillard reaction products during food processing. In *Amino-Carbonyl Reactions in Food and Biological Systems*, ed. M. Fujimaki, M. Namiki & H. Kato. Developments in Food Science, Vol. 13. Elsevier, Amsterdam, 1986, pp. 273–279.
183. Lingnert, H., Development of the Maillard reaction during food processing. In *The Maillard Reaction in Food Processing, Human Nutrition and Physiology*, ed. P. A. Finot, H. U. Aeschbacher, R. F. Hurrell & R. Liardon. Birkhäuser, Basel, 1990, pp. 171–185.
184. Chuyen, N. V., Utsunomiya, N., Hidaka, A. & Kato, H., Antioxidative effect of Maillard reaction products in vivo. In *The Maillard Reaction in Food Processing, Human Nutrition and Physiology*, ed. P. A. Finot, H. U. Aeschbacher, R. F. Hurrell & R. Liardon. Birkhäuser, Basel, 1990, pp. 285–290.
185. Aeschbacher, H. U., Anticarcinogenic effect of browning reaction products. In *The Maillard Reaction in Food Processing, Human Nutrition and*

Physiology, ed. P. A. Finot, H. U. Aeschbacher, R. F. Hurrell & R. Liardon. Birkhäuser, Basel, 1990, pp. 335–348.

186. Sugimura, T., Takayama, S., Ohgaki, H., Wakabayashi, K. & Nagao, M., Mutagens and carcinogens formed by cooking meat and fish: heterocyclic amines. In *The Maillard Reaction in Food Processing, Human Nutrition and Physiology*, ed. P. A. Finot, H. U. Aeschbacher, R. F. Hurrell & R. Liardon. Birkhäuser, Basel, 1990, pp. 323–334.

187. Nyhammar, T., Grivas, S., Olsson, K. & Jägerstad, M., Isolation and identification of beef mutagens (IQ compounds) from heated model systems of creatinine, fructose and glycine or alanine. In *Amino-Carbonyl Reactions in Food and Biological Systems*, ed. M. Fujimaki, M. Namiki & H. Kato. Developments in Food Science, Vol. 13. Elsevier, Amsterdam, 1986, pp. 323–327.

188. Jägerstad, M., Laser Reutersward, A., Olsson, R., Grivas, S., Nyhammar, T., Olsson, K. & Dahlqvist, A., Creatin(in)e and Maillard reaction products as precursors of mutagenic compounds: effects of various amino acids. *Food Chem.*, **12** (1983) 255–264.

189. Wakabayashi, K., Takahashi, M., Nagao, M., Sato, S., Kinae, N., Tomita, I. & Sugimura, T. Quantification of mutagenic and carcinogenic heterocyclic amines in cooked foods. In *Amino-Carbonyl Reactions in Food and Biological Systems*, ed. M. Fujimaki, M. Namiki & H. Kato. Developments in Food Science, Vol. 13. Elsevier, Amsterdam, 1986, pp. 363–371.

190. Jägerstad, M., Laser Reutersward, A., Oste, R., Dahlqvist, A., Grivas, S., Olsson, K. & Nyhammar, T. Creatinine and Maillard reaction products as precursors of mutagenic compounds formed in fried beef. In *The Maillard Reaction in Foods and Nutrition*, ed. G. R. Waller & M. S. Feather. ACS Symp. Ser. 215, American Chemical Society, Washington D.C., 1983, pp. 507–519.

191. Monnier, V. M., Sell, D. R., Miyata, S. & Nagaraj, R. H., The Maillard reaction as a basis for a theory of aging. In *The Maillard Reaction in Food Processing, Human Nutrition and Physiology*. ed. P. A. Finot, H. U. Aeschbacher, R. F. Hurrell & R. Liardon. Birkhäuser, Basel, 1990, pp. 393–414.

192. Sell, D. R. & Monnier, V. M., Structure elucidation of a senescence crosslink from human extracellular matrix. *J. Biol. Chem.*, **264** (1989) 21597–21602.

193. O'Brien, J. M. & Morrissey, P. A., The Maillard reaction in milk products. In *Heat-Induced Changes in Milk*, ed. P. F. Fox. *Bulletin of the IDF*, No. 238, 1989, 53–61.

194. Saltmarch, M. & Labuza, T. P., Nonenzymatic browning via the Maillard reaction in foods. In *Proceedings of a Conference on Nonenzymatic Glycosylation and Browning Reactions: Their Relevance to Diabetes Mellitus*, ed. C. M. Peterson. *Diabetes*, Vol. 31, Suppl. 3, Part 2 of 2, 1982, pp. 29–35.

195. Labuza, T. P. & Schmidl, M. K., Advances in the control of browning reactions in foods. In *Role of Chemistry in the Quality of Processed Food*, ed. O. R. Fennema, W.-H. Chang & C.-Y. Lii. Food & Nutr. Press Inc., Westport, Conn., 1986, pp. 65–95.

196. Eichner, K. & Ciner-Doruk, M., Formation and decomposition of browning intermediates and visible sugar-amine browning reactions. In *Water Activity: Influences of Food Quality*, ed. L. B. Rockland & G. F. Stewart. Academic Press, New York, 1981, pp. 567–603.

197. Saltmarch, M., Vagnini-Ferrari, M. & Labuza, T. P. Theoretical basis and application of kinetics to browning in spray-dried whey food systems. In *Maillard Reactions in Food*, ed. C. Eriksson. *Progress in Food and Nutrition Science*, Vol. 5 (1–6). Pergamon Press, Oxford, 1981, pp. 331–344.

198. McWeeny, D. J., Sulfur dioxide and the Maillard reaction in food. In *Maillard Reactions in Food*, ed. C. Eriksson. *Progress in Food and Nutrition Science*. Pergamon Press, Oxford, Vol. 5 (1–6) (1981) pp. 395–404.

199. Shu, C.-K. & Ho, C.-T. In *Thermal Generations of Aromas*, ed. T. H. Parliment, R. J. McGorrin & C.-T. Ho. ACS Symp. Ser. 409. American Chemical Society, Washington D.C., 1989, pp. 229–241.

Chapter 5

METAL-PROTEIN INTERACTIONS

R. D. GILLARD

Department of Chemistry and Applied Chemistry,
University of Wales College of Cardiff, Cardiff, UK

&

S. H. LAURIE
School of Chemistry,
Leicester Polytechnic,
Leicester LE1 9BH, UK

INTRODUCTION

It has long been recognised that a number of metallic elements are necessary for the maintenance of a healthy life form, be it animal or plant. Over the past decade increasing attention has been paid to the human nutritional aspects of these elements, as is evident from the number of recent publications in this area published by the Royal Society of Chemistry.[1-3] These publications serve to show that whilst our knowledge is improving it is still far from being complete. The nutritional importance of these elements is causing concern today with respect to modern food processing techniques, consumption of health foods, and the increasing tendency to vegetarian diets. There has been a recent lively debate over the adequacy of zinc in western diets.[4]

Essential to the many aspects of metal nutrition are the interactions between the metal ions and protein molecules.[5,6] It is the aim of this chapter to show the nature and range of these interactions. For this reason a number of the essential metal ions, i.e. Na^+, K^+ and Mg^{2+} are not included because their involvement with proteins is limited to a small

155

number of intracellular enzyme-activating roles (by relatively weak association with the proteins). Likewise cobalt is not included because, uniquely for a transition element, its occurrence and role is confined solely to that of one form, vitamin B_{12}, with no protein involvement. On the other hand, a number of the toxic metallic elements are included because of evidence pointing to some involvement with proteins.

METAL IONS AND HUMAN HEALTH

Essential, Beneficial and Toxic Elements

At present some 26 of the 90 naturally occurring elements are known to be required for animal and plant life. Eleven (C, H, O, N, S, Ca, P, K, Na, Cl and Mg) occur in high concentrations in the human body (Table 1) and are referred to as the major or macro-elements. Table 2 shows the levels of the macro-metals in more detail. The remaining 16 constitute the trace elements, i.e. Fe, Zn, Cu, Mn, Ni, Co, Mo, Se, Cr, I, F, Sn, Li, V, Si and As. Not all of these are required for human life and several others, like B, are of quite common occurrence in particular species. A trace element is regarded as constituting less than 0.01% of the total mass of an organism.

The trace elements may conveniently be subdivided into three categories: essential, beneficial and toxic. To define an *essential* element the following criteria, based on work by Mertz[7] and Cotzias,[8] are applied: (1) the element must be present in all healthy tissues; (2) there must be similar concentrations between species; (3) its withdrawal from the diet must induce reproducible physiological abnormalities regardless of the species type; (4) its re-addition to the depleted diet must reverse the abnormalities; and (5) its complete absence for a prolonged period would prove fatal. Within this rigid restriction the essential trace elements for *man* comprise Fe, Zn, Cu, Mn, Co, Mo and Se. The *beneficial* elements are those where life is possible in their absence but in a poorer form. From animal (not human) studies the elements V, Cr, Ni, F, Sn, Br and Sb come into this category. The *toxic* elements are those which have no recognised biological role and are deleterious to health, e.g. As, Ba, Be, Cd, Hg, Pb, Tl.

Classification of these elements has come from studies on animals other than human beings, so it is quite likely that changes in these categories as they apply to man will occur as research progresses. It could be that some of the present 'beneficial' elements may turn out to be 'essential', for

TABLE 1
APPROXIMATE ELEMENTAL COMPOSITION
OF A 70 kg MAN

Macro elements/g

O	43 000	C	16 000
H	7 000	N	1 800
Ca	1 000	P	780
S	140	K	140
Na	100	Cl	95
Mg	19		

Micro elements/mg

Fe	4 000	F	2 600
Zn	2 000	I	15
Cu	100	Se	15
Mn	15	Mo	10
Cr	15	Co	1

Other elements/mg

Ni	10	V	20
As	10	B	50
Sn	15	Al	70[a]
Br	200	Pb	12
Si	< 2 000	Cd	50

(Many other metallic elements also
found in mg and μg quantities)

[a]Improved analytical techniques point
to this figure being too high.

TABLE 2
METAL LEVELS IN ADULT HUMANS[a]

Metal	Whole blood ($\mu g/dm^3$)	Red Cells ($\mu g/dm^3$)	Liver ($\mu g/g$ dry wt)	Muscle ($\mu g/g$ dry wt)	Kidney ($\mu g/g$ dry wt)
Ca	61	3·23	100–360	140–700	400–820
Mg	24	57	590	900	630
Na	1 960	280	3 000	2 600–8 000	10 000
K	1 700	3 600	8 500	1 600	8 300

[a]M.N. Hughes, Chapter 62.1 in *Comprehensive Coord. Chem.*, ed. Wilkinson,
Gillard and McCleverty, Pergamon, U.K., 1987.

example Ni deficiency syndromes have been experimentally produced in several animal species, and so by implication these may be expected in man. The human body does in fact contain[9] approximately 10 mg Ni; this itself is not sufficient proof of its requirement, however, as a large number of elements with no known biological role are found[10] in trace quantities in man. In assigning an element to one of the above categories it must also be borne in mind that the response to an element is dose dependent. Even essential elements become toxic at high dose levels. Some of the toxic elements can indeed show beneficial effects at low doses (this may explain the therapeutic uses of elements such as As and Hg in earlier times). Se is an element long known for its toxicity but which in more recent years has been shown to be an essential constituent of glutathione peroxidase. Clearly, establishing the requirement for any element necessitates some exacting work, especially if the required dose level is very low.

An element shown to be essential in other animals is silicon[11], it is present at trace levels. The establishment of an interrelationship between Si and Al[12] is of considerable human interest because of the current concern over aluminium toxicity in humans (*vide infra*), so Si may well turn out to be another essential element in man.

Recommended Dietary Levels

It is the foods consumed rather than the water or the atmosphere that supply the major proportion of the total dietary intake of the trace elements. Public water supplies are 'normally' low in trace elements and generally supply less than 10% of the human intakes. Some useful data on the elemental contents of various foods have been compiled by McCance and Widdowson[13] and Southgate *et al.*[2] There are, however, some notable exceptions to the 'normal' situation where the concentrations of trace metals and of toxic metals can be considerably enhanced. For example, some water supplies can be rich in fluoride and so the water becomes the major dietary source of this element (and influences the uptakes of other trace elements). Water which is 'soft' can leach metal ions from plumbing materials, e.g. Cu and Pb. Dissolution from metal cooking vessels can also be a hazard, as with Cu in Indian Childhood Cirrhosis.[14] There is current concern over the possible release of metal elements and other compounds into the food chain as a result of manufacturing processes. A number of fatal diseases directly related to toxic materials having become incorporated into the human food chain are well documented, e.g. *itai-itai* disease from Cd poisoning,[15] Minimata disease from CH_3Hg^+ formation, and the large-scale Iraqi Seed poisoning fatalities also from organomercurials.[16] The high toxicity of organometallic compounds is reflected in current concern over the uses of tetraethyllead in petrol and organotin derivatives in marine paints, insecticides and plastics.

The above examples relating to toxicity emphasise the importance of the chemical form (species) of the metal to its dietary absorption. This is just one of several factors that contribute to the difficulties in defining quantitative dietary levels for each element. These factors have been discussed by a World Health Organisation Expert Committee and their findings published.[17]

One further complication is that amongst humans the requirements are not the same but can vary with age, sex and other human factors. The W.H.O. recommended daily intakes of iron and zinc (Table 3) illustrates this point. It is worthwhile noting that, in these days of increasing vegetarianism, meat is one of the best sources of the essential metals; this is not just a reflection of their high levels in meat but is attributed to interactions with meat protein enhancing their bio-availability. Published values of daily requirements of the various elements for adults are given in Table 4. In view of the foregoing discussion, these should be taken as approximate values only; there is no doubt that they will change as more information becomes available.

Table 5 shows effects of dietary imbalance in six metals. The normal or healthy range would be intermediate between the low levels giving rise to symptoms of deficiency, and the much higher levels giving rise to

TABLE 3
W.H.O. RECOMMENDED MINIMUM INTAKES[a] FOR IRON
AND ZINC/MG PER DAY

	Iron	Zinc
Infants: 0–4 months	0·5	1·25
5–12 months	1·0	1·1
Children: 1–12 years	1·0	1·6
Boys: 12–16/17	1·8	2·8
Girls: 12–16/17	2·4	2·6 –2·2
Adults	0·9	2·2
Menstruating women	2·8	—
Lactating women	1·0	5·45
Pregnant women	0·8–3·0	2·55–3·0

[a]Data taken from Ref. 17. These numbers need to be increased by a factor reflecting the bio-availability of the element in the diet.

TABLE 4
DAILY REQUIREMENTS OF ELEMENTS

Macro elements/g			
Ca	1·1	Na	4·4
K	3·3	Mg	0·3
S	0·9	P	1·4
Cl	5·1		

Micro elements/mg			
Fe	0·9	Mn	3
F	3	Cr	0·05–0·20
Zn	2·2	Cu	2
Co	0·3	Mo	0·2
I	0·1	Se	0·01

symptoms of toxicity. As two examples of the range of levels concerned, the gross effects of an excess of iron salt intake are of severity equal to that of insufficient iron intake.

For example, the well known 'bantu siderosis', supposedly caused by the ingestion of large quantities of beer brewed in iron pots at tribal ceremonies, causes problems of excess, whereas insufficient iron intake gives rise to the better known iron anaemias which can be aggravated by high levels of iron-binders in the diet, to the point of requiring iron supplementation for pregnant women. Of course, there are many other problems relating to metal levels caused by metabolic disfunction. One such is Cooley's anaemia, arising from faulty haemoglobin synthesis. Here, and in other related conditions, the therapeutic use of a metal chelating agent is required, and for Cooley's anaemia, desferrioxamine is used.

The medicinal use of iron is of course long-standing, Sydenham treated chlorosis in the late 17th Century with an iron tonic ('a syrup made by steeping iron or steel filings in cold Rhine wine'). Even earlier, the philosopher Melanpus cured the Thessalonian prince Iphiclus of impotence by causing him to drink the mixture of rust and wine made by scraping the rusty blade of a ram-castrating knife used by the prince's father years before.

In the same way, while copper is the third most abundant heavy metal in the body, and all cells seem to require copper, an excess is toxic. The toxicity toward humans is most commonly observed in cases of small

TABLE 5
RESULTS OF DIETARY IMBALANCE OF ESSENTIAL TRACE ELEMENTS[a]

Element	Deficiency	Excess
Iron	Widespread geographically; fatigue, anaemia	Danger in haemochromatosis, Cooley's anaemia, acute poisoning, Bantu siderosis
Zinc	Occurs in Iran, Egypt, TPN, genetic disease, traumatic stress; growth depression, delayed sexual maturation, skin lesions, depression of immunocompetence, change of taste acuity	Unlikely except from prolonged therapeutic use; can interfere with Fe and Cu metabolism
Manganese	Unknown	Toxicity by inhalation
Copper	Occurs in malnutrition, TPN, anaemia, neutropenia, skeletal and neurological defects, Menkes disease	Danger in Wilson's disease, Indian childhood cirrhosis
Cobalt	Vitamin B_{12} deficiency only, because of diet or failure to absorb	Unknown
Molybdenum	Faulty metabolism of xanthine	Gout-like syndrome in parts of Soviet Union

[a] Ref. 29, and II. J. M. Bowen *Environmental Chemistry of the Elements*. Academic Press, London, 1979.

children ingesting excess copper salts such as in the fatal Indian childhood cirrhosis.[14]

Aspects of Metal Metabolism
By far the greatest effort in what is called bio-inorganic chemistry has been devoted to understanding the functions and structures of the metal-loenzymes.[18] These are the catalytically active protein molecules in which the metal ions are tightly and irreversibly bound and are involved in the catalytic reactions of the enzyme. These represent the functional forms of the transition elements, of zinc and to a certain extent calcium. Outside this area of enzymology, however, are a large number of other metal-protein interactions of which our knowledge is still incomplete. These in-

clude the selective absorption and rejection of the elements, their transportation, and their storage.[19] In these cases the metal association with the protein molecules can vary from weak to exceedingly strong and the association is generally reversible.

A simple scheme showing the various facets of the metabolism of an element is shown in Fig. 1. It should be recognised that the different compartments in this scheme are not totally independent. Thus, while homoeostasis of an element is generally under the control of the absorption/ rejection stages, feedback from the storage form can occur to alter the control at these stages. A further complexity not shown in this scheme, and of vital importance in mineral nutrition, is the antagonistic interactions that occur between a number of the metallic elements. Table 6 shows these essential elements which are known to be depleted by an increase in concentration of some other element (the antagonist). Not included are non-essential elements, e.g. Cd, Pb, which can also act as antagonists when present at sufficiently high levels. In most cases the antagonism can be attributed to the interacting metal ions having similar coordination preferences (see p. 171) and therefore competing for the same protein sites. In other cases, the cause of the antagonism is not so evident and may occur at the initial dietary intake (e.g. Ca inhibition of Mn absorption).

The initial stages of food digestion and absorption from the gut involve chemical processes that are complex and still incompletely understood. The efficiency of these stages for metal ions as for other requisites is

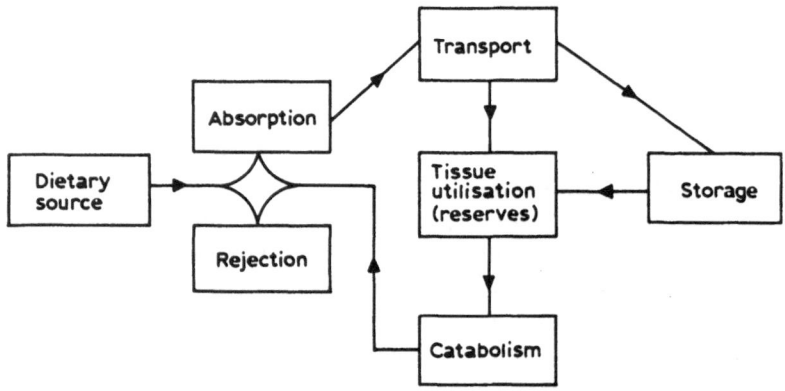

Fig 1. Schematic outline of the metabolism of a typical trace element (reproduced with permission from Ref. 19).

TABLE 6

ANTAGONISTIC INTERACTIONS BETWEEN THE ESSENTIAL
MICRONUTRIENTS

Target	Antagonist
Copper	Zinc, (molybdenum in ruminants)
Iron	Copper, calcium, manganese
Manganese	Copper, iron, phosphates
Molybdenum	Sulphur
Selenium	Sulphur analogues
Zinc	Copper, (iron), calcium, phosphate

strongly dependent on a number of dietary and physiological factors, which have been outlined by Bremner and Mills[20] and Lonnerdal:[21] these are summarised in Table 7. It is this range of factors that renders the determination of adequate dietary levels a difficult problem. One factor of particular importance here is the nature of the metal species present in the diet. Recent attempts have been made to determine the metal speciation in digested foods by computer simulation methods.[22] The amino acids obtained from protein digestion are strongly implicated in these simulations as aqueous soluble complexing agents for the micronutrients.[23] Such simulations are of value in assessing the nutritional status of elements in total parental nutrition applications.[24]

A protein of especial importance at the initial absorption stage in the gut is metallothionein. This is a low molecular weight protein comprising a single polypeptide chain of some 61 amino acids. One third of these are cysteine, which has the strong class b (vide infra: p. 178) thiol (—SH)

TABLE 7

FACTORS THAT INFLUENCE THE DIETARY UTILISATION OF ESSENTIAL NUTRIENTS[a]

1. Pre-existing tissue reserves
2. Anabolic demands: growth rate, age, pregnancy, lactation
3. Infection: stress increasing losses
4. Genetic variables
5. Physical nature of diet (e.g. fluid diets enhancing absorption)
6. Chemical composition of diet: form and content of the metal element; presence of ligands enhancing or decreasing absorption; presence of antagonists decreasing absorption or increasing tissue retention; fibre and phytate content.

[a]Adapted from Ref. 20.

metal-binding site.[25] The protein, widely distributed in mammalian tissue, strongly binds 7–10 copper or zinc ions, or toxic cadmium ions. The strong binding results entirely from the thiol residues. The level of metallothionein in gut mucosa is found to be under the control of the zinc concentration.

Transportation involves carriage of the metal ions via the portal bloodstream to the liver, and thence incorporation into the metallo-enzymes. The second phase of transportation is that from the liver to the various tissues. Proteins are prominent in the transport stage, particularly albumin which is present in large amounts in the blood-stream and can bind several ions of the metallic elements. An important feature of the metal binding to albumin is that, in contrast to that of metalloenzymes, it is reversible; hence[26] the metal ions can be readily released to the tissue where they are required.

The storage forms vary in kind and location between the elements. The storage form of iron is the one known in greatest detail: this is the widely distributed protein ferritin, (see p. 183) which can bind large concentrations of Fe(III). Calcium is stored in bone in a non-protein form while zinc stores occur in protein-bound forms in muscle tissue. For the other micro-nutrients, specific storage forms may not be necessary.[19] The high hepatic copper levels are often implicated as a storage form: however, the liver is also the site for the synthesis of a number of copper-containing enzymes. Copper is also found widely distributed among a number of tissues.[27] However, a storage form of this element *is* evident in pregnant and foetal animals. In the latter case, the very high hepatic copper levels account[28] for half of the total in the developing foetus. No locally concentrated forms of Cr, Co or Mn have yet been found, although this may be more a comment on the intensity of the search than a final truth, as is commonly the case for metal ions in nutrition.

The final metabolic stage, excretion, occurs mainly via the faeces for Cu, Mn and Zn, although significant zinc losses also occur in sweat. For these elements, excretion mostly involves unabsorbed dietary forms (e.g. estimated at approximately 30% of a normal copper diet). However, there appears to be a limited ability for the human body to excrete iron. Even during haemolytic crisis, when large amounts of iron are liberated in the body, less than 0.5% of the metal occurs in the faeces; the remainder is re-absorbed.[10] In contrast to these elements, chromium is excreted mainly via urine. It is unlikely that intact protein forms are involved in the excretory processes.

Analytical Techniques

Undoubtedly one of the most influential contributions to the increase in our knowledge of the metal food area over the past few years has been the development in analytical techniques. In the 1960s metal analysis was revolutionised by the availability of commercial atomic absorption (AA) spectrometers. Improvements in sample volatilisation methods and the use of emission rather than absorption, means that most metals can now be measured down to the ppb (1 in 10^9) concentration level. More recently, a related technique, inductively coupled plasma (ICP) spectrometry, has become available, allowing the automated and simultaneous determinations of many elements (approx. 20) in the same sample. The use of lasers for fluorescence and ionisation is predicted to lower the detection limits of these methods even further, down to fg (10^{-15} g) amounts.

Mass spectrometry (MS) is commonly associated with the analysis of volatile organic compounds. However, modern instruments can be used to detect polymers and other generally non-volatile compounds, including metal ions and their compounds. It is of particular significance that MS can resolve different isotopes of the same element.

The AA and ICP techniques analyse the *total* concentration of a metal in a matrix, they do not give information on the different forms in which the metal ion may exist within the matrix, i.e. the so-called speciation. A knowledge of the speciation is essential to our understanding of the behaviour of a metal ion in any biological organ, tissue or fluid. MS can be used for speciation determination and provides a powerful technique when combined with other techniques such as gas chromatography (GC-MS) and with ICP (ICP-MS). GC-AA is another combination which allows the analytical determination of species concentrations. Recent work on the bio-availability of metal ions and their compounds has made use of stable (non-radioactive) isotopes of the metals with analysis by MS.[29-32] These studies are providing information that is not obtainable by other means.

Another exciting development is the combination of a microprobe and analyser to give the location of metal ions within a tissue or organ. For example, scanning microscopy combined with energy-dispersive X-ray spectroscopy was used[33] to locate Al within the hippocampel neurones from patients' brain tissues. Other techniques include a combination of a laser probe with MS. As these newer techniques become more widely available we can look forward to a rapid gain in our understanding of metal behaviour in humans. An excellent and relevant review of the above techniques has been written by Savory and Wills.[34] Finally, we should mention that increasing attention is being paid to sample handling and processing, and newer methods (including robotics) are being studied.[34] As the ability to detect even lower amounts of metal increases, the concomitant problem of contamination becomes more serious and needs to be addressed. As an example of this, the levels reported of Al and Si in tissues has fallen over the years, not so much from the improvements in analytical techniques but more to improved sample handling. Obviously, with such ubiquitous elements, contamination was substantial in many of the earlier studies.

RELEVANT PROPERTIES OF METAL IONS

Metal ions are most familiar in such salts as sodium chloride, copper (II) nitrate or zinc sulphate. For almost all intake of a metal, essential or not, in diet, the metal ion will at one point or another be present as its solvated (aquated) ion dissolved in an aqueous medium. Hydrolytic equilibria (essentially involving acid dissociation from coordinated water, as shown in the equation) are well known. Fortunately, the common ions of diet and metabolism are,

$$[M(H_2O)_6]^{n+} \rightleftharpoons [M(H_2O)_5(OH)]^{(n-1)+} + H^+ \qquad (1)$$

within physiological and dietary ranges of pH, not greatly hydrolysed. However, for the sake of completeness, the possibility should be mentioned. For solutions containing such ions as iron (III), zinc (II), or additives like bismuth or titanium, if the pH rises much above 3.5, then *solid* oxo- and hydroxo-containing species may well form.

However, the property of a metal ion most relevant is its molecular interaction with the donor sites of protein and its consequent formation of complex compounds with the protein and its interaction with its environment. This complex formation is represented, in general terms, for a protonic ligand, H_XL, (which is here a protein) by the Eqn (3.2)

$$H_XL + M^{x+} \rightleftharpoons ML + xH^+ \qquad (2)$$

The feature of such equilibria (the displacement of protons from their conjugate base sites and the competitive binding of metal ions by those sites) will be treated separately, under the headings: stabilities (equilibria), rates (kinetics), oxidation and reduction (so-called 'redox' properties), classification of the metal ions and their favoured ligands (so-called class a versus b).

Thermodynamic Stabilities (Equilibria)

Most of the metal ions which are commonly observed to interact with proteins form their complexes with the proteins quite rapidly. That is, their equilibria are established quickly, so that thermodynamic treatments are sensible. Let us take the formation of a complex between a metal ion like copper (II) and an amino acid (Scheme 1), as a preliminary to a more detailed treatment later of the interactions of metal ions in general with poly-amino acids.

This example has been chosen because it represents some of the very earliest cases of deliberate complexing of a metal ion, and was due to Boussingault, the father of food chemistry. He used it to isolate glycine ('Leimzucker' to him) from protein hydrolysates.

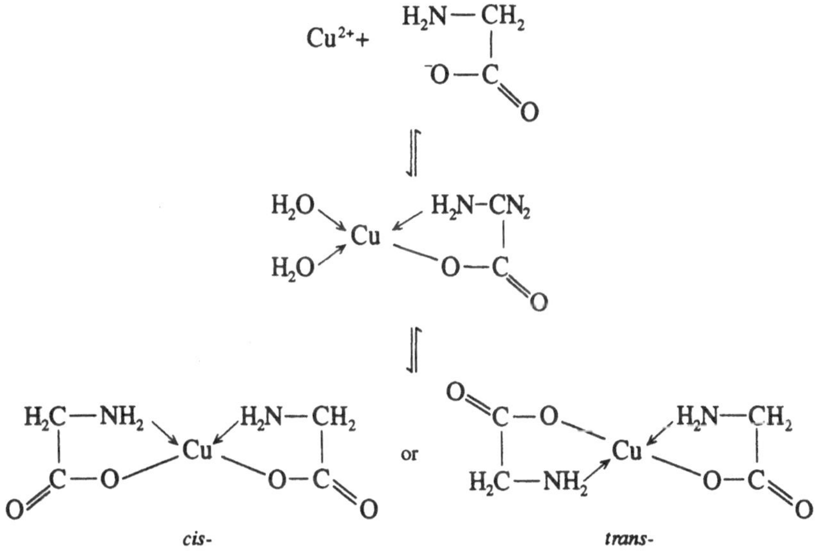

SCHEME 1

However, even in this rather simple system, complications soon ensue. In particular, the chelated 1:1 species can form only in one way, shown as the first stepwise equilibrium above.

This first formed 1:1 complex, as in Eqn (3) is

$$Cu(OH_2)_6{}^{2+} + gly-O^- \rightarrow [Cu(gly-O^-)(OH_2)_2]^+ \qquad (3)$$

a cation, still carrying a single positive charge.

Boussingault discovered that this charge can be neutralised by nitrate ion, giving the compound [Cu(glycinate)(nitrate)] as its monohydrate. Remarkably, this very easily prepared, very stable compound was not again made for 150 years. Its crystal and molecular structure show several unusual features, including glycinate bridging across two copper (II) ions, as in Fig. 2.

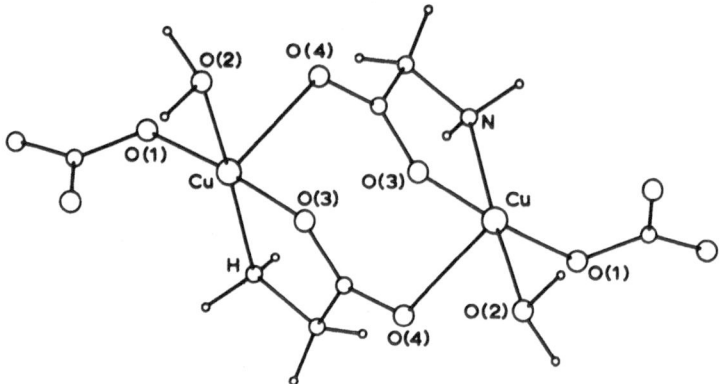

FIG 2. A view of part of the crystal structure of [Cu(glycinate)(nitrate) (Water)], showing the complete coordination sphere around two copper ions, and their connections through bridging glycinates; O(1) is from nitrate ion; O(2) is water; N is the NH$_2$ of chelating glycinate; O(3) is from the carboxylate of the same; O(4) is the other oxygen atom of the same carboxylate.

However, this 1:1 complex can then combine with a second amino-acidate ion in either of two ways, giving rise to the complexes which (in the common nomenclature of coordination chemistry) would be called *cis*- as against *trans*-species, 1:2 in composition. [These bis-complexes of copper (II) are, of course, essentially planar at the copper ion: the molecules in Scheme 1 lie essentially flat].

The convention has been adopted here that a bond from an uncharged donor site of a ligand (such as the oxygen atom of water, OH$_2$, or the

—NH_2 of glycinate or of any N-terminal amino-acid) is represented by an arrow from donor atom to metal ion. A primary chemical bond from a negatively charged site (such as the oxygen atom of a hydroxide ion, OH^- or the carboxylate group of an amino-acidate ion of a protein) is represented by a line. The nature of the electronic distribution within such bonds need not greatly differ: the metal-oxygen bonds to the two solvated ions of Eqn (1), a conjugate-acid and base pair ($M^{n+} \leftarrow OH_2$ as against M^{n+}—OH^-) are very similar in electronic distribution.

There has been a great deal of work on this simple system, and its analogues with amino acids other than glycine. With glycine as the amino acid, it was Boussingault who also first described the crystalline 1:2 species, the so-called bis-glycinatocopper (II). This solid complex, allowed to stand in its mother liquor, reverts from the Cambridge blue *cis*-species to the much more pleasing Oxford blue *trans*-species.

Now, for our present purpose of illustrating modes of combination of metal ions with protein constituents, the point of interest is that since the isomerisation shown in the Eqn (4)

$$cis\text{-}[Cu(gly-O)_2] \rightleftharpoons trans\text{-}[Cu[gly\text{-}O)_2] \qquad (4)$$

does not contain proton or hydroxide as a reactant, its equilibrium constant does not depend on pH. It is therefore not possible to deduce the isomeric composition of glycinate copper (II) (2:1) by using the common method of potentiometric titration (usually following the pH), followed by refined computer analysis of the results. Although this method will often give an accurate *stoichiometric* equilibrium constant, it is more and more necessary to confirm the nature of the species present (after achieving material balances) by some independent spectroscopic means.

The equilibrium constant:[5]

$$\beta_{120} = \frac{\{[Cu(gly-O)]_2\}}{\{[Cu^{2+}]\}\{[gly-O]\}^2} \qquad (5)$$

which describes the formation of this species is therefore called a 'macro' stability constant, because it describes the stoichiometry but not the detailed molecular nature of the product. To move to the 'micro' constants, exemplified in Eqns (6) and (7).

$$\beta_{120}(cis) = \frac{\{cis-[Cu(gly-O)]_2\}}{\{[Cu^{2+}]\}\{[gly-O]\}^2} \qquad (6)$$

$$\beta_{120}(trans) = \frac{\{trans-[Cu(gly-O)_2]\}}{\{[Cu^{2+}]\}\{[gly-O]\}^2} \qquad (7)$$

requires a knowledge of the equilibrium ratio of *cis:trans* species, which will require utilisation of a non-potentiometric method, such as spectrophotometry.

Note that in the equilibrium constant, β_{pqr}, (e.g. β_{120} above) the subscripts p, q, r refer to the exponents in the general Eqn (8)

$$pM^{n+} + qL^{x-} + rH^+ = [M_pL_qH_r]^{(np-qx+r)+} \tag{8}$$

There have been studies of this isomeric ratio using electron spin resonance spectra of frozen aqueous solutions of bis-glycinatocopper (II) and other amino acids. Because of the differing symmetries of the *cis* (non centric)- and *trans* (centric)-isomers, the fine structures of the spin resonance arising from ^{63}Cu (I = 3/2) and ^{65}Cu (I also 3/2) should differ. This difference has been used to suggest values of the equilibrium constant for isomerization.[35,36]

It is a great pity that the developing X-ray techniques for establishing molecular structural features and dimensions in aqueous solutions have not yet been successfully applied to this problem; they could confirm the reliability of the deductions from e.s.r. spectrometry. There has been one study of the radial distribution functions for X-rays in two aqueous solutions, containing chiefly the 1:1 complex $[Cu(gly-O)(OH_2)_2]^+$ and containing chiefly the 1:3 complex $\{Cu(gly-O)_3\}^-$, formed in very large excess of sodium glycinate (about 6.5 M in the structural study).

The peculiarities of these ostensibly simple systems may be illustrated by contrasting L-alaninate with glycinate. Whereas for solutions containing chiefly complex molecules of the stoichiometry $[Cu(gly-O)_2]$, the first crystallising solid is the *cis*-isomer (which, being metastable, will slowly convert to the solid *trans*-isomer), when glycine is replaced by L-alanine, the first form which crystallises is the *trans*-isomer and this readily converts to the more stable *cis*-solid!

There seems to be no means at present of predicting which isomeric environment for the copper ion (and presumably other metal) ions will be favoured. *Cycas circinalis* seed has been used as a foodstuff in Guam and elsewhere, and there have been suggestions[37,38] that this use (and some of the uses of the seed in traditional medicine) correlate with the incidence there of the amyotrophic lateral sclerosis (motor neurone disease).

The amino-acid L-α-amino-β-methylamino-propanoic acid (β-methylamino-L-alanine) though non-proteinaceous, has been found in all of a wide range of *Cycas* species.[39]

$$H_3C-\overset{+}{N}H_2-CH_2-\underset{\underset{\displaystyle NH_2}{|}}{\overset{\overset{\displaystyle H}{|}}{C}}-CO_2^-$$

The amino acid is abbreviated Cyc here. It forms extremely stable chelated complex compounds with doubly charged ions of the first transition series. The bis-complex of its racemic form with copper (II) perchlorate $[Cu(S\text{-}cyc)(R\text{-}cyc)](C10_4)_2$ is truly centric, with the *trans* structure about the copper (II) ion described above[40].

The scheme describing complexation of metal cations by amino acidates becomes very complicated if the amino acid is present in both enantiomeric forms. Fortunately, that is unlikely to be a concern in systems involving intact proteins.

A number of rules, usually empirical, relate equilibria of particular 'ligands' (a name used for any species which will bind the metal ion) and the sizes of equilibrium constants β_{pqr} to the metal ions available for combination. For ligands relevant to the present consideration of food proteins (that is small peptides, amino acids, and the polar atoms in their side-chains), the following general statements may be made. The extraordinarily specific metal-binding sites present in some proteins may make these generalisations less useful for proteins as ligands than for small molecules.

1. More highly charged cations bind more strongly than less highly charged cations: the stability (equilibrium) constant for a calcium ion combining with a given environment will be greater than that for a sodium ion in the same circumstances, despite the radii of these ions being about the same.

2. If there is no change of coordination environment (that is, if the number of binding sites for the metal ion and their relative distributions in space remain the same), then, within a like-charged series, the smaller the ion, the more stable the complex. So, for example, stability constants with a particular binding site would run in the order:

$$Ca^{2+} > Sr^{2+} > Ba^{2+}$$

3. The best known example of such an order is for the doubly charged

ions of the first transition series, which is called the Irving-Williams series:

$$Mn^{2+} < Fe^{2+} < Co^{2+} < Ni^{2+} < Cu^{2+} > Zn^{2+}$$

Here, superimposed on a size effect, arising from the d-electron contraction, there is an electronic effect (from the change in crystal fields).

In most equilibria between metal ions and proteins (or their constituents) the metal ions compete with protons for the binding sites on the proteins, so that, as in almost all aqueous equilibria including acid dissociations, pH will be of dominant importance. The division of metal cations between the aquo-species and protein-complex species will depend on pH, and much less metal ion should be complexed by proteins in acid regions, like the stomach, than in less acid regions.

Rates (Kinetics)

The speed of equilibration, that is to say the rates at which species containing metal ions attain equilibrium with their environment naturally varies with the metal ion and its oxidation state. Equilibria may generally be regarded as 'labile' (or rapidly attained) for the following metal ions: Na^+, K^+, Mg^{2+}, Ca^{2+}, Mn^{2+}, Zn^{2+}, Cu^{2+}. Conversely, slow attainment of equilibrium is more common for complex formation involving the following ions (less commonly involved in present views of food protein complexing): Cr^{3+}, Co^{3+}, Ni^{2+}.

The terms 'labile' and 'inert' here are defined empirically. Nevertheless, for protein–metal ion combination, examples falling in the grey region between the two are rare. The usual definition of a labile system is one in which tenth molar solutions of the metal ions and of the ligand come to full equilibrium in a minute or less after mixing, whereas inert systems take longer than a minute from mixing to come to equilibrium. In general, for metal ions like sodium and potassium, equilibrium is attained in nearly all circumstances in nanoseconds, for kinetically inert ions (which include, among the best studied, ions like cobalt (III) and chromium (III)) the half-lives for equilibration are usually hours rather than seconds. However, there are a few cases in which the very high stability of protein–metal combination, and the great complication of successive steps in altering the conformation of the ligand by unwinding it from around the metal binding site cause rates to slow down to levels much less than those for the simple coordination compounds of the metal ions.

Oxidation and Reduction

The stabilisation of one oxidation state of a metal ion rather than another is obviously most important, in the dietary context, in relation to the transition elements. For example, some fairly novel acid phosphatases have been isolated recently. Most of these from animal sources and some plants have been purple[41] and have contained iron. The sweet potato gives an apparently similar purple acid phosphatase. Despite the apparent similarity of this to the others, it actually contains manganese, and has an intense absorption around 515 nm. Both the oxidation state of this manganese ion and its nature are said to depend upon the particular variety of sweet potato from which the phosphatase is derived.[42]

No intact manganese phosphatase shows an electron spin resonance spectrum at room temperature: after denaturation in acid, the spectrum is characteristic of free manganese (II). However, this may not establish the oxidation state of the manganese when bound to the intact protein, since it is common experience (for iron and copper in particular) that chemical modification, and certainly denaturing, readily lead to the metal ion changing its original oxidation state. In the case of manganese, this is a particular danger, because, under most aqueous conditions, in the presence of most ligands, manganese (II) forms rather readily from the higher states such as Mn(III).

If the enthalpy change (ΔH_1) for the given combination with the oxygen species (or other most electronegative first row elements: fluorine among the halogens, or nitrogen among the pnictides) is greater than the corresponding enthalpy (ΔH_2) for complexing through the less electronegative second row analogue (sulphur among the chalcogens, phosphorus among the pnictides, or chloride among the halogens), the metal ion is said to belong to class (a). Conversely, if the metal ion has a higher affinity for the heavier donor atoms lower down the periodic table, exemplified by sulphur or phosphorus, then it is classified as belonging to class (b).

This distinction between classes of metal ions, although it is rather indefinite, corresponds to a true division of their chemical and biochemical behaviours. The transition elements and some others from the *p*-block do bind better to more polarizable ligands than do the *s*-block and other more electropositive metals. The strong binding of carbon monoxide to many heme-proteins is a good example of the effects of π-bonding. The stability of these Fe—CO bonds is in sharp contrast to the instability of the σ-bonded systems formed from carbon monoxide and Lewis acids: the prototype $(H—CO)^+$ is, of course, unstable in water. Another attempt to define these two classes of cations rests ultimately on comparisons not of enthalpies but of the kinetic properties of the ligands. Corresponding to the class (a) centres in this description are the 'soft' centres (whether metal ions, carbon-based, or other) which undergo reaction more rapidly with the more polarizable nucleophiles. The 'hard' centres react faster with the less polarizable nucleophiles.

In proteins, the thiophilic elements would be expected to have more affinity for sulfur and for aromatic nitrogen sites, whereas metals like magnesium and calcium will, in general, be attached to the class (a) sites, involving oxygen or nitrogen.

AMINO ACIDS, PEPTIDES AND PROTEINS AS LIGANDS

There are many classes of ligands, but in the context of metal–protein interaction, it is possible, at least initially, to ignore the π-bonding effects, and to concentrate on the ligands which attach to metal ions by electron donation through σ-bonding. Although it is nowadays used in many other contexts to signify some kind of attachment between two or more units, the word ligand in the present context merely means something which attaches to a metal ion. The naturally occurring ligands nearly all involve attachment to the metal ion through atoms of oxygen, nitrogen, or both. Indeed, almost the only simple ligand which is at all widespread in living cells attached to metal ions is water (or, of course, hydroxide, depending on the pH: water attached to metal ions becomes more acidic than free water). Other possible unidentate ligands, attached through one atom only are derivatives of sulfide and, in some special contexts, the halides.

In assessing the binding of proteins the attachment of a single ligand through more than one individual atom to the same metal ion, that is chelation, is more important. A chelated ligand is attached to the metal ion through a number of points, bidenentate referring to two points of attachment (as in a simple alpha-amino-acidate, like glycinate): terdentate (or sometime tridentate) refers to attachment through three points, and so on. The 'chelate' effect refers to the fact that formation of chelate rings is almost always favoured. That is, any such process as that shown in Eqn (9) is favourable: the equilibrium is positive and the free energy negative.

$$L_4M \overset{NH_3}{\underset{OH_2}{\big<}} + R-\overset{NH_3}{\underset{\underset{O}{\overset{\|}{C-O^-}}}{\big<}CH} \longrightarrow L_4M \overset{H_2N-CH \diagup R}{\underset{O-C \diagdown O}{\big<}} + NH_3 + H_2O \quad (9)$$

Let us now consider the points of attachment of proteins to metal ions.

Peptide Links and Side Chains
In the simple amino acids, like glycine or alanine (either α- or β-), the conjugate base (e.g. glycinate) has two points of attachment, the carboxylate and the amino group. (Other donor groups L will occupy the remaining coordination sites on the metal to complete its 'coordination sphere'; the number of atoms bound to a metal ion in a complex is its 'coordination number': in simple complex compounds this is commonly 4 or 6. In protein-metal centres, less regular arrangements often occur). In the corresponding dipeptides, the peptide link itself has, of course, two possible points of attachment, although it is important to realise that geometry would prevent the attachment of both the oxygen and the nitrogen of the peptide link to the same metal ion at the same time. For simple dipeptides like glycylglycine, or alanylglycine, the most coordination sites which could be occupied simultaneously is three. A typical partial structure is shown below:-

$$L_yM \overset{H_2N - CH_2}{\underset{O-C \diagdown O}{\underset{\big|}{\overset{\big|}{\longleftarrow N}}}} \overset{\overset{\diagup C-OH}{}}{\underset{CH_2}{\diagdown}}$$

Clearly, as the length of the peptide chain increases, and the folding and convolution of the chain develops, the number of potential binding sites which will be brought close to a single metal ion increases rapidly, and a single polypeptide chain of protein will be able to bind metal ions. Table 8 gives some binding constants for metal ions with peptides.

TABLE 8

BINDING[a] OF PEPTIDES $(H_X L)$[b] WITH METAL IONS, IN WATER AT 25°C, IONIC STRENGTH 0·1 mol/dm³, GIVEN AS $\log_{10} K$ FOR $M^{2+} + (H_n L)^{(x-n)-} \rightarrow [M(H_n L)]^{(2+n-x)+}$

Ion	Complex formed	Glygly $(H_3 L^+)$	Glyhis $(H_4 L^{2+})$	Glyglyhis $(H_5 L^{2+})$	Trigly $(H_4 L^+)$	Tetragly $(H_5 L^+)$	Pentagly $(H_6 L^+)$
Cu^{2+}	M(HL)	5·50	9·14	7·04^c			
	M(H$_2$L)				5·08		
	M(H$_3$L)					5·10	
	M(H$_4$L)						5·32^d
Zn^{2+}	M(HL)	3·44	3·65^c	3·31^c			
	M(H$_2$L)				3·18^d		
	M(H$_3$L)					3·14	—

[a]Abstracted from S. H. Laurie in *Comprehensive Co-ordination Chemistry*, 2 (1987) 739 (edited by G. Wilkinson, R. D. Gillard and J. A. McCleverty), Pergamon Press, Oxford.
[b]The subscript x indicates the maximum number of ionizable protons in the particular peptide, including those of the peptide links.
[c]at 37°C, I = 0·15M.
[d]I = 0.15M.

The discussion so far ignores the fact that many amino acids have potential σ-donor groups in their side chains. Such constituent amino acids of polypeptides as cysteine or glutamic and aspartic acids or lysine and histidine will contribute respectively sulfur donors, carboxylates and basic nitrogen-donors to the manifold of potential metal-binding sites, which may stabilise a metal ion in a particular environment on a particular protein.

A good example of the richness of possible binding arrangements, where amino acids have donors in their side-chains, is provided by the most readily isolated product of reacting cobalt (III) precursors with L-cysteine (calling the di-anion, deprotonated at carboxyl and thiol groups L-cys-O), in alkaline solution, i.e. $K_3[Co(L\text{-cys-}O)_3]$. The anion in this product has three chelate rings, which could each involve chelation by any two of the three available binding atoms of a single cysteinate (N, O, or S), as shown:

In fact, all three chelate rings are five-membered, being formed by N and S atoms, with the carboxylate not bonded to the cobalt ion. This is the (N,S)-linkage isomer.

There are two geometrically isomeric ways of arranging three such chelate rings octahedrally, as shown schematically:

facial   meridianal

Here, only the atoms directly attaching the cysteinate ion to the metal are shown. The facial isomer (which is the one actually formed in the present example) has the three nitrogen atoms as nearly adjacent as possible, on a face of the octahedral metal ion.

Finally, taking the facial, N,S-linked structure, there remains the possibility of optical isomerism at the metal ion, as shown in the structures.

The solid compound $K_3[Co(L\text{-cys-}O)_3]$, isolated by a reaction of a cobalt (III) precursor with L-cysteine in water, under defined conditions contains only one isomer.

Clearly, since protein molecules are ligands which exhibit marked capacity for such stereoselective combination, we would expect that in any case where a metal ion is bound to a protein by more than two or three binding sites, the combination will be specific, and the metal ion will find one of the many possible binding locations, with a specific environment of ligand attaching atoms, in the senses of linkage, geometric, and optical isomers.

The richness of potential sites plotted over the protein molecule as a whole is legion, but we could generalise that, in proteins with isoelectric points indicating the possession of many acidic side-chains, carboxylate binding to metal ions may be more important, and conversely, with proteins with high isoelectric points, the complexation may be through a high proportion of basic nitrogen sites.

The complex structure, and presence of multiple binding sites in proteins, gives rise to the possibility of their attachment of metal ions controlling the function and reactivity of that particular metal (or indeed of the metal–protein assembly) in ways which are often not yet mimicked by smaller coordination compounds. For example, the variation of the properties of the haem proteins with the nature of the environment of the iron is independent of the most subtle variations of the detailed protein structure. In more general terms, the protein may control the detailed geometry of coordination site (including the coordination number), the oxidation potential of the metal ion (which will in turn reflect its spin state) and the rate of oxidation or reduction of the transition metal centre in the protein environment. These are effects primarily at the metal centre, and all manner of subtle transmission of electronic influence, and of stereochemical modification of the control of reactivity at sites in the protein distant from the metal ion, are also common.

Among the particularly common ligands which actually bind metal ions in the proteins are the amino acid residues histidine and cysteine. In such a short compass, it is not possible to discuss in detail the sites typical of various metal ions, though the discovery of the beta-hydroxy aspartate residue as a component of some sites with high affinities for the calcium ion should be mentioned, since this amino acid had not previously been seen in protein. Another unusual amino acid residue gamma-carboxy-glutamate is found in some other calcium binding proteins, given the general name 'gla' proteins. They have widespread importance: for example,[43] 'gla' proteins inhibit the precipitation of calcium oxalate in urine. It is at least interesting that other proteins which inhibit precipitation of calcium salts from urine and the like have many proline and tyrosine residues, and this may be relevant in their calcium-binding capacities. One such is statherine.[44] With the low molecular weight of 5380, statherine contains many tyrosine residues, and is rich also in glutamic acid and proline residues. Its function is thought to relate to prevention of precipitation of basic calcium phosphate from the saliva, not by the thermodynamic effect of reducing the level of free calcium ions below that required to exceed the solubility product, but rather in some manner affecting the rate of nucleation, that is, a heterogeneous kinetic effect. The saliva is, of course, necessarily supersaturated with this calcium compound, in order to achieve protection of the dental enamel and recalcification.

In trying to relate the metal binding environment to activity in proteins (which of course include many enzymes of the highest importance), it is often difficult to make a correct deduction, even when the crystal structure and the environment of the metal is known at high resolution. This is the case, for example, with phosphoglyceratekinase, the enzyme catalysing, in the glycolytic pathway, the conversion of 1,3-diphosphoglycerate to 3-phosphoglycerate, and producing ATP from ADP. The magnesium-bound nucleotide attaches to the enzyme about a hundred times better than does the nucleotide itself. Nevertheless, from X-ray diffraction, the binding for this metal–nucleotide complex seems to be some 1 nm away from the phosphoglycerate binding site, which is at the N-terminal position. The high resolution diffraction study reveals[45] a single polypeptide chain, which contains two organised parts ('domains'), widely separated, but of almost equal size.

Typical Metal-Binding Sites
In discussing the ways in which metal ions are complexed by proteins, one problem is that, although a large number of crystal structures at medium resolution are now available, and an increasing number at high resolution, most studies have related to intact enzyme, and there are relatively few studies available on such important systems as calcium–protein units.

However, it is interesting to see how useful elementary concepts of coordination chemistry are in helping to anticipate and understand such facts from diffraction and similar studies. For example, among those enzymes with carbonic anhydrase activity, which occur so widely (in animals, plants, and some microorganisms) the human isoenzymes B and C have been studied by high resolution X-ray diffraction. Both these major isoenzymes have distinct amino-acid compositions, but there is in fact some two-thirds homology of sequences, and the general structures are similar. The molecular weight is 3000, and the proteins are ellipsoidal (roughly $5500 \times 4200 \times 4100$ pm) in shape. A cleft, some 1600 pm deep and reaching almost to the middle of the molecule, contains the zinc ion, at the bottom. This zinc ion, as might be guessed from the standard coordination chemistry of zinc, which is a class b ion, is attached[46] to three imidazole groups from the histidine residues, His-94, His-96, and His-119, with its tetrahedral coordination geometry completed by a water molecule, itself in turn hydrogen bonded to a threonine.

The environment of the zinc ion in the carboxypeptidases (which cleave the C-terminal amino acid residue from peptide proteins) released from inactive precursors in the pancreatic juice may similarly be guessed from simple models. A major study of these carboxypeptidases relates to the bovine enzyme: the zinc is again complexed by these side-chain groups of amino acid residues of the protein, namely two imidazole groups of His-69 and His-196, and a carboxylate group from Glu-72. The coordination environment is once more[47] a distorted tetrahedron, completed by a water molecule.

Just as the zinc ion has an apparent selectivity for imidazole function, so molybdenum has an apparent affinity for at least some sulfur-containing ligands. The molybdenum hydroxylases catalyse the 2-electron oxidation of a range of substrates, including purines and pyrimidines. The general reaction concerned is shown in Eqn (10):

$$RH + H_2O \rightarrow ROH + 2H^+ + 2e^- \qquad (10)$$

The best known of these molybdenum-containing enzymes is xanthine oxidase, which uses di-oxygen, O_2, as its terminal electron acceptor, and catalyses the hydroxylation of purines. The specific reaction is oxidation of xanthine to uric acid. In the resting form, the enzyme centre contains molybdenum(VI), but the catalytic cycle undoubtedly involves electron acceptance by the metal, forming transient lower oxidation states during the catalytic cycle. The oxidase has a molecular weight of 300 000; each of the two dimeric subunits contains one molybdenum, one FAD, and one each of two sorts of Fe_2S_2 units which make up the ferredoxins and similar non-haem iron proteins, isolable from spinach, parsley and other green plants. The coordination of the molybdenum in the resting enzyme involves[48] a double bond to a sulfur atom at about 225 pm (Mo=S), two further sulfur atoms, bonded at about 250 and 290 pm, and a soluble bonded terminal oxo group (Mo=O, with the distance around 175 pm), of the same type as is found in the molybdate(VI) ion.

Classification of Metal–Protein Interactions

Many metal ions occur naturally in foodstuffs and in the body, and nearly all are bound by proteins. The classification offered here for the way in which proteins and metals interact is no more than a very general summary of what is a very complex set of interactions: (i) Metalloenzymes; these are a sub-class of metalloproteins, where the protein is firmly bound to the metal ion, so that the assembly can be regarded as very stable and long-lasting. The metal ion becomes part of the overall protein structure, and can be removed from it only by severe conditions. As a typical example, we could consider carbonic anhydrase, where, despite the high turnover number, the rate of exchange of zinc with external aquated zinc ion in water is rather slow, in contrast with the very kinetically labile nature of zinc ion in its simpler complexes. (ii) Metal proteins (including enzymes); in such metal protein systems, the metal ion combines reversibly with the protein. The rates of exchange of such metal ions with the external free aquated metal ion are high. This makes for some difficulties in studying the nature of the binding site, as is generally the case even with simpler complexes of these kinetically labile metal ions. Despite such difficulties, the location of the binding site for metals in metal-activated protein is almost certainly just as much a specific site of fixed geometry, determined by the particular coordination groups, as it is in the metalloenzymes themselves.

However, finally, it may be worth pointing out the difficulty in distinguishing metal activated proteins, in terms of their stoichiometric behaviours, from the mere aggregation of protein with labile metal ions, at some or all of the many sites made available by oxygen and nitrogen atoms throughout the bulk of the protein. Whether these binding units are assembled into sites which give rise to particular reactivities is the distinction between mere metal binding, and metal activation of the protein, and it is very difficult to establish experimentally.

METAL-IONS IN VIVO

In this section specific examples of mainly metal–protein interactions of metabolic significance will be highlighted. Examples will be taken from both the macro and micro nutrients and from the toxic elements.

Calcium

This element[49] is commonly associated with its ionic form Ca^{2+}, as expected for a group IIA metal of class (a) type. Two salts of particular importance are the carbonate, $CaCO_3$, and the phosphate, $Ca_3(PO_4)_2$, both of these being highly insoluble and important biological structural materials. It is also capable of forming coordination-type compounds of high thermodynamic stability, particularly with O-donors or O- and N-polydentate donors. Such behaviour is found in the well known complex ion $[Ca(EDTA)]^{2-}$, where $EDTA^{4-}$ is the ethylenediaminetetraacetate ion, and in its association with a number of proteins.

The largest concentration of calcium is in bones and teeth, where in association with OH^- and PO_4^{3-} it has a structural role. The results on bone structure of deficiencies of calcium are well known.[50] Ca^{2+} ions are also found to form cross-links between the fibres of the protein collagen, the major constituent of the organic matrix of bones. Calcification of collagen may be an important factor in ageing as well as for other properties of the protein. Another calcifiable protein is found in tooth enamel. Apart from these familiar structural roles, calcium is also involved in a wide-range of other metabolic functions, mostly in association with proteins.

In contrast to its congener, magnesium, only low levels of calcium are found intracellularly. Although the higher level in the extracellular pool only accounts for less than 5% of the total body calcium, it is nevertheless an important fraction. The Ca^{2+} ions in both the intra- and extracellular pools are associated principally with proteins; in this way, the formation of the insoluble carbonate and phosphate salts by interaction with free Ca^{2+} ions is avoided. There appears to be an active pump mechanism controlling the calcium distribution between the two pools, which is driven by ATP hydrolysis brought about by a Ca-activated ATPase enzyme. One of the major functions of extracellular calcium is in the prevention of blood clotting, the clot being an insoluble matrix of blood proteins. The very small intracellular pool has a role in a number of cellular functions, e.g. nerve excitation and muscular contraction.

Muscle contraction is a complicated process in which a change in membrane potential triggers a rapid Ca^{2+} release which in turn affects muscle contraction; this calcium activation of muscle protein cannot be replicated by other divalent metal ions. Involved in the contraction mechanism is a calcium–troponin complex. Troponin-C, the major calcium binding component, is a protein of mass 18 000 Da which can bind 1–2 calcium ions. This protein contains a large number of dicarboxylic amino acid residues which act as the calcium-binding sites. The binding of calcium to troponiin-C induces a conformational change which, through other proteins, then induces the interaction between the myosin and actin proteins to give muscle contraction.

The conversion of trypsinogen to the active enzyme trypsin also involves calcium-binding. The enzyme is an important hydrolytic catalyst located in the pancreas. Calcium-binding to the inactive trypsinogen appears to be a requisite for conversion to the active form. The metal binding sites implicated are, again, the carboxylic groups from aspartic and glutmatic acid residues. Recently the binding site of calcium in γlactoalbumin has been identified[51] as involving five O atoms from the protein. This protein comprises some 15% of the total protein in human milk and is essential for lactose production; calcium binding of the protein is implicated in this process. Calcium binding to other proteins also has important metabolic effects, e.g. as with calmodulin.

One of the best food sources of Ca is milk, due not only to its relatively high Ca content but also to the greater bioavailability of the element from this source. This has been traced to two major components of milk, lactose and the protein casein. This protein has been shown[52] to cause the formation of intra-intestinal phosphopeptides, potent inhibitors against the formation of insoluble calcium phosphate, so keeping the Ca in a soluble and hence bioavailable form.

Iron[53,54]

Of the trace metals iron, is the best known and the most abundant (see Table 1), and more is known about its biochemistry.

Iron is found in most foods. Its bioavailability varies, however, from low values for most vegetables to relatively high values in meats where it occurs bound to proteins as iron-haems. Protein digestion products are also known to enhance the absorption of non-haem iron.[21] In milk the small iron concentration is associated with the protein lactoferrin, which is very similar to the human form, transferrin. Low dietary iron levels have been found to lead to an increased efficiency in iron absorption, as well as to increased absorption of other metals, both essential (cobalt, manganese, zinc) and toxic (cadmium, possibly lead). Conversely, high dietary iron can[55] seriously deplete copper levels in ruminants and zinc in humans.

Iron displays a wide variety of coordination behaviour in its bonding to proteins. This involves the haem prosthetic group, a variety of amino acid side chains, and iron-sulfur clusters linked to cysteine residues in the ferredoxin proteins of bacteria and plants. Information on the haem-proteins is available in Ledward's article (Chapter 6). Some of the more important iron-proteins and enzymes of the human body and their functions are listed in Table 9. In these examples the metal occurs in its two common oxidation status of II and III and cycling between these states is of importance in the redox catalyses.

More is known about the biochemistry of iron than for any other of the essential metals. The metabolism is shown in schematic form in Fig. 3.

TABLE 9

SOME IRON-PROTEINS AND METALLOENZYMES IN THE HUMAN BODY
AND THEIR FUNCTIONS

Protein	Function	Location
Transferrin	Fe transport	Blood, milk
Haemoglobin	O_2 transport	Blood
Myoglobin	O_2 storage	Muscle
Ferritin	Fe storage	Various tissues
Haemosiderin		
Lactoferrin	Fe storage (?)	Milk
Cytochromes	Electron transport	Mitochondria
Cytochrome oxidase (with Cu)	$O_2 + 4H^+ + 4e^-$ $\rightarrow 2H_2O$	Heart
Catalase	H_2O_2 decomposition	Blood

Apart from its functional forms, iron is also tightly bound to other proteins in its transport and storage forms.[139] This avoids the build-up of Fe(III) ions which, at physiological pH, would readily form the highly insoluble (and hence non-absorbable) hydroxide. Thus, transferrin, the transport protein of blood, has two strong binding sites for Fe(III) consisting of tyrosinate phenolic groups. These sites have only a weak affinity for Fe(II). The transferrins are in fact a group of glycoproteins that include lactoferrin (from milk), conalbumin or ovotransferrin (egg-white) and the serum transferrin; all functions as transport forms for iron.

Fe(III) is also found in the storage protein ferritin. The remarkable details of the storage forms are now well understood. The molecular structure consists of an outer protein shell which encases an Fe(III) hydroxy polymer which can contain up to 4500 iron atoms. Ferritin undergoes a partial stripping of the protein shell within lysosomes. The resulting Fe(III) complex haemosiderin is the second main form of iron storage in animals, but its molecular details are less well known.

Iron in haemoglobin and myoglobin is in the oxidation state (II). The haem ligands in these proteins stabilise this state against oxidation by oxygen to the state (III), thus ensuring that the O_2 binding is reversible. Fe(III)-haemoglobin is inactive towards O_2 binding.

Remarkably, the average daily iron dietary requirement is very low compared to the body's requirements. This reflects the relatively small excretory loss and the effective feedback mechanism that operate between the various compartments in the scheme of Fig. 3.

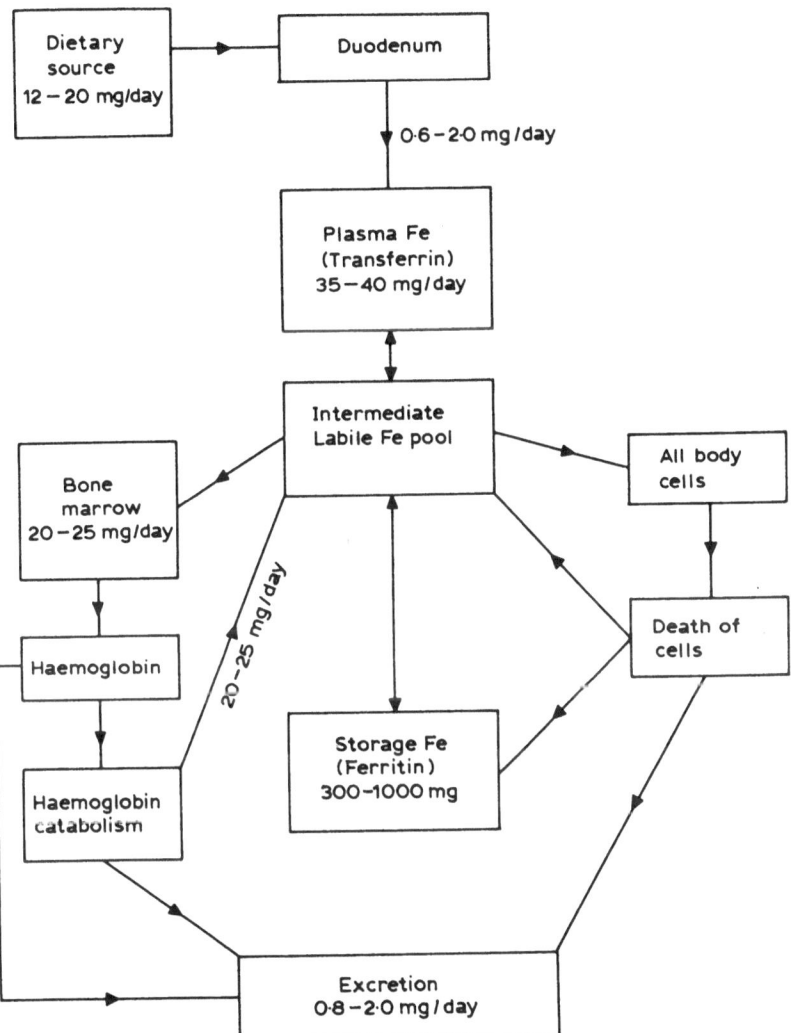

FIG. 3. Schematic outline of iron metabolism in an adult human; figures show
normal ranges.

Zinc

Some 200 enzymes are known to contain zinc[56,57] this is far more than is known for iron or any other metal. At present 30 of these enzymes have been identified in the human body; only a few of these have been well characterised at the molecular level.[56] As the examples in Table 10 show, the main function of the zinc enzymes is the catalysis of hydrolytic reactions; these involve zinc at the enzyme's active site acting as a 'superefficient' Lewis acid.

Zinc also has a structural role in stabilising protein structures, e.g. in insulin, liver alcohol dehydrogenase, alkaline phosphatase and superoxide dismutase. In the latter case, Cu(II) is the catalytically active metal ion. Zinc and copper also bind together to metallothionein and to serum albumin. The similarities between copper(II) and zinc(II) in their biological transport and storage and their mutual antagonism are not unexpected from their similar coordination chemistry. This similarity also extends to a common distribution in foodstuffs.

The intake of zinc is closely related to protein intake,[58] increasing with

TABLE 10

EXAMPLES OF COPPER AND ZINC ENZYMES IN THE HUMAN BODY AND THEIR FUNCTIONS

Enzyme	Function	Location
Copper		
Caeruloplasmin	Ferroxidase, transport, storage (?)	Blood
Cytochrome oxidase (with Fe)	$O_2 + 4H^+ + 4e^- \rightarrow 2H_2O$	Heart
Dopamine hydroxylase	Hydroxylation of dopamine	Adrenal gland
Amine oxidase	Oxidation of monoamines	Blood
Tyrosinase	Oxidation of tyrosine	Skin and liver
Lysyl oxidase	Oxidation of lysine side chain	Synovial fluid
Superoxide dismutase (with Zn)	Superoxide removal	Several tissues
Zinc		
Carbonic anhydrase	Catalysis of CO_2 equilibrium	Blood
Carboxypeptidase	Protein hydrolysis	Pancreas
Alcohol dehydrogenase	Alcohol oxidation	Liver
Alkaline phosphatase	Phosphate hydrolysis	Various tissues
Amino peptidase	Peptide hydrolysis	Kidney
Metallothionein	Homeostasis (?)	Various tissues
α_2-macroglobulin	Unknown	Blood

high protein meals and inhibited by cereals with a high phytic acid content. After dietary absorption, zinc is rapidly transported, as a zinc albumin complex, in the portal bloodstream to the liver. It then re-appears, mainly bound to the circulating proteins serum albumin (60–70% of the Zn) and α_2-macroglobulin (30–40%). This zinc transport mechanism is a facile exchange form for, after suffering trauma, there is a rapid redistribution of zinc within the body. After wound infliction, for example, the plasma zinc level rapidly falls whilst those of the liver and surrounding wound area increase. These changes implicate zinc as having a role in the biosynthesis of new proteins required for the wound-healing process.

Zinc homeostasis is maintained by regulation of intestinal zinc absorption and excretion; the efficiency of zinc absorption appears to increase with lowered zinc status.[59] The detailed mechanism of these regularity controls in man are not known as yet. There is no well-defined zinc storage form in the body, as there is for iron; however, there are zinc reserves in muscle and bone that appear to be utilisable. Appreciable zinc concentrations can also occur in the liver, particularly in the foetus; metallothionein appears to be involved in these cases. This protein (p. 8) has been implicated[20] in the control of uptake of zinc from the gut mucosa, although this has still to be proven.

The adverse effects of dietary zinc deficiency on factors such as growth, sexual maturity, skin conditions, and many others, are well documented. As mentioned earlier, concern about zinc deficiency in humans still abounds and is of particular concern in the use of synthetic infant foods, in intravenous nutrition and in food processing methods.[4] There is also evidence[20] that low zinc intakes can restrict the absorption of copper and iron from the same diets.

Copper

Because of its amenability to a range of magnetic and spectroscopic techniques, the binding of copper to a number of proteins has been studied extensively.[60,61] These studies show that, as for zinc, the major binding sites are the imidazole groups of histidine residues, followed by the thiol group of cysteine residues. Although similar to zinc in its coordination properties, copper is closer to iron in its biological functions; that is, it makes use of its accessible oxidation states of I and II (and possibly III) in catalysing redox processes. The functions of copper in its enzymes in the human and its distribution within the body are well established and documented.[27,29,60,61] Some of the better known copper enzymes are listed in Table 10.

Copper is present in a wide range of foods, the bioavailability from these sources is known to be dependent on a number of dietary and non-dietary factors; again, as with the other minerals, the food proteins are an important factor. A recent text surveys our present knowledge in the bioavailability of copper and its metabolism,[61] it also includes an extensive compilation of food copper contents. Like zinc, copper is transported after digestion to the liver via the portal bloodstream bound to albumin and transcuprein proteins. Within the liver most of the copper is incorporated into enzymes, in particular caeruloplasmin which then accounts for some 90–95% of the copper in the circulating bloodstream. Caeruloplasmin is something of an enigma: it is an α_2-glycoprotein of mass 130 000 Da and contains 6 copper atoms which are tightly bound. It has the intense blue colour associated with a number of copper enzymes and yet its role has escaped a definite assignment.[62] However, there is increasing evidence for a specific transport function, Harris and Percival,[63] for example, have shown that the copper atoms from caeruloplasmin are specifically incorporated into the superoxide dismutase located in the cytosols of K562 cells. Cell receptor sites have also been identified for caeruloplasmin.

Until fairly recently copper bound to albumin protein in blood was regarded as the transport form of the blood. However, this would appear to apply only after initial digestion in the portal blood stream: even there a second transport protein transcuprein is present.[64] This is a larger protein than albumin and binds Cu more strongly than albumin. The copper ions bound to these two proteins are readily exchangable. After incorporation into the liver a small amount of exchangable copper re-appears in the bloodstream, mostly bound to albumin. The serum albumin from a number of species, including man, has been shown to possess a single strong binding site for Cu(II). The site is located at the N-terminal end of the polypeptide chains and comprises[65] the terminal amino N atom, two deprotonated N atoms from the first and second peptide links, and an imidazole N atom from the thiol amino acid histidine. The four N atoms define an approximately square planar geometry around the metal. This type of binding, especially the use of the deprotonated peptide N atoms, is not found in any of the metal enzymes. Ni(II) can bind the same site but Zn(II) probably binds elsewhere along the polypeptide chain.

There is no evidence of a specific storage form of copper. The low dietary requirements and the absence of known cases of copper deficiency (not including the genetically linked Menkes' disease) would suggest that a storage form may not be necessary. The exception to this is the human foetus which accumulates relatively large concentrations of copper in the liver. This store is then used in the early stages following birth. In adults, the liver is still the largest source of copper (approx. 30 ppm), where it is distributed among several proteins, including metallothionein.

Toxic Metallic Elements

The ingestion of toxic metals can arise from a number of sources, principally from soil, sewage sludges, fertilisers, water and the containers and utensils used in various food processes.

The physiological disorders arising from the presence of these elements can be attributed to their binding to proteins and consequently, modifying the functions of the proteins or enzymes. Certainly for the heavier metals, lead and mercury, and their organic derivatives, there is abundant evidence of their ability to bind avidly to thiol groups of the cysteine residues of proteins, a reflection of the classical group b type behaviour of these metals (see p. 171). Toxicity can also arise from the displacement of essential metal ions from proteins, again altering the functional properties, e.g. Zn(II) by Cd(II), Fe(III) by Al(III). Other metals, e.g. beryllium, can form stable complexes with certain proteins which involve immune responses.

The more important members of the toxic metals group, with particular respect to their protein interaction, will now be considered.

Aluminium

This element is currently causing concern and controversy.[66,67] Claims to its toxicity range from the scaremongering to the dismissive (e.g. see Ref. 1, p. 272). It is the most abundant metal in the earth's crust and is widespread.

Aluminium occurs naturally in most foods: more recent analyses of common foodstuffs give a general concentration of less than 10 mg/kg. For milk it is less than 0.1 mg/kg.[68] These figures are more reliable and are lower than earlier results using less accurate procedures. Al is also used as an additive in baking powders, toothpastes and antacid remedies. It is also used as a flocculating agent during water purification: the recent inadvertent tipping of aluminium sulphate into a drinking water supply in the UK heightened the public awareness of the toxicity of this element. The leaking of Al from saucepans during cooking is a minor contribution to the amount of Al naturally consumed; the figure for the latter is put at approximately 6 mg/day but most of this is not absorbed.

The anxiety over Al toxicity arose from the lethal contamination of intravenous solutions for renal dialysis patients. It has also been proposed as an etiological factor in a variety of neurological disorders such as Parkinsonian dementia, Alzheimer's disease and motoneuron disease. Studies by Wills *et al.*[69] show that brain tissues exposed to aluminium lead to disturbances in the metabolism of cytoskeletal proteins. Other studies point to the inducement of skeletal disorders and disturbances of the haemopoietic system. In-vitro studies also show that Al^{3+} ions can promote the reaction between cytochrome *c* and succinic dehydrogenase and can act as a co-factor in adenylate cyclase activity. While these reactions may have no *in vivo* significance, there is some evidence that the toxicity of aluminium arises from its interference with enzymatic reactions.

Aluminium is widely found among human tissues; approximately 5 μg/l is found in blood, mostly bound to proteins, in particular transferrin.[70] This latter observation highlights the chemical similarity between Al^{3+} and Fe^{3+} ions; they are of similar size, strongly acidic, and of limited solubility at neutral pH. They are also both class *a* metal ions, as exemplified by their binding to transferrin which has a transport function (see p. 184). Lactoferrin has also been found to bind Al, and the iron-binding proteins may also be involved in the absorption of Al from the gut. This inter-relationship between Al and Fe obviously aids our understanding of the metabolism of Al but, as already pointed out pp. 14–25, there is also a further biological relationship with silicon.

Cadmium

Cadmium has long been recognised as a highly toxic element.[71,72] It is toxic to virtually every organ and tissue in the animal body. A serious outbreak of cadmium poisoning was observed in Japan in the late 1960s (*itai-itai* disease) as a result of industrial contamination of food and water supplies. The resultant depression of body growth and severe bone changes were shown to be exacerbated by a low-protein, Ca-deficient diet. During this crisis the daily Cd intake was 100–1000 μg compared with the W.H.O. recommended maximum of 50 μg. Contamination of soils and associated food crops is a problem concerning land adjacent to zinc processers and the use of fertiliser sludge from industrial sewage. In contrast to aluminium, cadmium is virtually absent from the human body at birth but accumulates with age up to approximately 50 years of age. Higher levels are found in smokers.

Cadmium metabolism in man features a long retention time of many years, a lack of homeostasis, and an accumulation in soft tissue. Highest levels are found in the kidney and liver where the cadmium is mainly bound to metallothionein protein. This has led to the hypothesis that metallothionein has a detoxifying role. Cadmium also has a powerful antagonistic effect on the divalent metal ions of copper, iron and zinc. Within the bloodstream, cadmium is principally associated with γ-globulin and albumin proteins; in adult humans the blood cadmium level is less than 1 μg/l. Undoubtedly, many of the toxic effects of cadmium can be related to its ability to replace zinc ions bound to proteins and enzymes, particularly when these involve thiol groups.

Lead

Long known to be toxic, lead[73] is still very much of current concern, particularly in its proposed deleterious effects on the mental health of young children. Lead is ubiquitous in the environment and in foods; fortunately, the levels are normally very low. In the U.K., the average diet is estimated to contain just 90 μg/kg food. Treated water supplies are generally also low in lead, but in prolonged contact with lead pipes, tanks or soldered joints the levels can rise above the EEC recommended maximum of 50 μg/l. Contamination of foods from contact with soldered joints in cans has been much reduced as a result of enforced legislative changes in the 1970s. Other sources of lead contamination are beverages stored in either inadequately glazed pottery or in crystal glass decanters. The normal intake from food and beverages is in the region 250–300 μg/day, only a small fraction of which is apparently absorbed.

Lead absorption is influenced by the dietary composition. A low protein diet, which usually decreases metal intake, results in an increase in lead absorption and retention. Of the 50–250 mg lead found in the body, more than 90% is located in the skeleton where it has a long residence time. Inorganic forms of lead whether given in the II or IV oxidation state are found in the body as Pb(II), mostly tightly bonded to protein fractions. Lead, like cadmium, can be shown *in vitro* to affect the properties of many enzymes and membranes although the *in-vivo* significance of these observations is not certain, as lead metabolism is not fully understood. However, lead has been shown to have a wide range of biochemical effects in its toxic reactions, many of which can be related to enzyme inhibition. Of particular importance are its poisoning effects on the cytochrome *P*-450 oxidase system and on the enzymes involved in haem synthesis. The lead inhibition of δ-aminolaevulinate dehydrase enzyme is a useful parameter for monitoring lead exposure in workers.

The comments above apply to both the inorganic and organic forms of lead. However, the latter form, mainly $(C_2H_5)_4Pb$, tetraethyllead from petroleum fuel, presents a more serious problem in that it has a stronger class b tendency and is far more lipid soluble. The latter property means that such forms can cross membrane barriers, e.g. the blood-brain barrier, which enhances their toxicity.

Mercury

The three common forms of mercury—the elemental form (a volatile liquid at room temperature), inorganic mercury (mainly Hg(II) salts), and organomercury (mainly RHgX, where R is an alkyl or aryl group and X a halide)—are all highly toxic.[74,75] They have different metabolic pathways and can interconvert in biological systems.

A great deal of information is now available on their biological behaviour, mainly as a result of two catastrophic incidents involving organomercury compounds, the Minamata outbreak in Japan, and the grain-poisoning outbreak in Iran. Both of these fatal incidents involved organomercury compounds. Like the organolead compounds, the organomercury compounds also exhibit increased lipid solubility. Furthermore the organo-forms have a much greater absorption from the gastrointestinal tract than the inorganic forms. Methylmercury is the most toxic of the various forms of mercury: it accumulates in the central nervous system with fatal results. The inorganic forms accumulate mainly in the kidney.

Both the inorganic and organic forms are good class b cations; their stability constants with thiol groups are some of the largest known for non-chelating ligands. Consequently, the binding with thiol groups, as with the cysteine groups of proteins for example, are not easily reversed. This is a property of considerable biological significance. Both forms are reabsorbed in the bloodstream by their binding to plasma proteins. CH_3Hg^+ shows a preference for binding to haemoglobin within the red blood cells. This fraction can be used as a guide to the extent of CH_3Hg^+ poisoning. A number of mercury compounds have also been shown to inhibit enzymes in the red blood cells. Extracellular protein binding depends to some extent on the dose level and the time lapse after receiving a mercury dose. Binding to albumin at different sites has also been noted. The highest concentrations of mercury are found in the kidney where it is again associated with proteins, the major one of which is identical to metallothionein. CH_3Hg^+ has a much greater effect on neuroreceptors than the inorganic forms. This is ascribed to its binding to membrane thiol groups, blocking transmembrane flux of Ca^{2+}. Physiologically, this is reflected in the muscular weakness that is associated with poisoning by methylmercury but not normally by the inorganic forms. Chemical signs of organomercury poisoning in adults appear at doses above just 300 μg/day. Children are even more sensitive.

CONCLUSIONS

Interactions of metal ions with proteins are quite well understood, and there is intense activity in several aspects of metal metabolism. In the context of food-protein studies, the situation must be described as less well developed. There have been a number of studies of the interference by proteins with metal metabolism, and though there is some knowledge of cases where particular food additives inhibit transport of metal ions, much remains to be established.

Perhaps an example will clarify this. Aluminium has been discussed earlier. The tea bush *Camellia sinensis* requires acid conditions of pH for successful growth, and this mobilises aluminium ions from the soil into solution and thence into the plant. Consequently, the raised level of Al in tea means that this is a major contributor[76] of aluminium at least to the British diet, and this has given rise to some concern, because of the recent demonstration of conditions related (not necessarily causally) to high aluminium levels (p. 190).

The effects of metals like aluminium, which are essentially a recent additive to human diet, will need careful evaluation. At a recent discussion meeting[77] studying the incidence of senile dementia in its relation to dietary aluminium intake, Perl commented that 'aluminium appears to do most damage when calcium and magnesium are absent. It is then that aluminium can bind with proteins, cross cell walls and disrupt biological processes'.

With this kind of development in mind, the need for the establishment of rates and equilibria of metal binding by dietary protein is clear, and many more studies of this kind are to be encouraged. Understanding the details of metal metabolism will prove helpful, and enormous efforts are already going into this kind of biochemical study aided by the newer analytical techniques outlined at the beginning of this chapter. The competition between metal ions for similar sites is fairly predictable, but an area which may well require considerable further research is the reverse, the competition between ligands for metal ions, which will manifest some striking effects. Thus, if a metal ion is an essential requirement for healthy metabolism, and an additive (such as a highly proteinaceous foodstuff) complexes it all, successfully competing with its proper receptor area, how can such an outcome be predicted and the effects avoided?

REFERENCES

1. Coultate, T. P., *Food, The Chemistry of its Components,* Royal Society of Chemistry, London, 1989.
2. *Nutrient Availability: Chemical and Biological Aspects,* ed. D. A. T. Southgate, I. T. Johnson and G. R. Fenwick, Royal Society of Chemistry, London, 1989.
3. *Trace Elements in Health and Disease,* ed. A. Aitio, A. Aro, J. Jarvisalo and H. Varnio. Royal Society of Chemistry, London, 1991.
4. See, Bryce-Smith, D., *Chem. Br.,* **25** (1989) 783 and **26** (1990) 24 and reply by H. T. Delves *et al., Chem. Br.,* **25** (1989) 1207.
5. Williams, R. J. P., *Chem. Br,* **19** (1983) 1009.
6. Crichton, R. R. & Mareschal, J.-C., in *Inorganic Biochemistry,* ed. H. A. O. Hill, Vol. 3. Royal Society of Chemistry, Specialist Periodical Report, London, 1982, p. 78.
7. Mertz, W., *Proc. Fed. Am. Soc. Exp. Biol.,* **29** (1970) 1482.
8. Cotzias, G. C., in *Trace Subst. Environ. Health,* in Proceedings Univ. Missouri 1st Annual Conf., ed. D. H. Hemphill. Columbia, 1967, p. 5.
9. Schroeder, H. A. & Nason, A. P., *Clin. Chem.,* **17** (1971) 461.
10. Underwood, E. J., *Trace Elements in Human and Animal Nutrition,* 4th edn, Academic Press, New York, 1977.
11. *Biochemistry of Silicon and Related Problems,* ed. G. Bendz and I. Lindqvist, Plenum, London, 1978.
12. Birchal, J. D. & Chappell, J. S., *Lancet,* i (1989) 953.
13. McCance, R. A. & Widdowson, E. M., in *The Composition of Foods,* eds. A. A. Paul and D. A. T. Southgate. *H.M.S.O.,* London, 1978.
14. O'Neill, N. C. & Tanner, M. S., *J. Ped. Gastroenterol. Nutr.,* **9** (1989) 167.
15. Kobayachi, J., *Proc. Int. Water Pollution Res. Conf. 5th,* 1970, p. 1.
16. Craig, P. J., *Organometallic Compounds in the Environment,* Longmans, Harlow, UK, 1986.
17. *Trace Elements in Human Nutrition,* W. H. O. Technical Report Series No. 532, W. H. O., Geneva, 1973.
18. Hughes, M. N., *The Inorganic Chemistry of Biological Processes,* 2nd Edn, Wiley, Chichester, 1981.
19. Laurie, S. H., *J. Inher. Metab. Dis.,* **6** (1983) 9.
20. Bremner, I. & Mills, C. F., *Phil. Trans. R. Soc. London,* **294** (1981) 75.
21. Lonnerdal, B., in Ref. 2, p. 131.

22. Barnett, M. I., Duffield, J. R., Evans, D. A., Findlow, J. A., Griffiths, B., Morris, C. R., Vesey, J. A. & Williams, D. R., Ref. 2, p. 97.
23. Robb, P., Williams, D. R. & McWeeny, D. J., *Inorg. Chim. Acta,* 125 (1986) 207.
24. Duffield, J. R., Hall, S. B., Williams, D. R. & Barnett, M. I., in *Progress in Medicinal Chemistry,* Vol. 28, ed. G. P. Ellis and G. B. West. Elsevier Science Publishers, Amsterdam, 1991, p. 175.
25. Vasak, M. & Kagi, J. H. R., in *Metal Ions in Biological Systems,* ed. H. Sigel, Vol. 15. Marcel Dekker, New York, 1983, p. 213.
26. Sarkar, B., *Chem. Scr.,* 21 (1983) 101.
27. Fisher, G. L., *Sci. Total Environ.,* 4 (1975) 373.
28. Williams R. B., McDonald, I. & Bremner, I., *Br. J. Nutr.,* 40 (1987) 377.
29. Serfass, R. E., Thompson, J. J. & Houk, R. S., *Anal. Chim. Acta,* 188 (1986) 73.
30. Dalgarno, B. G., Brown, R. M. & Pickford, C. J., in Ref. 2, p. 31.
31. Eagles, J., Portwood, D. E., Fairweather-Tait, S. J., Gotz, A. & Heumann, K. G., in Ref. 2, p. 35.
32. Fairweather-Tait, S. J. *et al.,* in Ref. 2, p. 45.
33. Perl, D. P., Gajdusek, D. C., Garrato, R. M., Yanagihara, R. T. & Gibbs, C. J., *Science,* 217 (1982) 1053.
34. Savory, J. & Wills, M. R., in Ref. 3, p. 5.
35. Goodman, B. A. & McPhail, D. B., *J. C. S. (Dalton),* (1987) 1717.
36. Goodman, B. A., McPhail, D. B. & Powell, H. K. J., *J. C. S. (Dalton),* (1981) 822.
37. Garruto, R. M. & Yase, Y. M., *Trends Neurol. Sci.,* 9 (1986) 368.
38. Whiting, M. G., *Econ. Bot.,* 17 (1963) 271.
39. Dossaji, S. F. & Bell, E. A., *Phytochemistry,* 12 (1973) 143.
40. Hursthouse, M. B., Motorelli, M., O'Brien, P. & Nunn, P.B., *J. C. S. (Dalton),* (1990) 1985.
41. Antanaitis, B. C. & Aisen, P., *Adv. Inorg. Biochem.,* 5 (1983) 111.
42. Fujimoto, S., O'Hara, A. & Uehara, K., *Agric. Biol. Chem.,* 44 (1980) 1659.
43. Nakagowa, Y., Kaiser, E. T. & Coe, F. L., *Biochim. Biophys., Res. Commun.,* 84 (1978) 1038.
44. Schlesinger, D. H. & Hay, D. I., *J. Biol. Chem.,* 252 (1977) 1689.
45. Banks, R. D., Blake, C. C. F., Evans, P. R., Haserd, R., Rice, W., Hardy, G. W., Merrett, M. & Phillips, A. W., *Nature,* 279 (1979) 773.
46. Kammon, K. K., Notstrand, B., Fridborg, K., Lovgren, S., Ohósson, A. & Petef, M., Proceedings of the National Academy of Sciences (New York), 72 (1975) 51.
47. Quiocho, F. A. & Lipscomb, W. N., *Adv. Protein Chem.,* 25 (1971) 1.
48. Bordas, J., Bray, R. C., Garner, C. D., Gutteridge, S. & Hassein, S. S., *Biochem. J.,* 191 (1980) 499.
49. *Calcium Binding Proteins and Calcium Function,* ed. R. H. Wassermanm, R. A. Corradino, E. Carafoli, R. H. Kretsinger, D. H. Maclennan, and F. L. Seigel. Elsevier, New York, 1977.
50. Cooper, C., *Chemistry and Industry,* (1986) 445.
51. Stuart, D. I., Acharya, K. R., Walker, N. P. C., Smith, S. G., Lewis, M. & Phillips, D. C., *Nature,* 324 (1986) 84.

52. Naito, H., Gunshin, H. & Noguchi, T., in Ref. 2, p. 253.
53. *Iron in Biochemistry and Medicine*, ed. A. Jacobs & A. P. Worwood, Vol. 1 and Vol. 2, Academic Press, New York, 1974, 1980.
54. Bothwell, T. H., Charlton, R. W., Cooke, J. D. & Finch, C. A., *Iron Metabolism in Man*. Blackwell Scientific, Oxford, 1979.
55. Mills, C. F., *Annu. Rev. Nutr.,* **5** (1985) 173.
56. Vallee, B. L. in *Zinc Enzymes,* ed. T. G. Spiro. Wiley, New York, 1983.
57. Chelbowski, J. F. & Coleman, J. E., *Metal Ions in Biological Systems,* ed. H. Sigel, Vol. 6, Marcel Dekker, New York, 1976, p. 1.
58. Sandstrom, B., Arvidsson, B., Cederblad, A. & Bjorn-Rasmussen, E., *Am. J. Clin. Nutr.,* **33** (1980) 739.
59. Bosworth, C. M., Bacon, J., Bremner, I. and Aggett, P. J., in Ref. 2, p. 213.
60. *Metal Ions in Biological Systems*, ed. H. Sigel, Vols, 12 and 13, Marcel Dekker, New York, 1981.
61. *Copper Bioavailability and Metabolism,* ed. C. Kies. Plenum Press, New York, 1990.
62. Laurie, S. H. & Mohammed, E. S., *Coord. Chem. Rev.,* **33** (1980) 279.
63. Harris, E. D. & Percival, S. S., in Ref. 61, p. 95.
64. Goode, C. A., Dinh, C. T. & Linder, M. C., in Ref. 61, p. 131.
65. Iyer, K. S. N., Lau, S., Laurie, S. H. and Sarkar, B., *Biochem. J.,* **169** (1978) 61.
66. Allfrey, A. C., in *Trace Elements in Human and Animal Nutrition,* 5th edn., ed. W. Mertz, Vol. 2. Academic Press, New York, 1986.
67. *Aluminium in Food and the Environment,* ed. R. Massey and D. Taylor. Royal Society of Chemistry, Special Publication No. 73, London, 1989.
68. Sherlock, J. C., Ref. 67, p. 68.
69. Wills, M. R., Hewitt, C. D., Savory, J. and Herman, M. M., in Ref. 3, p. 227.
70. Tapp, G. A., *Life Sci.,* **33** (1983) 311.
71. *Chemistry, Biochemistry and Biology of Cadmium,* ed. M. Webb. Elsevier, Amsterdam, 1979.
72. Kostial, K. Ref. 66, p. 319.
73. Quaterman, J., in *Trace Elements in Human and Animal Nutrition,* Vol. 2, 5th edn., ed. W. Mertz. Academic Press, New York, 1986, p. 281.
74. Suzuki, T., in *Toxicology of Trace Elements,* ed. R. A. Goyer and M. A. Mehlman. Wiley, New York, 1977, p. 1.
75. Craig, P. J., in Ref. 16, p. 65.
76. Coriat, A. M. & Gillard, R. D., *Nature,* **321** (1986) 570.
77. Pearce, F., *New Scientist,* (1985) issue of 25th April, p. 7.

Chapter 6

HAEMOPROTEINS IN MEAT AND MEAT PRODUCTS

D. A. LEDWARD

Food and Agricultural Chemistry Department, The Queen's University of Belfast, Newforge Lane, Belfast BT9 5PX, UK

INTRODUCTION

Since publication of the original review in 1984, there has continued to be significant research effort in the general field of haemoprotein chemistry. This has resulted in further understanding of the nature of many of these coloured pigments in meat and meat products. However, in some areas little further advance has been made. Thus this review will assume the reader already has access to this 1984 work[1] and will keep to the original format but condense those areas in which little or no advance has been made and modify or expand those in which we have furthered our understanding. Thus the areas concerned with the nature of the haemoproteins in fresh, cooked, cured, irradiated and dried meat are significantly modified as also is the section concerned with the relationship between lipid and haemoprotein oxidation.

OCCURRENCE AND FUNCTION

Although the haemoproteins only constitute about 0·5% of the wet weight of red meats such as beef and lamb, and several times less than this of the white meats such as pork, they are of paramount importance in determining meat quality because, to a large extent, they govern the colour of the product. In addition, their presence is essential in contributing to the high iron bioavailability of meat and may well affect the stability of the lipids present in the product.

In most meats, the haemoprotein in greatest concentration is the muscle pigment myogloblin, although the blood pigment haemoglobin is also present in significant concentrations. Even though the absolute concentration of haemoprotein in meat will obviously affect its colour, the haemoprotein rich meats such as beef and lamb being much darker than the haemoprotein poor meats such as pork and chicken, and may also affect the bioavailability of the iron,[2] the ratio of haemoglobin to myoglobin has little effect since their reactivities are similar. As the reactivity of these two haemoproteins are similar it is not surprising that early workers found different relative concentrations.[3] For example, Rickansrud & Hendrickson[4] found that in beef, haemoglobin constituted about 20 and 25% of total haemoproteins of the longissimus dorsi and biceps femoris muscles, respectively, whilst Warriss & Rhodes[5] found that, in these muscles, haemoglobin constituted less than 10% of the total haemoproteins, a result confirmed by Hazell.[6] Warriss & Rhodes[5] also demonstrated that the ratio of haemoglobin to myoglobin in beef varied with the anatomical position of the muscle and the age and sex of the animal. Although differences in haemoglobin levels are seen between different samples of meat, the major reason for the wide differences in haemoprotein contents found in muscles is the inherent variations in the concentration of myoglobin.[7]

Myoglobin is apparently distributed uniformly throughout muscle and its role appears to be that of facilitating the diffusion of oxygen from the capillaries to the intracellular structures where the oxygen is used in oxidative processes.[8] In general it would appear that high levels of muscular activity lead to higher concentrations of myoglobin[7] reflecting, in this respect, differences due to species (the muscles of the hare are richer in myoglobin than those of the rabbit), breed, sex[9] (the muscles of the bull contain more myoglobin than those of the cow), age (the muscles of the steer are richer in myoglobin than those of the calf), type of muscle (the leg muscles in the chicken are richer in myoglobin than those of the little used breast), and training (the muscles of stall fed animals generally contain less myoglobin than those of their free range counterparts). Diet can also significantly affect the level of myoglobin in meat animals and it has recently been shown that in pigs a restricted feeding regime leads to a significant increase in the haemoprotein content of the longissimus dorsi.[9]

Although it is possible to rationalise, on the above lines, the wide range of myoglobin concentrations found in different muscles, it is rather more difficult to explain the variability very occasionally encountered within a specific muscle. For example, it has been claimed that in a given muscle the myoglobin concentration may vary several hundredfold over distances of a few centimetres.[7]

Although myoglobin is undoubtedly the major haemoprotein in red meats it is readily apparent that white meats must contain a greater proportion of their haemoproteins as haemoglobin. For example, in chicken meat haemoglobin may account for well over 50% of the total haemoproteins.[6,10] As well as haemoglobin and myoglobin, other haemoproteins (e.g. cytochromes) are present in meat. However, these proteins are present at such low concentrations that they have, except in very special circumstances such as their role in curing, minimal effects on the quality of meat and meat products.

The concentration of the different iron containing compounds present in several muscle and organ meats are shown in Table 1 and it is seen that in red meats the haemoproteins are the major source of iron but in liver and white meats non-haematin iron compounds (predominantly haemosiderin and ferritin) are the major sources.

TABLE 1

IRON CONTENT, AND PERCENTAGE PRESENT AS HAEMOPROTEINS, OF MEATS (FROM REFS. 6, 10–13)

Meat	Mean iron content mg $100 g^{-1}$ wet wt	% Haematin iron
Beef	2·44–2·61	62–73
Beef liver	6·30	30–40
Lamb	1·64–1·90	57–59
Lamb liver	8·40	—
Pork	0·69–1·00	47–49
Pork liver	9·50	—
Chicken breast	0·53	17
Chicken leg	0·88–1·83	23–28
Chicken liver	20·15	13–17

STRUCTURE OF MYOGLOBIN AND HAEMOGLOBIN

Myoglobin

As explained in the previous review, sperm whale myoglobin consists of a single polypeptide chain of 155 amino acid residues complexed to a haematin moiety. The protein chain consists of eight α-helical segments, ranging in length from 7 to 24 residues, separated by non-helical regions. The helical regions constitute about 80% of the molecule.

The haematin moiety, in both myoglobin and haemoglobin consists of the protoporphyrin IX ring system (Fig. 1) in the centre of which an iron atom is attached to the four nitrogen atoms of the pyrroles. However, the iron atom (whether ferrous or ferric) is coordinated in an octahedral-type environment so that it is capable of accepting two further ligands at right angles to the haematin. In myoglobin one of these ligands is the nitrogen in the imadazole ring of a globin histidine residue, whereas the other may be any molecule of the correct electronic configuration, small enough to occupy the haematin pocket in the protein. The nature of this sixth ligand, and the oxidation state of the iron atom will affect the electronic arrangement of the electrons of the iron, which will in turn affect the electronic arrangement of the d electrons of the iron and consequently the spectral absorption characteristics of the molecule, hence its colour.[1] The haematin moiety itself fits into a space in the hydrophobic interior with both propionic acid side chains being hydrogen bonded to amino acid side chains of the globin. The propionic acid side chains extend through the interior with the carboxyl group on the hydrophilic surface whereas the vinyl groups are buried in the hydrophobic interior. This interaction between the haematin and globin serves to stabilise the whole molecule.[14]

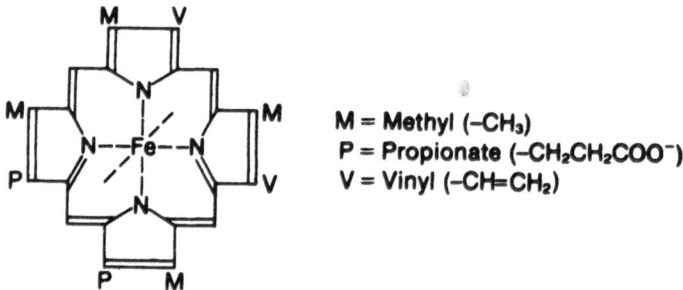

M = Methyl (–CH₃)
P = Propionate (–CH₂CH₂COO⁻)
V = Vinyl (–CH=CH₂)

M = Methyl ($-CH_3$)
P = Propionate ($-CH_2CH_2COO^-$)
V = Vinyl ($-CH=CH_2$)

FIG. 1. The iron protoporphyrin IX group of myoglobin.

Although the precise structure of sperm whale myoglobin is well established, the structures of myoglobins extracted from meat producing animals have not all been fully elucidated. It is apparent, though, that the basic structures of all myoglobins are similar.[15] However, the subtle differences that do exist between myoglobins from different species may be important in determining the visual appearance and colour stability of different meats. Thus Satterlee & Zachariah [16] found significant differences in the stability and properties of bovine, ovine and porcine myoglobins. These workers claimed that, although ovine and bovine myoglobins were very similar in their stability to both heat and acid, porcine myoglobin behaved differently, being more susceptible to acid denaturation, although its stability to heat was similar to that of the other myoglobins. The amino acid composition (and isoelectric point) of porcine myoglobin was also very different to that of both bovine and ovine myoglobin, which although possessing similar compositions, had themselves significantly different amino acid sequences. To further complicate the picture it has been reported that myoglobin extracted from pale, soft, exudative (PSE) pork is very different from that obtained from normal pork.[17] PSE pork is produced from stress susceptible animals when the carcase goes into rigor very rapidly, so that the pH of the muscle decreases from an *in-vivo* value of over 7 to one of 5·5 or less whilst the temperature is still high. Thus, this within species variation may be an artefact created by the unusual pH/temperature environment the molecules experience after the slaughter of stress susceptible animals.[1]

Haemoglobin

Haemoglobin, the transporting pigment of mammalian serum, although closely related to myoglobin, is more complex as it consists of four individual chains each of which possess a haematin moiety. The individual chains are not covalently linked and can be separated relatively easily. The two pairs of identical chains that constitute the intact haemoglobin molecule are called the α- and β-chains. In human haemoglobin the α-chain has a molecular weight of about 15 130 and the β-chain a molecular weight of about 15 870. The intact tetramer ($\alpha_2 \beta_2$) has a molecular weight of about 65 000. The α- and β-chains have different physical and chemical properties and the four oxygen binding sites in haemoglobin are not equivalent.

Bovine haemoglobin is slightly less stable to heat than bovine myoglobin.[18] The relative stabilities of the haemoglobins from other meat animals are not known but they are unlikely to vary to any significant extent.

Although differences do exist in the structure and reactivity of different myoglobins, and haemoglobin may behave differently from myoglobin, the differences are of degree rather than kind. For this reason the discussion in this review will be concerned primarily with the myoglobins.

ROLE OF HAEMOPROTEINS IN DETERMINING MEAT QUALITY

Colour

The visible absorption characteristics of myoglobin depend on the electronic configuration of the haematin group, which in turn is dependent on the nature of the ligands attached to the 5th and 6th coordination positions of the iron atom, the oxidation state of the iron and the conjugated double bond system of the protoporphyrin molecule.

How the electronic distribution around the ferrous or ferric ion affects the colour of the haemoproteins has been discussed in detail elsewhere.[1,19]

Fresh Meat

When meat is freshly cut the myoglobin is in the purple reduced state (Mb) in which the iron is in the high spin ferrous state and the sixth coordination site is unoccupied.[20] On exposure to air myoglobin, because of its great affinity for the oxygen molecule, combines rapidly and reversibly with oxygen to form the bright red oxymyoglobin (MbO_2) and it is this pigment that the consumer associates with freshness. Because of its great affinity for oxygen the meat surface 'blooms' to the red colour within minutes of exposure to air and, with time, the small layer of MbO_2 spreads downwards into the meat. The depth to which the oxygen diffuses depends on the activity of the oxygen utilising enzymes, i.e. the oxygen consumption rate of the meat, the temperature and external oxygen pressure.[7] After 2 h exposure to air at 0°C the MbO_2 layer in different beef muscles was 1–3 mm thick and increased to 7–10 mm after 7 days.[21] The gaseous oxygen diffuses through the aqueous environment and enters the hydrophobic haematin cleft to occupy the sixth coordination site. Oxygenation-induced conformational shifts are believed to bring a histidine residue, known as the distal histidine to differentiate it from the proximal histidine which occupies the fifth coordination site, within interacting range of the liganded oxygen, thereby stabilising the complex. Recent years have seen some controversy regarding the electronic distribution around the haematin in the oxymyoglobin molecule. The covalent structure ($Fe^{2+}O_2$) shown in Fig. 2 is now considered to be very improbable and a great deal of evidence does suggest that the complex is best represented as a low-spin superoxide ferric complex, i.e. $Fe^{3+}O_2^-$. Gidding[19] reviewing the available evidence suggests that the single best electronic representation of the iron–oxygen complex lies between the completely covalent and ionic extremes and that the dioxygen iron model

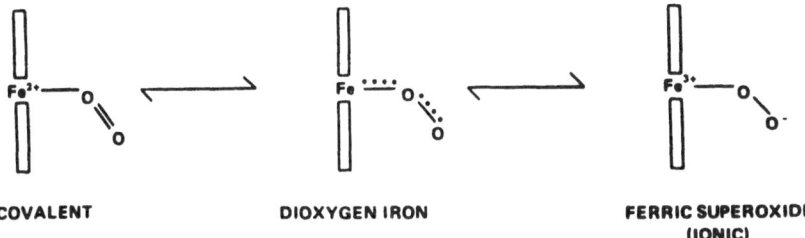

COVALENT   DIOXYGEN IRON   FERRIC SUPEROXIDE
                                                                  (IONIC)

FIG. 2. Three possible resonance forms of the haematin–oxygen bonding in oxymyoglobin.

(Fig. 2) with substantial, but less than complete charge transfer to oxygen, is most representative.

As might be anticipated 'blooming' is more efficient under conditions which increase oxygen solubility and discourage enzymic activity, i.e. at low temperatures and low pH values. The dark colour of meat of high ultimate pH, e.g. DFD (dark, firm, dry) beef is to some extent due to the high activity of the enzymes that minimise oxygen penetration so that the purple reduced myoglobin is visual at or just below the surface. In addition meat that has been held (aged) for several weeks *in vacuo* prior to exposure to air blooms more rapidly than fresh meat[21] due to some loss of activity of the oxygen utilising enzymes during storage.

The commercial exploitation of electrical stimulation (ES) in the USA was to some extent helped by the fact that ES-treated beef yielded a better, i.e. redder, 'colour' when the longissimus dorsi was used to grade the carcase 24 h after slaughter. This was due to increased formation of MbO_2.[22] Electrical stimulation is a process which effectively accelerates post-mortem glycolysis so that the muscle pH is reduced to its ultimate value within a few hours of slaughter, rather than the 20–24 h necessary at chill temperatures in unstimulated beef carcases. The lower pH thus permits more oxygenation of the haematin. Although a better colour is seen 24 h after slaughter there is little difference in the rate or extent of MbO_2 formation in stimulated and unstimulated carcases 48 h after slaughter,[22] presumbaly because the protein environments are now identical.

Although electrical stimulation of carcases generally improves the 24-h colour of the longissimus dorsi (loin) muscle, it can have adverse effects on the initial colour of the deep seated muscles that cool far more slowly.

For example the semi-membranosus (topside) muscles of low voltage stimulated beef carcases can be significantly paler than their unstimulated controls.[23,24] In low voltage systems the current is applied immediately after slaughter so that high temperatures and low pH values may co-exist. A major contribution to this paleness is due to partial unfolding of the myofibrillar proteins, which consequently lose some of their ability to hold water[24] so that the meat becomes pale and watery. A similar phenomenon occurs in PSE pork.[25] Very rapid chill-freezing will ameliorate this problem.[24] There is no evidence that the haemoproteins themselves are affected by low voltage electrical stimulation although the work of Bembers & Satterlee[17] suggests that in PSE pork structural modification of the myoglobin may occur.

Whatever factors affect the rate and extent of MbO_2 formation the complex readily oxidises to metmyoglobin (metMb) which is brown and unable to form an oxygen adduct. This brown pigmentation, which downgrades the quality of fresh meat, is formed by the removal of a superoxide anion (or its conjugate acid, HO_2) from the haematin[19] and its replacement by a water molecule to produce a high spin ferric haematin. The water molecule attached to the ferric iron may ionise so that a hydroxyl group occupies this 6th site to yield a low spin, red coloured complex. However, this transition has a pK value above 8[26] and is of little significance in meat and meat products.

In pure solution the autoxidation of MbO_2 is first order with respect to unoxidised myoglobin although the actual reaction mechanism is still the subject of some controversy.[19,27] In spite of the actual mechanism still being the subject of considerable debate the factors which affect the rate are well documented. The rate is maximal at oxygen tensions that correspond to half saturation of the iron,[28,29] i.e. at oxygen pressures of about 1–1·4 mmHg. At oxygen tensions above 30 mmHg the rate is independent of the partial pressure of oxygen. In addition, the autoxidation is very pH and temperature dependent. It has been estimated that at saturated oxygen pressures the half-life of the autoxidation is 2·8 h at pH 5 and 25°C whilst at 0°C and pH 5 the half-life is about 5 days. At pH 9 the values are 7 days and 1 year respectively.[30] Thus, under saturating oxygen tensions the activation energy is about 26·5 cal/mol (Q_{10} 5·3) although at well below saturated oxygen pressures the activation energy is significantly reduced. It is also well established that metal ions such as Cu^{2+} and Fe^{3+} accelerate the reaction.[31]

The rate of autoxidation does vary to some extent with the type of myoglobin, fish myoglobins being most susceptible and those of mammalian origin least reactive.[27]

However, the results found for myoglobins in solution cannot be applied indiscriminately to meat as, in meat, a reducing system is present which is capable of reducing metmyoglobin to the ferrous state, and a catalytic process also operates. Although it was originally believed that the reduction of metmyoglobin in meat only took place in the absence of oxygen[32] it is now established that it can occur under aerobic conditions[33] which would agree with the known properties of, for example, beef heart metmyoglobin reductase[34] since in vitro this enzyme is unaffected by the presence or absence of oxygen and has optimum activity at about 37°C. The pH optimum for the enzyme is at about 6·5 and the activity at this pH is about twice that at pH 5·6. Obviously, the reduction of metmyoglobin in meat may be more complex than the studies with the purified enzyme suggests since more than one reaction may be responsible for the reduction observed in meat systems, as several possible substrates and intermediates are present.[1]

As both oxidation and reduction of the haematin pigments may occur in meat it is not surprising that, in intact meat pieces, the kinetics of metmyoglobin formation are complex. However, consideration of the factors known to affect the autoxidation of myoglobin and enzymic reduction of metmyoglobin do enable the effect on the rate of formation to be understood. Thus, sterile beef or pork stored aerobically at low temperatures (−1 to 10°C) accumulate increasing concentrations of metmyoglobin at their surface until a 'pseudoequilibrium' level is attained as the rates of the oxidation and reducing reactions equalise (Fig. 3). However, on prolonged storage, as the enzymic reducing system becomes exhausted, the pseudoequilibrium is lost and further metmyoglobin accumulates at the surface (Fig. 3). Also, as the autoxidation is maximal at low partial pressures of oxygen and the enzymic reducing system is believed to be oxygen independent, when meat is stored in air for a day or so at refrigerated temperatures, a brown layer of metmyoglobin forms a few millimetres below the surface where the oxygen concentration corresponds to that required for maximum formation. Increasing the oxygen tension in the outside atmosphere causes the metmyoglobin layer to occur at greater distances below the meat surface and in atmospheres containing 60% oxygen meat pieces 2 cm thick do not normally possess such a layer.[32] Conversely, storage at low oxygen pressures causes the layer to move closer to the surface until at about 7 mmHg (~ 1% oxygen at STP), it actually occurs at the surface and visibly discolours the meat (Fig. 4).[33,36] If meat can be stored in oxygen containing atmospheres that prevent the formation of this brown metmyoglobin layer or cause it to occur well below the surface, (> 1 cm) then browning at the surface is much

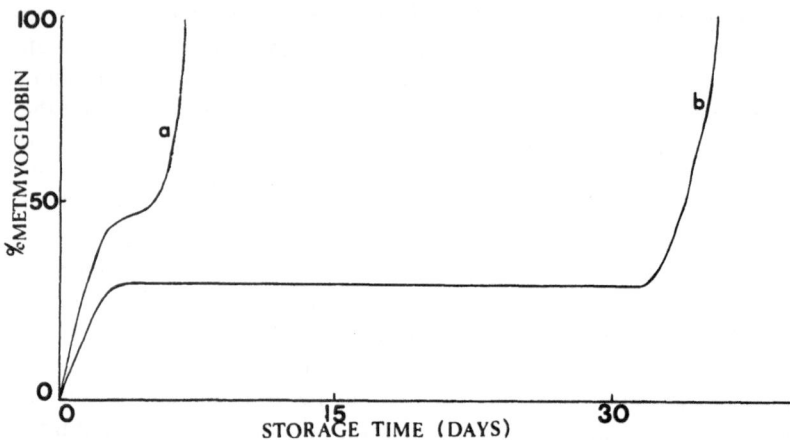

FIG. 3. Schematic representation of typical time courses for the formation of met-myoglobin at the surface of fresh meat during aerobic storage at 0–2°C. Curve (a) is for a colour labile muscle and (b) for a colour stable muscle (Adapted from Refs. 33, 35–37, 39).

reduced;[33,37] however, if the layer is allowed to form within a few mil-limetres of the surface, the formation of metmyoglobin at the surface appears to be largely independent of external oxygen pressure, i.e. from 5 to 20% (Fig. 4).[35] This complex dependence on oxygen pressure has led

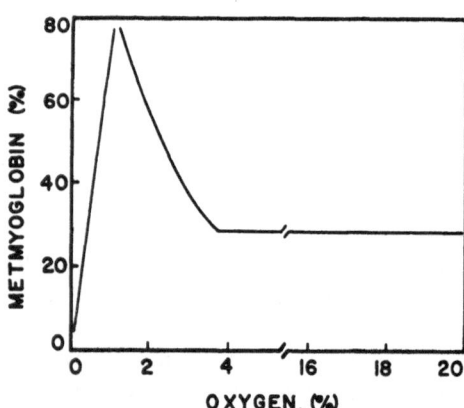

FIG. 4. Relationship between the 'equilibrium' level of metmyoglobin (metMb) formed at the surface of fresh, sterile beef semitendinosus muscle stored at 7°C and the oxygen concentration of the surrounding atmosphere (Redrawn from Ref. 35. Copyright © by the Institute of Food Technologists).

to the development of systems for the transport and storage of fresh meat at high (> 60%) oxygen partial pressure. It is usual in these systems to incorporate carbon dioxide (20%) to retard bacterial growth.

At the rate of the autoxidation reaction increases with decreasing pH whilst the enzymic reduction is far less effective at low pH, it is not surprising that, in general, muscles of low ultimate pH discolour more rapidly than those of high ultimate pH.[38] However, it must be remembered that, in meat of high pH, due to the high oxygen consumption rate, the metMb layer will be near to the surface and thus visual discoloration may occur more rapidly than anticipated (unpublished observations). Also, as the temperature dependence of the autoxidation[29] is far greater than that of the enzymic reduction[34] and since oxygen penetration will be less at higher temperatures, the observation that a 3–5°C increase in temperature in a display cabinet may double the rate of discoloration is not unexpected.[33] Meats aged in the absence of the oxygen for extended periods brown relatively rapidly on exposure to air, presumably due to the depletion of essential reactants in the reducing system during storage,[21,23] i.e. although aged meat blooms rapidly it subsequently forms metMb more rapidly than unaged meat.

The ability of different muscles to maintain a low equilibrium level of surface metmyoglobin during storage varies greatly and to a very large extent depends on the anatomical location within the carcase;[39–41] in beef the longissimus dorsi and semitendinosus muscles are relatively colour-stable whilst the psoas major discolours very rapidly. O'Keefe & Hood[21,42,43] extensively studied the factors affecting the formation of met-myoglobin in different beef muscles and claim that the inherent differences are not due to different metmyoglobin reducing activities of the muscles but primarily to variations in the rate of myoglobin oxidation and the ability of the muscle to consume oxygen, i.e. muscles with high activities of oxygen utilising enzymes, which allow little penetration of oxygen into the tissue, tend to discolour most rapidly.

More recent studies have also shown a significant positive correlation between oxidative activity (oxygen consumption rate) and colour instability in three 'fresh' beef muscles.[44] In addition there was no significant correlation between metmyoglobin reducing activity and colour stability. However, as aged muscles, of low oxygen consumption rate, discolour very rapidly,[23] this conclusion must be viewed with caution. It has also been shown that in very fresh meat, 1 h post-slaughter, the metmyoglobin reducing activities of 4 muscles were *inversely* proportional to their colour stabilities.[45]

Further insight into the factors responsible for the formation of metMb in meat (especially beef) has been afforded by the numerous studies carried out on electrically stimulated beef. It is now generally accepted that in muscles allowed to cool relatively rapidly, such as the longissimus dorsi, then the effects of rapidly decreasing the muscle pH by electrical stimulation has little effect on the colour stability.[23,24] This is presumably because the rapid decrease in temperature protects any temperature sensitive reactants in the oxidative and/or reducing systems. However if muscles cool more slowly then there is some controversy as to whether rapidly decreasing the pH by stimulation improves or worsens the colour stability of the muscle. Sleper et al.[46] found that, at low oxygen pressures, stimulated longissimus dorsi muscles could form metMb more rapidly than non-stimulated muscles whilst other workers have claimed that no significant difference exists between such muscles.[47] In contrast, Claus et al.[48] observed that hot-boned muscles tend to produce more metMb when freshly sliced than do muscles which have undergone electrical stimulation prior to hot-bonding. However, Ledward et al.[49] found that low voltage electrical stimulation decreased the colour stability of beef semi-membranosus muscles.

The proposed rationale to explain these apparently conflicting results is as follows.[23] In fresh meat, exposed to air 24 or 48 h after slaughter the rate of metMb formation is primarily governed by the oxygen consumption rate of the muscle. Thus, exposing the meat to high temperatures and low pH, as occurs in stimulation, may decrease the activity of the oxygen utilising enzymes and thus improve the colour stability of the muscle.[48] However, the oxygen consumption rate of meat decreases exponentially with time at 2°C[50] and thus in meat exposed to air several days post-slaughter the rate of enzymatic reduction, which decreases only slowly with time[23] largely dictates the rate of metMb formation. Faustman & Cassens have recently demonstrated that in beef held in 1% O_2 for from 2 to 6 days the reducing ability, when measured aerobically, apparently increases quite significantly.[51] This supports the contention that the oxidative ability decreases far more rapidly than the reducing ability over this period. Thus in 'aged' beef any damage to the enzymatic reduction system, as may occur by exposure to high temperature and low pH, will diminish the colour stability.[23] It has been shown that the activity of a reducing system present in beef of normal pH decreases by a factor between 0·3 and 2 for each hour it is held at 42°C rather than 2°C.[52]

The rapid rate of decline of the catalytic, oxidative process and the slower loss of enzymatic reducing ability will explain why beef muscles have a maximum colour stability when displayed for sale about 5–7 days after slaughter.[23]

A highly significant correlation ($r = 0.72$) has been reported between the oxygen consumption rate and metmyoglobin reducing activity[23] of beef muscles.

Although there is still some debate about the relative roles of the reducing and oxidising systems in determining the rate and extent of metmyoglobin formation in meat, studies have shown that the rate of formation can be inhibited by feeding α-tocopherol (Vitamin E)[54] to the animals or by treatment of the meat with either Vitamin C or Vitamin E.[55] However, if Vitamin E is added to meat post-slaughter the effects are minimal, suggesting that feeding enables this antioxidant to become incorporated into the tissue in such a way as to be able to act as an antioxidant towards both myoglobin and lipid.[54] Further studies have confirmed the low effectiveness of Vitamin E (6 ppm) compared to Vitamin C (500 ppm) when added to meat, although a combination of the two antioxidants was most effective.[55]

As well as the factors discussed above radiant energy, relative humidity and bacterial load can affect the rate of metmyoglobin formation. If meat is dried to an a_w of 0.98 then the rate of metmyoglobin formation is increased;[39] however, darkening due to concentration of the haem pigments easily overrides the visual impact of this effect.

The effect of bacterial load on the formation of metmyoglobin is thought to be due to depletion of oxygen at the surface;[56] in fact very high ($> 10^8$ cm^{-2}) levels of *pseudomonas* sp. effectively convert the metmyoglobin back to the purple reduced form since the meat surface becomes essentially anaerobic.[56] Although thermal oxidation to metMb is of major concern photo-oxidation may be important at low temperatures. For example, during storage at $-18°C$ photo-oxidation contributes to the formation of metMb since packaging meat in opaque, oxygen permeable films inhibits its formation when stored under fluorescent light.[57] However, in chilled meat the effect of visible light is negligible[7,58] although ultra-violet light is a powerful pro-oxidant.[7,59] Thus at pH 5.4 and 0°C the relative rates of photo-oxidation at 546, 366 and 254 nm are 1, 10 and 4700, respectively.[59] The relative rates of metMb formation from oxymyoglobin, due to thermal and photo-oxidation at different wavelengths, are shown in Fig. 5.

Although the formation of the brown metmyoglobin is the major colour problem facing the processor of fresh (or frozen) meat it is possible, under certain conditions, for reactions leading to modification of the porphyrin structure to occur, giving rise to other marked colour changes. Of these changes reactions leading to saturation of one of the methene bridges and consequent disruption of the conjugation (resonance) around the porphyrin ring are of most importance as these usually lead to the

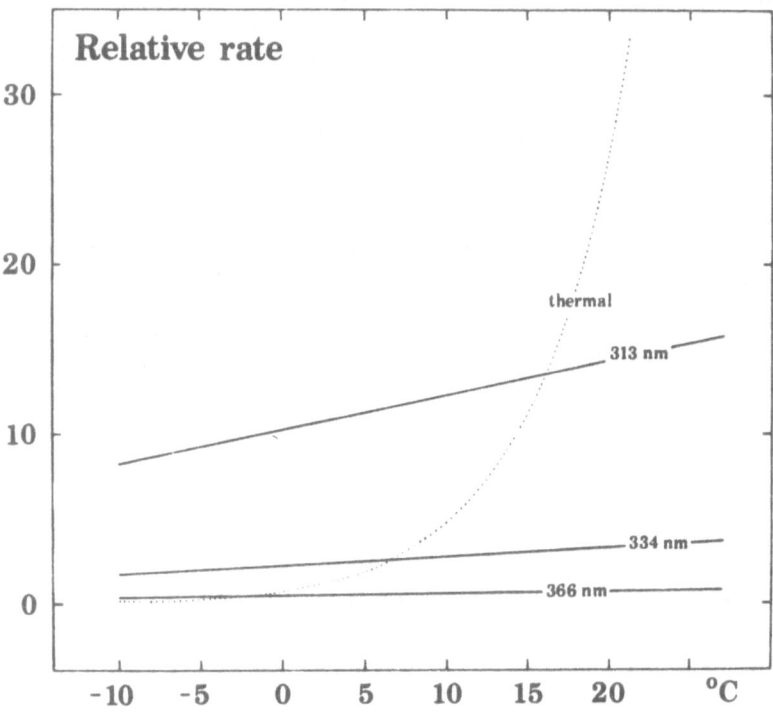

FIG. 5. Rate of competitive thermal and light-induced autoxidation of MbO_2 (relative to thermal oxidation at 1°C, pH 5·5), for model aqueous oxymyoglobin solutions illuminated by monochromatic light of indicated wavelengths (From Ref. 57, reproduced by permission).

development of green colorations. For example an H_2O_2 induced oxidation may bring about a hydroxyl substitution of the ring system which may undergo further oxidation to form *choleglobin*, which is believed to be a complex of biliverdin (an oxidation product of the porphyrin,

$C_{33}H_{34}O_6N_4$), iron and globin. However, in fresh meat catalase and perhaps peroxidase minimise the H_2O_2 content and thus the formation of choleglobin is not usually a problem in these systems, although, as will be discussed later, it may give rise to unacceptable discolorations in cured meats in which the salt present effectively destroys any catalase activity. Of more concern, especially with the increasing use of vacuum packaged meat, is the ability of H_2S to oxidise the porphyrin ring to green sulphmyoglobin.[60] In this compound it is known that each molecule of haematin contains one sulphur atom which is bound to the ring system; however, the precise nature of the complex is not clear[61,62] though it is likely to possess a hydroxythiol or episulphide structure. Present evidence suggests that in sulphmyoglobin the globin is still in its native form and the iron is in the ferrous state as denaturation with alkali yields the denatured protomyochrome and oxidation with ferricyanide or molecular oxygen yields the dull red metsulphmyoglobin pigment, which may be reduced back to the green sulphmyoglobin with dithionite.[60]

Sulphmyoglobin (and haemoglobin) formation has been a problem in undrawn (New York dressed) poultry for many years due to the production by the gut microflora of H_2S which subsequently reacts with the haematin pigments present in the tissue. However, in the late 1960s it was observed that stored, vacuum packed beef of high pH tended to 'green' with time and this was shown to occur under conditions of high pH (> 6) and low oxygen tensions (1–2%), conditions under which the bacteria were able to produce H_2S. Originally it was thought that the spoilage organism responsible belonged to the *Pseudomonas* spp. but recent studies have shown that the organism primarily responsible is *Altermonas putrefaciens*.[63,64] Australian workers have also identified a number of lacto-bacilli-type organisms that may produce H_2S on meat of pH below 6 and which are probably responsible for the greening occasionally observed in meat of normal pH.[63,64]

The mechanism for the formation of sulphymyoglobin in meat, has never been elucidated but it may involve a 'ferryl' ion as an intermediate. In a 'ferryl' complex the iron atom has an apparent valency of 4 and has been described as having only four 3d valence electrons and an apparent spin of 1.[1] This quadrivalent iron is complexed to an oxygen atom (i.e. it may be called ferrimyoglobin peroxide). This type of complex has been implicated in the haem catalysed oxidation of lipids in meat (p. 226) and as a product of the irradiation of myoglobin (p. 221). In the laboratory sulphmyoglobin is usually prepared via the ferryl complex[61] and a possible mechanism for its formation in meat is as follows.

$$Mb^{2+}O_2/Mb^{3+}O_2^- \rightleftharpoons Mb^{3+}H_2O + O_2^-$$

oxymyoglobin   metmyoglobin +
                                      superoxide anion

$$\downarrow {\scriptstyle H_2O_2}$$

$$Mb^2SH \xleftarrow{\quad H_2S \quad} Mb^{4+}O_2$$

                        from bacteria

Sulphmyoglobin   ferrimyoglobin peroxide

The H_2O_2 required for production of the ferryl complex has been detected in ground meat and may in fact be a product of the oxidation of oxymyoglobin to the met form or a product of bacterial or enzymic action, peroxidising lipids or flavins (see Flavour Considerations, p. 226).

If further oxidation of the porphyrin ring occurs then a green verdohaem complex may form and ultimately the haematin pigments may degrade to yellow or colourless pyrrole fragments (bile pigments) and globin.[7] This formation of bile pigments though is rarely a problem in fresh meat though it may be significant in dried and semi-dried meat products (p. 224).

In view of the relative instability of the red oxymyoglobin pigment several attempts have been made to stabilise the red colour associated with fresh meat by the formation of other, more stable, low-spin haematin complexes. Carbon monoxide binds tightly to myoglobin forming the bright red carboxymyoglobin which is more stable than the corresponding oxy derivative.[65] Thus the colour of refrigerated ground beef patties stored in atmospheres containing 1% CO at 2°C was still perfectly acceptable after 6 days of storage whilst similar samples stored in air were discoloured within 3 days.[65]

As well as carbon monoxide other strong field ligands, that can be accommodated at the sixth coordination site of the haematin group to yield a stable, red, low-spin complex, have been suggested as means of preserving the colour of fresh meat. Hines[66] patented the use of exposure of the fresh meat to ammonia gas prior to packaging to form a stable, red ammonia complex whilst Tarladgis[67] claimed that dipping meat in solutions containing appropriate concentrations of heterocyclic nitrogen containing compounds such as imidazole, led to the formation of stable red complexes. However, none of these ideas are being used commercially.

Cooked Meat

As the haemoproteins are so important in determining fresh meat colour it is not surprising that the response of these pigments to heat is of major importance in determining cooked meat colour. Myoglobin is one of the most heat stable of the proteins present in meat although its actual stability does depend on the source (see. p. 201), the nature of the ligand at the sixth coordination site and the oxidation state of the iron.[68]

There has recently been some controversy over the actual nature of the ligands attached to the 5th and 6th coordination sites in the cooked meat haemoproteins, although it is well established that the haematin iron is usually in the ferric state.[18,69,70] Studies on solutions of haemoglobin suggest that the reversible and irreversible denaturation of the haemoglobin gives rise to complexes in which both ligands are supplied by denatured globin.[71] Depending on conditions, these low-spin complexes have nitrogen (from the imidazole of histidine) at both sites or a nitrogenous base at the fifth position and a sulphur atom from methionine or cysteine at the sixth position.

However, this information cannot be transferred directly to the behaviour of the haemoproteins in meat since several other complexing agents, other than globin, are present.

Tappel[69] suggested that the haemoproteins in cooked meat were mixed denatured globin nicotinamide hemichromes, whilst Tarladgis[70] argued that the reflectance spectra of the meat were typical of high-spin ferric complexes and suggested that the fifth and sixth positions were occupied by a carboxylate group from denatured globin and water respectively. However, Ledward[18] found that the ESR spectra of cooked meat indicated that the complexes present possessed no high-spin character at $-196°C$, the temperature of measurement. However, at $20°C$ the reflectance spectra do suggest that high-spin complexes may be present. He suggested therefore that, if a nitrogen atom of histidine or possibly a sulphur atom of a methionyl or cysteine residue was one of the ligands and water the other, this could give rise to mixed high–low spin equilibrium, being about 100% low-spin at $-196°C$, the proportion of high-spin complex increasing with temperature.[18] This type of equilibrium is well established for other ferric haematin complexes. However, if water is one of the ligands it would be expected that on increasing the pH this water molecule would ionise to a hydroxyl group with the formation of a low-spin complex. This does not occur:[72] rather, the reflectance spectra develop more high-spin character at pH values of 9 and above, suggesting that at these high pH values lysine becomes one of the ligands.[72,73] These subsequent studies[72] suggested that, at neutral pH values, the major products are ferric di-imidazole complexes formed by the fifth and sixth positions being bound to histidine residues of the bound protein, further stabilisation being afforded by salt-linkages and hydrogen and hydrophobic bonds between the haematin and denatured protein. Giddings[19] has criticised this suggestion by arguing that, on steric grounds, it is unlikely that the haematin can accommodate two protein bound imidazole groups. However, as several haemoproteins do possess two protein-bound imidazole residues, as also do some forms of denatured haemoglobin, this criticism does not in itself invalidate the di-imidazole model.[74] Thus, at the present time, it is not possible to define unequivocally the nature of the cooked meat haemoprotein and as the reflectance spectrum of cooked meat is not unique[18] it may well be that more than one type of complex is present. However, recent work with freeze-dried beef lends further support to the hypothesis that the major haematin pigment in cooked meat contain two protein-bound imidazole groups (see p. 224).

Another unusual feature of myoglobin 'denaturation' in meat is that it tends to occur, as evidenced by precipitation, at temperatures of 60°C and above, whilst in pure solution at pH 5·5–6·0 marked denaturation and precipitation only occurs at temperatures of 75°C and above.[75] Thus Draudt[68] concluded that myoglobin in meat coprecipitates with the other proteins present and part of the thermal denaturation behaviour of myoglobin has to be attributed to the denaturation of the other proteins present in meat. It is thought that myoglobin undergoes a small conformational change at some temperature below its denaturational temperature[76,77] and it has been argued that this conformational change permits some of the other, less stable proteins present in meat to attack the partially exposed haematin following, or during, their denaturation,[18,74] this attack displacing the globin which spontaneously denatures.

If this is correct, a whole range of haematin-denatured protein complexes are formed in which the bound protein is unlikely to be globin. Studies on muscle extracts have suggested that the concentration and type of bound protein is dependent on the thermal history of the meat. For example, muscle extracts held at 52°C, a temperature at which several of the sacroplasmic proteins in meat will denature and precipitate but at which myoglobin will be largely unaffected, was found to form 41% cooked meat haemoprotein on subsequent heating to 60°C whilst samples heated directly to 60°C formed 61% cooked meat haemoprotein.[18] The effectiveness of several of the muscle proteins in forming 'cooked meat haemoprotein' was dependent on the conditions employed. Thus at 65°C most of the formation was due to reaction with the soluble proteins of molecular weight about 100 000 whilst at 60°C those of molecular weight about 80 000 were most effective. The temperature dependence presumably relates to the different thermal stabilities of the proteins present in meat, as differential scanning calorimetry has shown that although, in general, there is minimal interaction between proteins prior to, and during, thermal denaturation, in the presence of an excess of protein of relatively low thermal stability (bovine serum albumin) myoglobin can denature and coprecipitate with that protein.[78] This observation agrees with the above hypothesis that, during heating, specific interactions take place between the denatured protein and the haematin of conformationally altered, but undenatured, myoglobin.

The rate at which the reactions leading to these colour changes take place will depend on the rate of heating. Palombo and Wygaards[79,80] have shown that the colour changes are essentially complete within 5 min at 100°C and 10 min at 80°C but at lower temperatures (50 and 60°C) the colour is still changing after 3 h. These authors have also shown that increase in lightness of the meat on heating is, at least after the first few minutes, mirrored by decreases in the hue and chroma of the sample. However small decreases in hue and chroma appear to take place after the lightness has achieved a steady value.[80] Since general protein precipitation is responsible for the lightening in colour of the meat, whilst the hue and chroma are primarily due to the nature of the haemoproteins, this work further verifies the relative stability of myoglobin. These results were reported in meat of normal pH and it is established that this coprecipitation is pH and salt dependent.[81] Salt accelerates the denaturation of myoglobin, whilst increasing pH in the range 5–7 stabilises the pigment. Thus meat of high ultimate pH is more colour stable with respect to both heat denaturation and oxidation to metmyoglobin.

It would also appear that on heating meat there is some cleavage of the porphyrin ring and subsequent release of the iron, so that the total haemoprotein content decreases.[11] This should also cause some change in the hue and chroma of the meat. The amount of iron released depends on the severity of heating, there being an apparently linear increase with time, and the presence of other constituents. Ascorbic acid accelerates the release whilst nitrate inhibits it.[82] This release of iron, as well as contributing to the colour changes observed on prolonged heating[79,80] may have some effect on the bioavailability of the iron (p. 228) and the stability of the lipids on storage (p. 226). Though the breakdown is only slight in full moisture meat it may be very significant in dried meats (p. 224).

Whatever the reactions responsible for, and the nature of, cooked meat haemoproteins, their formation is used to assess the degree of 'doneness' of red meat. For example, beef cooked to an internal temperature of 60°C is considered rare, that cooked to 71°C medium-done and that to at least 77°C as well done.[1] Some white meats such as turkey and pork have a persistent pink coloration even after heat treatment at 95°C.[83] This may be due to the formation of carbon monoxide or nitric oxide type pigments generated in gas ovens.[1] However, more recent work suggests it may be associated with the stable reduced cytochrome present at low concentrations in meat.[83]

It appears that the unreacted myoglobin in rare beef, even after prolonged heating at low temperatures is in the red, oxymyoglobin form,[69] but the reasons for this are not clear as, at the elevated temperatures and low oxygen pressures assumed to be present in the centre of the meat one might expect the pigment to be present as the brown oxidised metmyoglobin (p. 202) or possibly the purple reduced myoglobin. It is known though that when aqueous muscle extracts are held at temperatures up to 60°C the undenatured haemoproteins remaining in solution, although initially oxidising to the met form, are subsequently reduced back to a red pigment

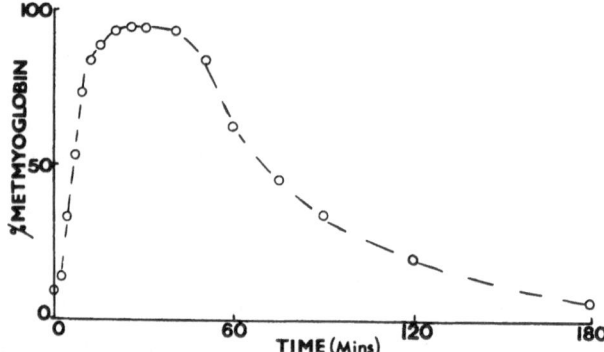

FIG. 6. The formation and subsequent reduction of metmyoglobin as a percentage of the total haemoproteins present in an aqueous muscle extract during heating at 60°C (0). The pH of the solution was 5·86 at time zero, 5·94 after 50 min and 5·96 after 180 min. Due to precipitation the haemoprotein content of the extract decreased with time, from 0·137 mM at time zero to 0·088 mM after 50 min, however, after 180 min the concentration was 0·081 mM indicating that the decreased concentration of metmyoglobin observed between 50 and 180 min was due to preferential precipitation of metmyoglobin.[1]

with the same spectral characteristics as oxymyoglobin. A typical time course for the reaction at 60°C is shown in Fig. 6, and the reduction appears to be first order with respect to metmyoglobin concentration; the half life being about 30 min at pH 5·95 and 60°C. Thus at these temperatures a reductive mechanism may also operate in meat to convert the undenatured pigment to the ferrous state and thus yield the red colour consumers associated with 'rare' beef. γ-Irradiation of raw poultry can give rise to a pink/red colour throughout the tissue which has spectral characteristics similar to oxymyoglobin (p. 221) and there may well be similarities involved in the formation of such a pigment in the two systems. It is possible that at these temperatures the oxygen utilising enzymes are inactivated, and thus oxygen solubility increases so that oxygen can permeate to the centre of the meat.

Cured Meats

In *uncooked* cured meat products the predominant pigment is the low-spin nitric oxide myoglobin (nitrosomyglobin), which is spectrally similar to oxy- and carboxymyoglobin but in which nitric oxide, rather than oxygen or carbon monoxide, occupies the sixth ligand position. Possible electronic structures have been discussed previously.[1]

It is well established that nitric oxide can combine with both ferric and ferrous haematin iron but the ferric complexes of both myoglobin and haemoglobin auto-reduce to yield a partial ferrous nitrosyl (Fe^{2+}–N^+O) configuration by partial transfer of the odd nitrogen electron to iron.[1]

Although nitric oxide is the essential reactant in the formation of the desirable nitric oxide myoglobin it is, in practice, produced by the action of the nitrite present in the curing solution with the myoglobin. The initial reaction of nitrite with haemoglobin is to produce both nitric oxide haemoglobin and methaemoglobin, although with myoglobin, metmyoglobin appears to be the only significant product.[7] However, in the presence of reducing systems present in meat, nitric oxide myoglobin is formed. Several schemes have been proposed for the reduction of metmyoglobin in the presence of nitrite[1,38] and that proposed by Walters *et al.* in 1967[84] is still widely recognised. In this scheme nitrite oxidises myoglobin to the met form and also converts ferrocytochrome *c* to nitrosoferricytochrome *c*. The nitroso group is then transferred from nitrosoferricytochrome *c* to metmyoglobin by NADH cytochrome *c* reductase action to yield nitric oxide metmyoglobin. This in turn is reduced to nitric oxide myoglobin by enzyme systems of the mitochondria. However, recent work has suggested that nitric oxide metmyoglobin is only a transient intermediate[85] as it rapidly auto-reduces to a nitric oxide radical cation (Fig. 7).

In this proposed mechanism nitrite is reduced to nitric oxide which reacts with the metmyoglobin to form either nitric oxide metmyoglobin (which rapidly auto-reduces to the ferrous nitric oxide myoglobin radical cation) or undergoes a conjugated reaction involving simultaneous NO coordination and auto-reduction to the radical cation. This radical cation reduces to the nitric oxide myoglobin; this latter reduction is relatively slow in model systems containing only the protein and nitric oxide[85] but may be relatively rapid in meat.

Once formed, nitric oxide myoglobin is stable in the absence of oxygen although in its presence it does lose its colour (fades) as it oxidises to met-myoglobin. This occurs because, although the nitric oxide pigment is far more stable than the oxy derivative, the oxygen is invariably present in much higher quantities than the nitric oxide and on dissociation will oxidise the released nitric oxide to higher oxides that are not capable of binding to the haematin. Since oxygen cannot react directly with bound nitric oxide, the stability of the pigment is largely dependent on the rate of dissociation of the nitric oxide from the haematin. In the absence of light, this dissociation is very slow so that the colour of uncooked cured meats, even in the presence of oxygen, fades only slowly with time when kept in the dark. However, the pigment is photolabile and thus in the presence of both light and oxygen the colour fades relatively rapidly,[1,19] although, as indicated above, in the absence of either, it is relatively stable. Wiltshire-cured bacon that has been oxidised will recover its colour during subsequent anaerobic storage indicating that the reducing system is still active in these products for a considerable period after processing.[86]

On heating cured meats the nitric oxide myoglobin converts to a bright red pigment in which the iron is in the low-spin ferrous state.[87] Nitric oxide undoubtedly occupies one of the positions axial to the porphyrin ring but there is some controversy regarding the nature of the ligand, if any, at the other axial position. Early workers concluded that denatured globin supplied the other axial ligand but the suggestion that protein supplies one of the ligands is now thought to be unlikely. As early as 1962 Tarladgis proposed that in the cooked cured meat pigment both axial ligands were nitric oxide.[87] This structure has received support from the work of Lee & Cassens[88] who found that 2 mol of nitric oxide was bound to each mole of myoglobin in the formation of this pigment and Wayland & Olsen[89] who provided evidence of a dinitric oxide complex formed from a model ferrous porphyrin compound, tetraphenylporphyriniron.

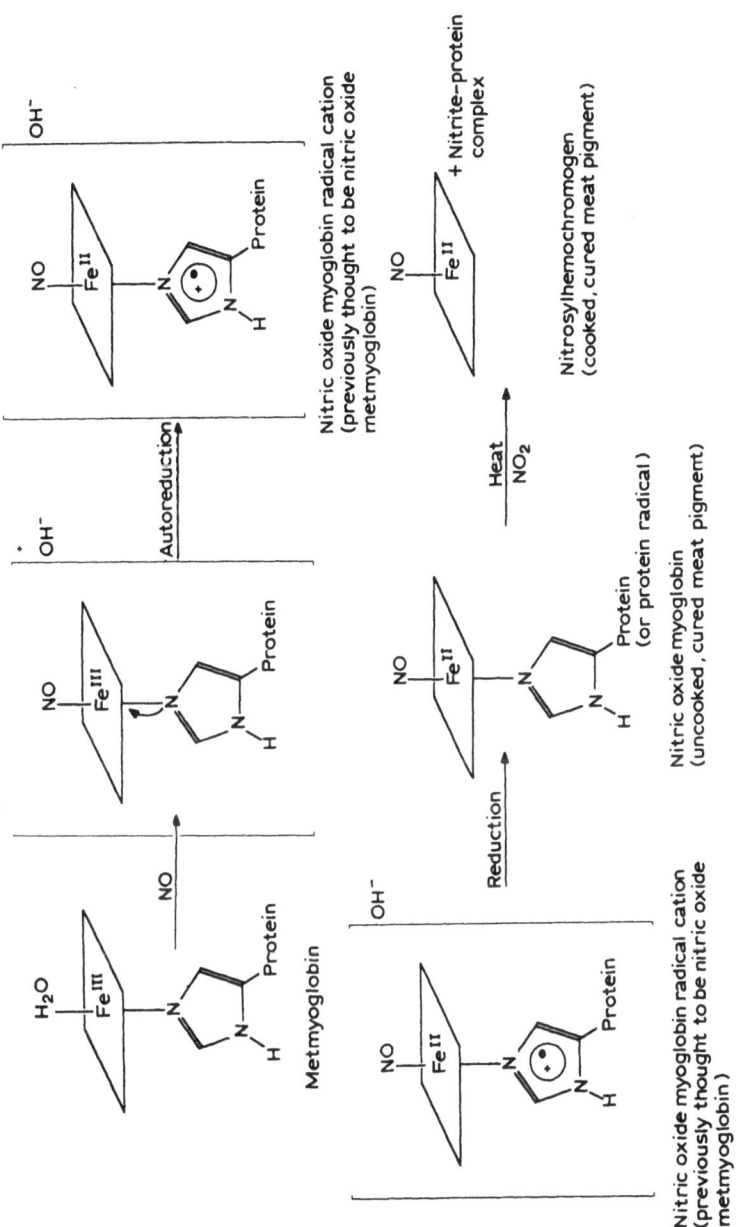

FIG. 7. Proposed mechanism for the formation of the cured, cooked meat pigment (Adapted from Ref. 85).

Sebranek & Fox[90] also support this model. However, the cured meat pigment from cooked, cured beef when characterised by infra red and visible spectroscopy was found to be identical with a synthetic compound that was identified by fast atom bombardment mass spectrophotometry as the mono nitric oxide species nitrosyl (Fe^{2+}) protoporphyrin, a five coordinated haematin compound.[85] These authors postulate that a further molecule of nitrite is bound to the denatured protein which would thus explain the taking up of two molecules of nitrite in forming this pigment.[88] The postulated mechanism is shown in Fig. 7.

Although further work is necessary to elucidate fully the nature of the haematin pigments present in cooked, cured meat, it is well established that they are stabler than the uncooked pigments. However, the mechanism of fading (oxidation) appears to be similar to that of the uncooked pigment. Thus pasteurised ham stored in the dark maintains its colour even in the presence of oxygen, but when illuminated rapidly fades.[91]

When vacuum packed ham is displayed under normal illumination it fades until the residual oxygen in the pack is consumed by biological and/ or microbiological activity at which time the red colour starts to regenerate, although it never recovers its original quality.[91] The fading can be prevented by holding the vacuum packed ham in the dark for a few days, until the residual oxygen is consumed prior to retail display.[91] Interestingly the photo-oxidation of the nitro pigments does not appear to be wavelength dependent, unlike the photo-oxidation of oxymyoglobin which is very dependent on the wavelength (p. 210).

Although greening in fresh meats is relatively rare it occurs more frequently in cured meats and is invariably due to breaking of the conjugation around the porphyrin ring. Thus, as mentioned on p. 210, hydrogen peroxide producing bacteria will oxidise the porphyrin ring to the green choleglobin as the salt present in cured meats prevents catalase activity from rapidly destroying the hydrogen peroxide produced.[7] In cured meats green discolorations can also occur due to the use of excess nitrite in the curing solution ('nitrite' burn). It has been claimed that this discoloration, which occurs about twenty times faster with bovine haemoglobin than bovine myoglobin, involves nitrosylation of the porphyrin ring to form a nitrihaematin complex,[92] i.e. the conjugation around the ring system is interrupted by the binding of a nitro group. Chemical analysis of the complexes suggest that only one such group is bound to the haematin.[92] In these complexes the iron is apparently in the ferric state and the globin may still be attached to the haematin. On reduction with dithionite a green ferrous pigment is formed. As the spectral characteristics of this ferrous complex are the same whether or not excess nitric oxide, which should form a nitrosyl pigment with the reduced haematin, present Fox & Thompson[92] tentatively assume that the chemical modification taking place in the formation of the nitrihaematin complexes involves both the porphyrin ring and iron. This type of modification renders the iron 'unavailable' for reaction with other ligands.

Over recent years concern has been expressed regarding the use of nitrite in meat products[1] and research effort has been devoted to developing alternative, stable cured meat pigments. Numerous suggestions have been made regarding possible substitutes and an extensive study of over 300 compounds indicated that the best pink colour was formed by reaction with 3-acylpyridines.[93] Isoquinoline, pyrazine and imidazole also formed stable pigments. However, as nitrite in meat not only modifies the colour and flavour of the products but also prevents the growth of *Clostridium botulinum* it seems unlikely that safe and effective substitutes for this compound will be found.

Irradiated Meats

In the 1940s research was directed towards the use of ionising radiation to sterilise and pasteurise food. This work included the effect of irradiation on the haemoproteins in meat. Giddings & Markakis[94] have summarised this earlier work. However, there was little opportunity for commercial exploitation of radiation preservation until recently when irradiation as a means of preserving specified foods was permitted in several countries. Thus research into the effect of ionising radiation on the haemoproteins present in meat has been stimulated.

At the low doses envisaged to 'pasteurise' meat (γ-irradiation of less than 10 K Gy) no denaturation of myoglobin occurs[94] and thus most work has been devoted to the changes taking place at the haematin. Several workers have investigated the nature of the haematin pigments formed during low-dose γ-irradiation of haemoproteins. However, there is uncertainty as to the nature of some of the products formed during irradiation of solutions of myoglobin in the absence of oxygen.[1] It has been suggested that the product is oxymyoglobin or a mixture of reduced myoglobin, ferrimyoglobin peroxide and unreacted metmyoglobin.[1]

On irradiation in the presence of oxygen, which scavenges hydrated electrons and thus inhibits reductive processes, aqueous solutions of metmyoglobin change in colour from brown to red as the ferrimyoglobin peroxide is formed by the action of H_2O_2, generated during radiolysis, on the ferric haematin.[94] However, some workers have claimed that no such colour change is observed in oxidised (brown) stored beef due to the presence of scavengers, such as catalase, which remove H_2O_2 as rapidly as it is produced.[94] Also, when fresh well oxygenated meat is irradiated it has been claimed that no significant change in colour occurs, suggesting that the oxymyoglobin pigment remains unchanged. Any oxymyoglobin becoming oxidised is rapidly reverted to the ferrous form by the reducing systems naturally present in fresh meat.[19] However, other work has complicated the issue. Thus when irradiated at doses of 2–3 K Gy, or higher, poultry takes on a pink/red coloration, whether oxygen is present or not.[92,96] Pork loins also beome redder when irradiated at 3 K Gy[97] and a red coloration has also been reported in irradiated beef muscle.[98–100]

It is now generally accepted that γ-radiation, in the range 2–10 K Gy, encourages the formation of a red pigment in raw meat.[7,95] However, it forms far more easily in poultry, and to some extent in pork, than in beef. If at a given dose and dose rate the formation is dependent on the myoglobin concentration then the relative lack of reactivity of beef muscle would be explained. It is possible though that other differences between the species account for the relative ease of formation of this red pigment in different meats.

Very low doses of irradiation (< 2 K Gy) favour the formation of metmyoglobin[95] and this appears to be the initial reactant in subsequent formation of the red or pink pigment. Spectrally the pigment is similar to oxymyoglobin and in fact this could well be the pigment concerned since it could be formed, even in the absence of oxygen, by reaction involving ferrylmyoglobin as an intermediate,[99] i.e.

$$\text{met Mb} \xrightarrow[\text{or } H_2O_2]{\cdot OH} \text{ferrylmyoglobin}$$

$$\downarrow H_2O_2 \text{ or } HO\cdot$$

$$\text{oxymyoglobin} + H_2O \text{ or } HO\cdot$$

Alternatively, as Giddings & Markakis[94] suggest, oxymyoglobin may form from reduction of metmyoglobin by hydrated electron, with subsequent oxygenation with residual oxygen or oxygen produced during irradiation by such reactions as:

$$\cdot OH + HO_2 \cdot \longrightarrow O_2 + H_2O$$

$$HO_2 \cdot + Fe^{3+} \longrightarrow H^+ + O_2 + Fe^{2+}$$

$$2HO_2 \cdot \longrightarrow H_2O_2 + O_2$$

Another possibility, discussed by these authors is the capture of hydrated electrons by oxygen (produced as above) to form the superoxide anion, O_2^- which could complex with metmyoglobin to form the oxypigment. At very high irradiation doses a green pigment forms[98] which, though not identified in the original publication, is spectrally identical to sulphmyoglobin. It is known that sulphmyoglobin forms by reaction of H_2S (produced from protein breakdown at high doses of radiation) with ferrylmyoglobin (p. 211) suggesting that ferrylmyoglobin is one product of irradiation and lending tentative support to the first mechanism. However, as discussed later this pigment may itself be the end product.

Whatever the mechanism, the formation of the oxypigment throughout the tissue, not just at the surface, could account for the pinkness observed, especially in white meats.

However, arguments against the pigment being oxymyoglobin include the observations that during storage and cooking the pink colour is relatively stable.[7] It persists in chicken cooked to an internal temperature of 82–85°C,[95] a temperature at which unirradiated chickens take on the normal cooked appearance. In addition, at the low oxygen tensions present in the interior of the stored meats one might expect rapid oxidation to the met form; this is not observed as the pink colour decreases only slowly with time at chilled temperatures.[95] Also, the pigment appears to differ in absorption characteristics to oxymyoglobin in the Soret region of the spectra[100] and $\cdot OH$ scavengers have no noticeable effect on the perceived colour.[100]

Thus the pink compound may well be a stable pigment other than oxymyoglobin and a likely candiate is a ferryl myoglobin species. The ferryl myoglobin could form by reaction of metmyoglobin with H_2O_2 or $HO \cdot$ as discussed previously.

When vaccum packed *cooked* beef is sterilised by ionising radiation a pink coloration is produced which, on exposure to the atmosphere, reverts to the normal brown colour of 'cooked meat haemoprotein'.[101] The nature of this pink pigment has not yet been established, but it may be the reduced counterparts of the denatured 'cooked meat haemoproteins', formed by reduction of the ferric haematin to the ferrous state by reaction with a hydrated electron or a ferryl derivative. If oxygen is present during the irradiation step then the formation of the reduced 'pink' complex appears to be inhibited in beef.[99,101] However, when cooked chicken is irradiated in the presence of oxygen a pink colour develops throughout the muscle[95] which persists for some time. This, again, may be the reduced counterpart of cooked meat haemoprotein or a denatured ferryl pigment.

Although the magnitude of the colour changes undergone by meat following irradiation are dependent on several parameters such as dose rate (high rates favouring the formation of the pink pigment), pH, temperature and salt concentration[95,101] they appear to be most obvious in the poorly pigmented meats such as chicken where pinkness/redness is not a desirable attribute.

Dried and Intermediate Moisture Meats

Dehydrated and intermediate moisture meat products are being developed both as foodstuffs in their own right[102] and as functional ingredients for use in meat-based products.[103] Intermediate moisture meat products are prepared by equilibrating meat in solutions of low water activity, such that moisture diffuses out of the sample and the humectants (e.g. salt and glycerol) diffuse in to yield a resultant product of decreased water activity and consequent increased microbial stability.[102] An antimycotic, such as potassium sorbate, is usually incorporated in the humectant solution to inhibit mould and yeast growth.

As summarised previously there has been some debate in the literature regarding the nature of the pigments present in dehydrated (freeze-dried) meats.[1] However, analysis of the reflectance spectra of beef samples treated to contain all the haemoproteins in one pigment form have clarified the situation.[104] These results suggest that little or no change occurs in the haematin environment on freeze-drying fresh, oxymyoglobin-rich meat or cooked meat.[104] Although the oxygen molecule may be displaced from the haemoprotein during the vacuum drying the reduced myoglobin so formed rapidly regenerates oxypigment on exposure to air. On the other hand, dehydration of metmyoglobin in meat results in the water molecule at the sixth coordination site being replaced by a protein bound imidazole group.[104] The spectra of this compound is very similar to that of cooked meat, which suggests the pigments in both meats, are di-imidazole complexes (p. 213). On rehydration the imidazole group at the sixth coordination site may be replaced by water to yield metmyoglobin but the degree of conversion depends on the rate of rehydration. Slow absorption of water into the meat apparently enables the protein chain to stabilise its coordination to the ferric iron and prevent displacement by water. Thus, a mixture of metmyoglobin and the freeze-dried pigment may coexist following rehydration.[104]

The process of freeze drying itself brings about little oxidation of oxymyoglobin.[104,105]

With glycerol-desorbed intermediate moisture meats there is no marked change in the nature of the haemoproteins following processing. Thus samples prepared from raw meat have their pigments predominantly in the oxy form and, if the desorption process involves heat, the resultant pigments are typical of those found in cooked meat.[1,102]

The haemoproteins in both freeze-dried and intermediate moisture meats degrade to bile pigments during aerobic storage, due to oxidation of the porphyrin ring.[1,102] Not unexpectedly the rate of degradation is very temperature dependent; thus during aerobic storage of raw, freeze-dried beef at 38°C about half the porphyrin was degraded within a week whilst at −12°C over 20 weeks storage was needed to cause 50% destruction of the porphyrin.[1] Fishwick found that, in freeze-dried turkey, 36% destruction occurred at 0°C and 83% at 37°C after 116 days storage.[106] It seems likely that the agent(s) responsible for the breakdown is one or other of the products of the peroxidised lipids present in these systems.[1] Although the haemoproteins in freeze-dried meat are susceptible to oxidation those present in blood[1] and liver[10] are relatively stable during prolonged storage in air; it is not clear why the meat environment should be conducive to such breakdown and liver and blood so immune. It may be related to the types of lipids and phospholipids and/or natural antioxidants present in different systems.

The haemoproteins present in glycerol/salt, or salt only, desorbed intermediate moisture meats are also very susceptible to degradation, the spectra typical of cooked meat being lost within 3 weeks at 38 or 30°C.[102,107] However, peroxidising lipids may not be the agents responsible for the degradation in these products and it has been suggested that peroxides and/or aldehydes present in most glycerol samples,[92] or sorbate oxidation products,[108] may be the essential reactants. Present evidence suggests sorbate oxidation products may be the major reactants.[108] There is no loss of haemoprotein character during anaerobic storage.[102,107]

The rate of degradation of the haemoproteins in these systems (both freeze-dried and intermediate moisture) is greater in the cooked than in the raw products.[1,102] This is not unexpected since lipid (and presumably sorbate and glycerol) oxidation occurs far more rapidly in cooked, than in raw meat.

There does not appear to be any marked loss of haemoprotein character in stored, cooked meat of normal water content although limited breakdown may occur during extended storage at 38°C.

Flavour Considerations—Relationship Between Lipid and Haemoprotein Oxidation

Obviously the postulated interaction of peroxidising lipids with the haemoproteins of dried and intermediate moisture meats (p. 225) is an example of a relationship between lipid and haemoprotein oxidations. However, this type of reaction, in which the haematin is oxidised to non-porphyrin compounds, does not occur to any significant extent in full moisture meats.[1] Rather, such meats involve oxidation of the haemoproteins to the met form (p. 204) and simultaneous oxidation of the lipids to rancid products. Thus, although haemoproteins *per se* do not affect the flavour of meat and meat products, as certain derivatives are thought to be potential pro-oxidants (and in certain circumstances antioxidants) of unsaturated lipids, they may catalyse the development of rancid odours and flavours in meats.[109-111]

However, the interaction between the haemoproteins and unsaturated fats is not clear-cut. For example in model systems, at lower lipid:haematin ratios than are usually found in meat, the haematin compounds may stabilise peroxides and/or free radicals and thus exert a marked antioxidant effect. Results obtained by Johns *et al.*[111] for washed muscle fibres containing added haemoglobin are shown in Fig. 8. This observation may be important in products artificially enriched with rich sources of haemoproteins, such as blood,[112] but at the relative levels found in fresh meat the haematin pigments present are at levels that should catalyse the oxidation.

Recent studies have shown that one of the major catalysts of lipid oxidation in raw meat is the ferryl ion (+4) formed by reaction of H_2O_2 with metmyoglobin.[111,113,114] Thus in washed muscle fibres pure metmyoglobin had very little catalytic effect but a mixture of oxy and methaemoglobin was a powerful catalyst.[111] This was claimed to be due to the production of H_2O_2 in these systems by oxidation of the oxypigment to produce superoxide anions which on dismutation and protonation generate H_2O_2.[111,113]

However, other workers believe the major catalysts of lipid oxidation in meat are low molecular weight ferrous compounds and that haem compounds are of limited importance.[115,116] This hypothesis has been criticised on the grounds that the model systems used by these authors would not permit uniform distribution of potential catalysts, the concentration of catalysts used were not at levels appropriate to meat and, perhaps most importantly, no H_2O_2 for haemoprotein activity was present in the haem containing system.[111] It is known that H_2O_2 is present in post-rigor meat.[117] Although the evidence that the ferryl derivative is the major catalyst of lipid oxidation in raw meat is very persuasive, catalyst(s) and lipids need to be able to interact and thus ferrous iron located in the membrane may well play a significant role.[118]

If the concentration of metmyoglobin is the rate limiting step in the oxidation of the lipids, meat of high metmyoglobin content should be relatively unstable. However, if the concentration of H_2O_2 needed to form the catalytic ferryl radical, is rate limiting then there may be no dependence on metmyoglobin content. The situation is further complicated since the concentration of H_2O_2 may be governed by the kinetics of the oxidation of oxymyoglobin to the met form.[113] Such considerations may, to some extent, explain the apparent confusion in the literature regarding the role of haematin compounds in lipid oxidation.

Although rapidly peroxidising lipids may lead to increased rates of metmyoglobin formation in some model systems there is little evidence that in raw meat such an effect is of any real practical significance.[110] A

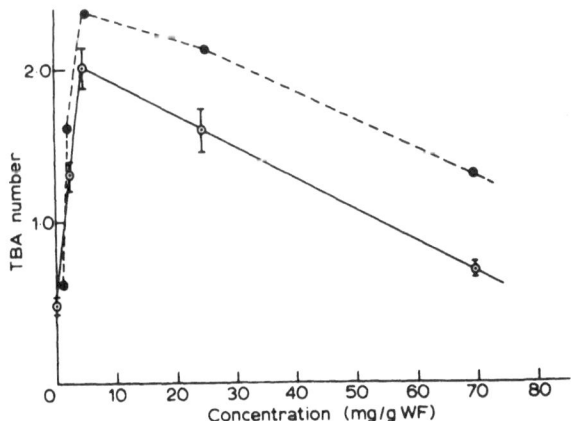

FIG. 8. Effect of haemoglobin (Hb) concentration on the TBA number (an index of the extent of lipid oxidation) of washed muscle fibres after 4 (0) and 7 (●) days storage at 2°C (From Ref. 111, reproduced by permission).

possible relationship between the two oxidations is that rapidly peroxidising lipids deplete the available oxygen around the haemoprotein to such a level that increased formation of metmyoglobin occurs.[1] The lipids in raw and cooked cured meats are relatively stable and it has been claimed this is due to the nitric oxide pigments being non-catalytic,[119,120] or this may be because the cured meat pigments cannot react with H_2O_2 to produce catalytic ferryl derivatives.

The lipids in cooked meats are less stable than those in the raw products and this could be because the heating process *per se* accelerates the oxidation, inorganic iron is released from the haematin environment on heating, and/or the haematin pigments are converted to a catalytic (+4) state. All three factors may in fact contribute to the observed catalysis. It is now generally accepted that inorganic iron is released from the haematin environment on heating[115,116] but whether in sufficient quantities to bring about the increased rate is open to question. In well cooked meat the ferric cooked meat haemprotein is brown/grey in colour but under certain circumstances, e.g. following irradiation or an increase in pH, it can be converted to a red pigment. The nature of this red pigment is not known but if it is a ferryl type derivative (p. 211) it could, if present at low concentration, contribute to the rapid increase in rate of lipid oxidation in cooked, uncured, meats. However, potential sources of H_2O_2, needed to bring about oxidation to the ferryl pigments are not so obvious in cooked as in raw meat, although the H_2O_2 utilising enzymes catalase and peroxidase will presumably be inactive after cooking, allowing any H_2O_2 generated to accumulate. As with the raw meats only traces of the activated ferryl pigments would be needed for effective catalysis.

Nutritional Significance of Haemoproteins

It is well established that meat is an important dietary source of iron, and numerous studies involving the measurement of haemoglobin absorption have been devised to give some insight into the digestion/absorption mechanism, e.g.[2,21,123] These studies, involving isolated haemoglobins, have led to the general belief that the superior nutritional availability of meat iron, compared with that from plant products, is due to a special mechanism involving the uptake into, and possible transfer of, intact haematin across the intestinal mucosa;[2,122,123] the haematin complex being a digestion product of myoglobin and haemoglobin, which is maintained in a soluble form by chelation to the appropriate amino acid/polypeptide residues formed by digestion of the accompanying protein.[121,122] Studies have shown that several amino acid residues can bind to isolated haematin at neutral pH and thus prevent the aggregation and subsequent precipitation of the haematin that occurs in their absence.[72,121,122] At neutral pH, histidine and to some extent cysteine appear to be the amino acid residues with the greatest affinity for the haematin.[72]

Although it has been confirmed that haematin complexes are the major products of the digestion of isolated haemoglobin[124] it has also been demonstrated that the major iron-containing compounds formed during the simulated in-vivo digestion of meat are low molecular weight (< 5000) non-haematin iron complexes.[124,125] In-vivo studies showed that these compounds were well absorbed[125] and it was suggested that it is not the haemoproteins per se which are responsible for the high availability for absorption of iron in meat, but rather the nature of their degradation products formed within the meat environment.

Consideration of the results obtained from in-vitro and in-vivo digestion studies led Hazell et al.[125] to propose the following mechanism for iron absorption from the haemoproteins in meat:

(i) On entering the stomach the undenatured haemoproteins in meat dissociate to yield acid haematin and denatured globin; heat denatured proteins may or may not dissociate under these conditions. At the low pH of the stomach the haematin is relatively stable although some digestion of the accompanying protein occurs.

(ii) On entering the duodenum the haematin chelates to some of the protein fragments and continued digestion of the protein takes place.

(iii) When the average molecular weight of the 'haemoproteins' reaches a critical size (~ 10 000) the haematin becomes accessible to the environment and degrades to liberate the iron. At this stage though the duodenum contains high concentrations of small polypeptides, formed by digestion of the proteins, and some of these chelate the iron and protect it from any inhibitors present in the partially digested food. These polypeptide/amino acid–iron complexes diffuse to the mucosa where they are readily transferred across.

Recent work has suggested that iron, whether derived from haematin or not, is chelated primarily by the carboxylic acid groups of the amino acids or small peptides produced by digestion of the proteins in the diet.[126] It must be noted, though, that although the above mechanism is plausible it is not universally accepted and some workers still favour the mechanism involving the uptake of intact haematin complexes by the mucosa.[127]

FUTURE RESEARCH NEEDS

Perhaps the most significant advance in our understanding of the reactivity of haemoproteins in meat and meat products since the original review in 1984 is the apparent key role the 'ferryl' pigments may play in various reactions. If hydrogen peroxide (or ·OH radicals) are available metmyoglobin certainly appears able to form such a pigment and its formation may be critical in the formation of, for example, the green sulphmyoglobin in vacuum packaged beef; it may also be the major catalyst involved in the development of rancid odours and flavours in meats. Even though their lifespan in such meats may be very short, it is apparent we need to know more about their nature and chemical reactivity. Such pigments may form during radiation processing and if this becomes an accepted method of meat preservation it gives further impetus to the need for research into the chemistry of these compounds.

Although our understanding of the factors affecting the formation of metmyoglobin has increased in recent years we still do not have an acceptable technology to preserve the bright red colour of some meats during extended display at chilled temperatures. Modified atmospheres containing elevated oxygen levels are reasonably successful but consumers are increasingly turning away from sophisticated packaging systems, which are seen to be environmentally unfriendly. Simple packaging or the development of edible films with characteristics of maintaining appropriate colour and microbial quality have possibilities, but we need to further understand the complex reactions that control metmyoglobin formation in meat before slaughtering, processing and packaging systems can be optimised to maximise colour stability.

Research and development to find new 'processed or cured meats' to meet the needs of developing countries or the increasing health conscious consumer of the developed world will continue. The success of such products will depend to some extent on the 'appeal' and stability of their colour and thus there is a need for a better understanding of the reactions undergone by the haemoproteins in these 'novel' meat products.

Finally, as discussed previously,[1] the widespread incidence of anaemia in all parts of the world makes it imperative that more work be carried out on the nutritional availability of iron from all foods. In this respect it is

highly desirable that the fate of the haemoproteins in meat during digestion be fully understood as these compounds, in a meat environment, are
known to be excellent sources of iron. The information so gained would
be invaluable in ensuring that new meat products, or meat replacers, are
developed along lines that ensure they are nutritious sources of iron, as
well as protein.

REFERENCES

1. Ledward, D. A. In *Developments in Food Proteins—3*, ed. B. J. F. Hudson.
   Elsevier Applied Sci. Publ., London, 1984, p. 33.
2. Jacobs, A., *Proc. Nutr. Soc.*, **35** (1976) 159.
3. Warriss, P. D., *J. Sci. Food Agric.*, **28** (1977) 457.
4. Rickansrud, D. A. & Henderickson, R. L., *J. Food Sci.*, **32** (1967) 57.
5. Warriss, P. D. & Rhodes, D. N., *J. Sci. Food Agric.*, **28** (1977) 931.
6. Hazell, T., *J. Sci. Food Agric.*, **33** (1982) 1049.
7. Lawrie, R. A., *Meat Science*, 4th edn. Pergamon Press, Oxford, 1985.
8. Antonini, E. & Brunori, M., *Haemoglobin and Myoglobin and their Interactions with Ligands*. Elsevier, New York, 1971.
9. Warris, P. D., Brown, S. N., Adams, J. M. & Lowe, D. B., *Meat Sci.*, **28**,
   (1990) 321.
10. Bogunjoko, F. E., *The Bioavailability of Iron from Chicken Meat and Liver*,
    PhD Dissertation, University of Nottingham, 1982.
11. Schricker, B. R., Miller, D. D. & Stouffer, J. R., *J. Food Sci.*, **47** (1982) 740.
12. Jacobs, A. & Greenman, D. A., *Br. Med. J.*, **1** (1969) 673.
13. Cook, J. D. & Monsen, E. R., *Am. J. Clin. Nutr.*, **29** (1976) 859.
14. Breslow, E., Kochler, R. & Girotti, A. W., *J. Biol. Chem.*, **242** (1967) 4149.
15. Atassi, M. Z., Tarlowski, D. P. & Paull, J. H., *Biochim. Biophys. Acta*, **221**
    (1970) 623.
16. Satterlee, L. D. & Zachariah, N. Y., *J. Food Sci.*, **37** (1972) 909.
17. Bembers, M. & Satterlee, L. D., *J. Food Sci.*, **40** (1975) 40.
18. Ledward, D. A., *J. Food Sci.*, **36** (1971) 883.
19. Gidding, G. G. *CRC Crit. Rev. Food Sci. Nutri.*, **9** (1977) 81.
20. Trautwein, A., Zimmermann, R. & Harris, F. E., *Theor. Chim. Acta*, **36**
    (1974) 67.
21. O'Keefe, M. & Hood, D. E., *Meat Sci.*, **7** (1982) 209.
22. Orcutt, M. W., Dutson, T. R. Cornforth, D. P. & Smith, C. G., *J. Anim.
    Sci.*, **58** (1984) 1366.
23. Ledward, D. A., *Meat Sci.*, **15** (1985) 149.
24. Hector, D. A., Brew-Graves, C., Hassan, N. & Ledward, D. A., *Meat Sci.*,
    31 (1991) 299.
25. Stabursvik, E., Fretheim, K. & Froystein, T., *J. Sci. Food Agric.*, **35** (1984)
    240.
26. George, P. & Hanania, G., *Biochem. J.*, **52** (1952) 516.
27. Livingstone, D. J. & Brown, W. D., *Food Technol.*, **35** (1981) 244.

28. George, P. & Stratmann, C. J., *Biochem. J.*, **51** (1952) 418.
29. Brown, W. D. & Mebine, L. B., *J. Biol. Chem.*, **244** (1969) 6696.
30. Gotoh, T. & Shikama, K., *J. Biochem. (Toyko)*, **80** (1976) 397.
31. Synder, H. E. & Skrdlant, H. B., *J. Food Sci.*, **31** (1966) 468.
32. Stewart, M. R., Hutchings, B. K., Zipser, M. W. & Watts, B. M., *J. Food Sci.*, **30** (1965) 487.
33. Ledward, D. A., *J. Food Sci.*, **37** (1972) 634.
34. Hagler, L., Coppes, R. I. Jr. & Herman, R. H., *J. Biol. Chem.*, **254** (1979) 6505.
35. Taylor, A. A. & MacDougall, D. B., *J. Food Technol.*, **8** (1973) 453.
36. Ledward, D. A., *J. Food Sci.*, **35** (1970) 33.
37. Ordonez, J. E. & Ledward, D. A., *Meat Sci.*, **1** (1977) 41.
38. Owen, J. E. & Lawrie, R. A., *J. Food. Technol.*, **10** (1975) 169.
39. Ledward, D. A., *J. Food Sci.*, **36** (1971) 138.
40. Hood, D. E., *Proc. 17th Europ. Meeting Meat Res. Workers, Bristol 1971*, pp. 677–684.
41. MacDougall, D. B. & Taylor, A. A., *J. Food Technol.*, **10** (1975) 339.
42. O'Keefe, M. & Hood, D. E., *Meat Sci.*, **5** (1980–81) 27.
43. O'Keefe, M. & Hood, D. E., *Meat Sci.*, **5** (1980–81) 267.
44. Renerre, M. & Labas, R., *Meat Sci.*, **19** (1987) 151.
45. Echevarne, C., Renerre, M. & Labas, R., *Meat Sci.*, **27** (1990) 141.
46. Sleper, P. S., Hunt, M. C., Kropf, D. H., Kastnel, C. L. & Dikeman, M. E., *J. Food Sci.*, **48** (1983) 479.
47. Orcutt, M. W., Dutson, T. R., Cornforth, D. P. & Smith, C. G., *J. Anim. Sci.*, **58** (1984) 1366.
48. Claus, J. R., Kropf, D. H., Hunt, M. C., Kastner, C. L. & Dikeman, M. E., *J. Food Sci.*, **49** (1984) 1021.
49. Ledward, D. A., Dickinson, R. F., Powell, V. H. & Shorthose, W. R., *Meat Sci.*, **16** (1986) 245.
50. Bendall, J. R. & Taylor, A. A., *J. Sci. Food Agric.*, **23** (1972) 707.
51. Faustman, C. & Cassens, R. G., *J. Food Sci.*, **55** (1990) 1278.
52. Billinge, A. & Ledward, D. A., Refrigeration Science and Technology International. Inst. Refrigeration Commission C-2, Bristol, 1973, p. 55.
53. Faustman, C., Cassens, R. G., Schauffer, D. M., Buege, D. R., Williams, S. N. & Scheller, K., *J. Food Sci.*, **54** (1989) 858.
54. Okayade, T., *Meat Sci.*, **19** (1987) 179.
55. Mitsumat, M., Faustman, C., Cassens, R. G., Arnold, R. N., Schaeffer, D. M. & Scheller, K. K., *J. Food. Sci.*, **56** (1991) 194.
56. Ledward, D. A., *Proc 19th Meat Workers Conference 1973*, Paris, **1**, 259.
57. Andersen, H. J., Bertelsen, G. & Skibsted, L. H., *Meat Sci.*, **25** (1989) 155.
58. Solberg, M. & Franke, W. C., *J. Food Sci.*, **36** (1971) 990.
59. Bertelsen, G. & Skibsted, L. H., *Meat Sci.*, **19** (1987) 243.
60. Nicol, D. J., Shaw, M. K. & Ledward, D. A., *Appl. Microbiol.*, **19** (1970) 937.
61. Nicholls, P., *Biochem. J.*, **81** (1961) 374.
62. Nichol, A. W., Hendry, E., Morrell, D. B. & Clezy, P. S., *Biochim. Biophys. Acta*, **128** (1968) 97.

63. Egan, A. F., Shay, B. J. & Stanley, G., *Meat Res. CSIRO (Ann. Rep.)* (1980) 30.
64. Egan, A. F. & Shay, B. J., *Meat Res. CSIRO (Ann. Rep.)* (1981) 28.
65. Gee, D. L., & Brown, W. D., *J. Agr. Food Chem.*, **26** (1978) 273.
66. Hines, L. R., *U.S. Patent 3, 023, 109,* 1962.
67. Tarladgis, B. G., *U.S. Patent 3, 360, 381,* 1967.
68. Draudt, H. N., *Proc. 22nd Ann. Reciprocal Meat Conf. 1969,* Am. Meat Sci. Assoc., Chicago, p. 180.
69. Tappel, A. L., *Food Res.*, **22** (1957) 404.
70. Tarladgis, B. G., *J. Sci. Food Agric.*, **13** (1962) 481.
71. Peisach, J. & Blumberg, W. E., *Proc. 1st Inter. American Symp. Haemoglobins,* Caracas, Karger, 1969, Basel, 199.
72. Ledward, D. A., *J. Food Technol.*, **9** (1974) 59.
73. Kaziro, K. & Tsushima, K., In *Haematin Enzymes,* ed. J. E. Falk, R. Lembery & R. K. Morton. Pergamon Press, Oxford, 1959, p. 173.
74. Ledward, D. A., *J. Food Sci.*, **43** (1978) (iii).
75. Bernofsky, C., Fox, J. B. & Schweigert, D. D., *Food Res.*, **24** (1959) 339.
76. Awad, E. S. & Deranleau, D. A., *Biochemistry,* **7** (1968) 1791.
77. Atanasova, B. P., Derzhanski, Al. & Georgieva, A., *Biochim. Biophys. Acta,* **160** (1968) 255.
78. Ledward, D. A., *Meat Sci.*, **2** (1978) 241.
79. Palombo, R. & Wygaards, G., *J. Food Sci.*, **55** (1990) 601.
80. Palombo, R. & Wygaards, G., *J. Food Sci.*, **55** (1990) 604.
81. Trout, G. R., *J. Food Sci.*, **54** (1989) 536.
82. Schricker, B. R. & Miller, D. D., *J. Food Sci.*, **48** (1983) 1340.
83. Girard, B., Vanderstoep, J. & Richards, J. F., *J. Food. Sci.*, **55** (1990) 1249.
84. Walters, C. L., Casselden, R. J. & Taylor, A. McM., *Biochim. Biophys. Acta,* **143** (1967) 310.
85. Killday, K. B., Tempesta, M. S., Bailey, M. E. & Metral, C. J., *J. Agric. Food Chem.*, **36** (1988) 909.
86. Cheah, K. S., *J. Food Technol.*, **11** (1976) 181.
87. Tarladgis, B. G., *J. Sci. Food Agric.*, **13** (1962) 485.
88. Lee, S. H. & Cassens, R. G., *J. Food Sci.*, **41** (1976) 969.
89. Wayland, B. B. & Olson, L. W., *J. Am. Chem. Soc.*, **96** (1974) 6037.
90. Sebranek, J. G. & Fox, J. B., *J. Sci. Food Agric.*, **36** (1985) 1169.
91. Andersen, H. J., Bertelsen, G., Boegh Soerensen, L., Shek, C. K. & Skibsted, L. H., *Meat Sci.*, **22** (1988) 283.
92. Fox, J. B. & Thompson, J. S., *Biochemistry,* **3** (1964) 1323.
93. Dymicky, M., Fox, J. B. & Wasserman, A. E., *J. Food Sci.*, **40** (1975) 306.
94. Giddings, G. G. & Markakis, P., *J. Food Sci.*, **37** (1972) 361.
95. Blythe, K. M., *MSc Dissertation,* Queen's University Belfast, 1990.
96. Coleby, B., Ingram, H. J., Shepherd, H. J. & Thornley M. J., *J. Sci. Food Agric.*, **11** (1960) 678.
97. Lebepe, S., Molins, R. A., Charoen, S. P., Farrar, IV, H. & Skowronski, R. P., *J. Food Sci.*, **55** (1990) 918.
98. Ginger, I. D., Lewis, V. J. & Schweigert, B. S., *J. Agric. Food Chem.*, **3** (1955) 156.

99. Bernofsky, C., Fox, J. B. & Schweigert, B. S., *Arch. Biochim. Biophys.*, **80** (1959) 9.
100. Satterlee, L. D., Mari, S. W. & Barnhart, H. M., *J. Food Sci.*, **36** (1971) 549.
101. Kamarei, A. R., Karel, M. & Wierbicki, E., *J. Food Sci.*, **44** (1979) 25.
102. Ledward, D. A., In *Developments in Meat Science—2*, ed. R. A. Lawrie. Applied Sci. Publ., Barking, 1981, p. 159.
103. Hamm, R. In *Developments in Meat Science—2*, ed. R. A. Lawrie. Applied Sci. Publ., Barking, 1981, p. 93.
104. Ledward, D. A., *Meat Sci.*, **21** (1987) 231.
105. Penny, I. F., *Food Process Packag.*, **29** (1960) 363.
106. Fishwick, M. J., *J. Sci. Food Agric.*, **21** (1970) 160.
107. Zapatta, J. F. F., Ledward, D. A. & Lawrie, R. A., *Meat Sci.*, **27** (1990) 109.
108. Ledward, D. A., *Food Additives and Contaminants*, **7** (1990) 677.
109. Watts, B. M. In *Symposium on Foods: Lipids and their Oxidations*, ed. H. W. Schultz. Avi Publ. Co., Westport, Conn., 1962, p. 202.
110. Verma, M. M., Parajape, V. & Ledward, D. A., *Meat Sci.*, **14** (1985) 91.
111. Johns, A. M., Birkinshaw, L. H. & Ledward, D. A., *Meat Sci.*, **25** (1989) 209.
112. Oellingrath, I. M. & Slinde, E., *J. Food Sci.*, **53** (1988) 967.
113. Kanner, J. & Harel, S., *Lipids*, **20** (1985) 625.
114. Rhee, K. S., *Food Technol.*, **42**(6) (1988) 127.
115. Sato, K. & Hegarty, G. R., *J. Food Sci.*, **36** (1971) 1098.
116. Love, J. D. & Pearson, A. M., *J. Agric. Food Chem.*, **22** (1974) 1032.
117. Harel, S. & Kanner, J., *J. Agric. Food Chem.*, **33** (1985) 1186.
118. Kanner, J., Hazan, B. & Doll, L., *J. Agric. Food Chem.*, **36** (1988) 421.
119. Igene, J. O. & Pearson, A. L., *J. Agric. Food Chem.*, **27** (1979) 838.
120. Pearson, A. L. & Gray, J. I., *The Maillard Reaction in Foods and Nutrition*, ed. G. R. Waller & M. S. Feather. *ACS Symposium Series 215*, American Chemical Society, Washington D.C., 1983, p. 287.
121. Conrad, M. E., Cortell, S., Williams, H. L. & Foy, A. L., *J. Lab. Clin. Med.*, **68** (1966) 659.
122. Conrad, M. E., Weintraub, L. R., Sears, D. A. & Crosby, W. H., *Am. J. Physiol.*, **211** (1966) 1123.
123. Wheby, M. S., Suttle, G. E. & Ford, K. T., *Gastroenterology*, **58** (1970) 647.
124. Hazell, T., Ledward, D. A., Neale, R. J. & Root, I. C., *Meat Sci.*, **5** (1980–81) 397.
125. Hazell, T., Ledward, D. A. & Neale, R. J., *Br. J. Nutr.*, **39** (1978) 631.
126. Shears, G. E., Ledward, D. A. & Neale, R. J., *Int. J. Food Sci. Technol.*, **22** (1987) 265.
127. Wheby, M. S. & Spyker, D. A., *Am. J. Clin. Nutr.*, **34** (1981) 1686.

Chapter 7

MODIFICATION OF FOOD PROTEINS BY NON-ENZYMATIC METHODS

Frederick F. Shih

US Department of Agriculture, Southern Regional Research Center, New Orleans, Louisiana, USA

INTRODUCTION

The effective utilization of proteins in food systems is dependent on tailoring the protein's functional characteristics to meet the complex needs of the manufactured food products. Many food proteins, particularly those from plant sources, require modification to improve such functional properties as solubility, whippability and emulsification activity. Chemical treatments used for modification can be classified into two main groups: enzymatic and non-enzymatic methods. This chapter concerns chemical methods that modify food proteins non-enzymatically. For enzymatic modification, the reader should refer to Chapter 8 of this book. Even though enzymatic modifications are considered safer for food uses and therefore more desirable, many non-enzymatic methods have been proven safe. Indeed, because of their simple approach, ready availability and great effectiveness, non-enzymatic methods are always desirable and often irreplaceable. Reviews on food protein modifications for improved functionality are available in the literature.[1-3] This review covers only non-enzymatic modifications with an emphasis on current developments.

Functional properties of protein are closely related to size of the protein, structural conformation and level and distribution of ionic charges. Chemical treatments that could cause alteration of these properties include reactions which either introduce a new functional group to the protein or remove a component part from the protein. Therefore, reactions such as succinylation, acetylation, phosphorylation, limited hydrolysis and specific amide bond hydrolysis (deamidation) have been used to impart improved functional properties to the protein.

235

ACYLATION

Protein acylation concerns the reaction of a nucleophile such as the amino or hydroxyl group of the protein with the carbonyl group of an acylating agent, and results in the addition of a new functional group to the protein. The tyrosine phenolic group and serine and threonine hydroxyl groups are weak nucleophiles and they are not easily acylated in aqueous solution.[4] Most acylating agents, including various acyl halides and anhydrides, react more readily with the amino groups. Acylation has been useful in protein characterization and modification. Free amino groups in a protein molecule, especially the ε-amino group of lysine, are often marked by acylation for structure and composition analysis or are blocked to protect them from other reactions. The blocking is reversible if reagents such as maleic anhydride and citraconic anhydride are used and the acylated products can be easily deacylated.[5-8] In food processing, protein acylation produces derivatives with new functional groups and, depending on the acylating agent used, alters the physical and chemical properties of the protein. As succinic anhydride and acetic anhydride are the most often used acylating agents, we will focus our discussion on the modification of various food proteins with these two reagents. The term acylation used in the following paragraphs therefore refers to succinylation and acetylation unless noted otherwise.

The ease of protein acylation can be demonstrated by the treatment of soy protein with succinic anhydride or acetic anhydride.[9] In about 1 h, up to 95% of the available amino groups can be acylated by gradually adding the acylating agent to the protein (1:1, w/w) in water. During the reaction, the pH is maintained between 7 and 8 with NaOH. The product can be purified by dialysis or ultrafiltration. Both succinylation and acetylation have been utilized to improve functional properties for various food proteins from sources including soybean,[8-12] milk,[13,14] oat,[15,16] wheat,[17,18] cottonseed,[18,20] egg,[21-23] rapeseed,[24-27] sunflower,[28-30] sesame,[5,31] fish,[32] corn,[33] canola,[34] pea,[35] castor bean,[36] winged bean,[37] arachin,[38] and field bean.[39] Acylation has also been used to separate protein (rapeseed and navy bean proteins) from phytic acid and to prepare low phytate isolates with good functional properties.[40] The functional properties improved by acylation include solubility, surface hydrophobicity, emulsification and fat binding capacity. The improvement of a specific aspect of functionality depends on reaction conditions, particularly the type and extent of acylation. For instance, high levels of acetylation are effective in masking lysine residues, exposing hydrophobic interiors and causing subunit dissociation. Kim & Rhee[10] reported a twofold increase in surface hydrophobicity for glycinin at 95% acetylation and, at 65% or higher acetylation, the enthalpy dropped from 3·6 cal/g native protein to 0 cal/g protein, implying complete denaturation.

Succinylation converts the cationic amino groups in protein to anionic residues. The increase in net negative charge produced by succinate anions alters the physico-chemical character of the protein, resulting in enhanced aqueous solubility and subsequent changes in emulsifying and foaming capacity. Succinylation of over 90% of the available amino groups in soy protein has been reported to shift the isoelectric point from 4·5 to 4·0 and to improve solubility above pH 4.[9] Kim & Kinsella[11] reported that succinylation of glycinin at 25–98% of available amino groups progressively increased hydrophobicity, viscosity and exposure of aromatic amino acids. Maxima in surface pressure, film-yield stress, film elasticity and foam stability were observed at 25% succinylation of glycinin. Excessive succinylation (> 50%) reduced these parameters because the high net negative charges may have impaired protein–protein interactions in films. Increased solubility and decreased surface hydrophobicity at alkaline pH were reported for canola protein as the level of succinylation increased.[34] In contrast, the intrinsic viscosity of sunflower[30] and rapeseed[26,30] proteins was reported to increase sharply at succinylations of 50–65% and higher. Cottonseed isolates have been reported[20] to display the sharpest increases of water solubility and emulsion capacity at succinylation of greater than 95 and 60%, respectively.

Much attention has been given to the effects of acylation on protein digestibility and nutritional value. Siu & Thompson[13] reported that both in-vivo and in-vitro digestibilities of lysine, cysteine, methionine, and threonine were greatly reduced at high levels of succinylation. True protein digestibility, reflecting in-vivo amino acid digestibility, was 3–6% lower for the highly succinylated protein as compared with that of the control. Siu & Thompson[41] also reported that whey concentrate with 37% succinylation had a net protein ratio (NPR) higher than that of casein, but the NPR was lower at higher succinylation. Rats fed with the 37% succinylated whey had high urinary N excretion, but little of the succinylated lysine, cysteine and threonine were recovered as bound amino acids in the urine. Apparently these amino acids were partially catabolized and no longer suitable for digestion. It is obvious that acylation interferes with normal biological activities in the utilization of protein. Further studies are needed to evaluate the metabolic and toxic nature of the acylated proteins and to assess whether protein acylation is suitable for food use.

PHOSPHORYLATION

Phosphorylation is another effective way to increase negative charges in a protein molecule and thereby to improve functionality, particularly solubility. Inorganic phosphate (P_i) can be transferred to proteins by either O- or N-esterification reactions. In an O-esterification reaction, P_i reacts with the primary or secondary hydroxyl on serine or threonine, respectively; or with the weakly acidic hydroxyl on tyrosine, forming a —C—O—P_i bond (Fig. 1). In N-esterification, P_i combines with the ε-amino group of lysine, the imidazole group of histidine, or the guanidino

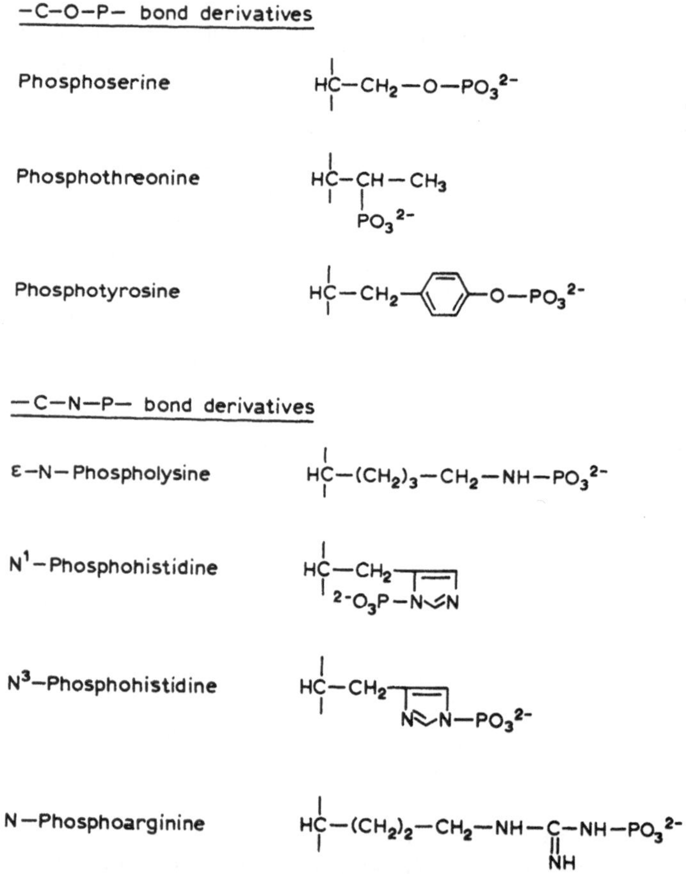

−C−O−P− bond derivatives

Phosphoserine   HC—CH$_2$—O—PO$_3{}^{2-}$

Phosphothreonine   HC—CH—CH$_3$
                              |
                            PO$_3{}^{2-}$

Phosphotyrosine   HC—CH$_2$—〈benzene〉—O—PO$_3{}^{2-}$

—C−N−P− bond derivatives

ε−N−Phospholysine   HC—(CH$_2$)$_3$—CH$_2$—NH—PO$_3{}^{2-}$

N^1−Phosphohistidine   HC—CH$_2$
                              $^{2-}$O$_3$P—N≷N

N^3−Phosphohistidine   HC—CH$_2$
                              N≷N—PO$_3{}^{2-}$

N−Phosphoarginine   HC—(CH$_2$)$_2$—CH$_2$—NH—C—NH—PO$_3{}^{2-}$
                                              ||
                                              NH

Fig 1. Phosphorylated amino acid derivatives in proteins.

group of arginine, forming an —C—N—P$_i$ bond (Fig. 1). The nitrogen-bound phosphates are acid labile and are readily hydrolysed at pH values at or below 7.[42] Proteins containing oxygen-bound phosphate are acid stable and are the modification of choice for food proteins[43] since the pH of most food systems is 3–7.

Non-enzymatic methods for the phosphorylation of proteins have been reviewed by Frank[44] summarizing the reaction conditions involved and detailing experimental procedures for the most important methods. Matheis & Whitaker,[43] in their review of phosphorylation for food proteins, discussed the reagents used, the nature of the phosphate linkages involved, and the effects of phosphorylation on the functional and biological properties of the proteins. Food proteins can be phosphorylated by phosphorus oxychloride (POCl$_3$), sodium trimetaphosphate (STMP), phosphorus pentoxide/phosphoric acid, and many other reagents. According to Matheis & Whitaker,[43] of the phosphorylating agents surveyed, only POCl$_3$ and STMP appear to be suitable for large-scale phosphorylation of food proteins. These two reagents will be the focus of our discussion in the following paragraphs.

Woo et al.[45] reported that when a 1% aqueous solution of β-lactoglobulin was treated with solutions of POCl$_3$ in n-hexane, carbon tetrachloride or mineral oil and the pH was kept at 8·5 during the reaction by the addition of NaOH, the incorporation of 14 mol of P$_i$/mol of protein could be achieved. Based on [31]NMR spectral data, phosphorylation occurred at the ε-amino group of lysine residues and the imidazole group of histidine residues. As expected of a nitrogen-bound phosphate, the phosphorylated β-lactoglobulin was acid labile, and it dephosphorylated rapidly under acidic pH conditions. Using somewhat different reaction conditions, Hirotsuka et al.[46] reported that up to 140 mol of P$_i$/mol of protein could be incorporated into a soy isolate, whereas Matheis et al.[42] incorporated 7·4 and 6·2 mol P$_i$/mol of casein and lysozyme, respectively. The phosphate in phosphorylated casein was exclusively bound to the hydroxyl oxygen whereas the majority of the phosphate in phosphorylated lysozyme appeared to be bound to nitrogen. These workers also observed that protein cross-linking occurred during phosphorylation. Sung et al.[47] used STMP to modify serine and lysine residues in soy isolate and found that the reaction was dependent on pH, temperature, and STMP concentration. At pH 11·5, 35°C, and 1% STMP, 30% of the serine residues in soy isolate were phosphorylated. However, Matheis et al.[42] were unable to detect any covalently bound P$_i$ in soy protein and lysozyme using the method of Sung et al.[47]

Phosphorylation with $POCl_3$ improved the gel-forming properties of casein[42] and β-lactoglobulin,[48] particularly in the presence of Ca^{2+}. Both phosphorylated casein and lysozyme[42] adsorbed more moisture than the corresponding control proteins, but the emulsifying capacity of phosphorylated casein was lower than that of control casein. Soy protein phosphorylated with $POCl_3$ showed improved emulsifying ability and gel-forming ability even at the pH of its isoelectric point (pH 4·5).[46] However, solubility of the protein decreased at pH < 5·5 due to protein cross-linking. On the other hand, soy protein treated with STMP-catalysed phosphorylation displayed increased solubility and emulsifying properties, particularly under acidic conditions.[47]

Huang & Kinsella[49] phosphorylated yeast protein with $POCl_3$ at a protein to reagent ratio of 2:1 and found the treatment effective in the removal of nucleic acids (to < 3%). The phosphorylated yeast protein showed improved emulsifying activity and produced stable but weak foams at neutral pH.[50] Other investigators used STMP for the same purposes.[51,52] Sung et al.[51] reported that yeast cells, treated with a ten times by weight of 1% STMP solution at pH 12 and 40°C for 8 h, yielded a protein isolate with significantly reduced nucleic acid content and enhanced solubility and emulsifiability. Similarly, Giec et al.[52] reported that when yeast homogenate was phosphorylated at pH 11·0, 38°C, and 3% STMP, a protein isolate was obtained with decreased nucleic acid (by 60%) and markedly improved solubility, emulsifying activity, foaming capacity and fat and water absorption.

Matheis et al.[42] studied the digestibility of phosphorylated casein by enzymatic hydrolysis with trypsin and α-chymotrypsin. They found that the initial rate of hydrolysis was slower for the phosphorylated casein than the control but the hydrolysis after 24 h was the same with or without phosphorylation. *Tetrahymena thermophili* grew as well on the phosphorylated casein as on the control. These studies indicated that very little inhibition of digestibility, *in vitro* and *in vivo*, had been suffered by the casein in the phosphorylation. Enzymatic hydrolysis of phosphorylated soy proteins by pepsin, pancreatin and pronase E[47] were studied. The results indicated that the digestibility value of the soybean proteins tested was not reduced to a significant extent by the phosphorylation.

In addition to differences in functionality of proteins phosphorylated with $POCl_3$ and STMP, there are other advantages and disadvantages for the use of these reagents in food systems. $POCl_3$ can be used in aqueous or non-aqueous systems. However, when it is added directly to an aqueous protein solution, it reacts rapidly with water to produce P_i and HCl vapor. The low pH and heat generated could lead to denaturation of the protein. In order to minimize these problems, $POCl_3$ is often added in small portions to aqueous protein solutions or dissolved in organic solvents such as carbon tetrachloride. The latter, however, could cause handling problems in a food processing facility. In addition, $POCl_3$ promotes protein cross-linking and its site of phosphorylation is non-specific, as both O- and N-esterification occur. Phosphorylation of protein with STMP can avoid many of the problems associated with $POCl_3$. For example, STMP does not promote protein cross-linking, and hydrolysis in water produces only harmless P_i. Also, STMP is an FDA-approved food additive.[53] However, phosphorylation with STMP requires alkaline pH conditions, more so than with $POCl_3$, which could lead to undesirable reactions, including the formation of lysinoalanine. Both $POCl_3$ and STMP show potential for food protein modification on a commercial scale, but further research is needed to avoid alkaline pH conditions and to enhance O-esterification while reducing N-esterification.

HYDROLYSIS

Chemical hydrolysis remains one of the most popular forms of protein modification. Peptide bond hydrolysis, by acid or base, may be used to yield smaller protein products with more uniform molecular size. Due to the decrease in size and the increase of ionizable groups of α-NH_4^+ and α-COO^-, the resulting peptides are more water soluble and more amenable to further chemical modification. However, excessive peptide bond hydrolysis is undesirable because it could release bitter and off-flavor peptide components. More importantly, because retaining macromolecular characteristics is essential for the protein ingredient to function effectively in food systems, too much decrease in molecular size could result in reduced functionality.

To prevent excessive hydrolysis, reaction conditions have to be carefully controlled and the hydrolysis preferentially limited to less than 4% of the peptide bonds in the protein. It should be noted that both peptide and amide bonds are hydrolysed during protein hydrolysis. Under mild conditions for limited hydrolysis and with very low level peptide bond hydrolysis, both amide bond hydrolysis and its effect on the functional properties of the protein could be significant. Indeed, mild acid treatment of food protein, so often used for enhancement of functionality, may be due primarily to deamidation. With deamidation recognized as an important factor in the development of protein functionality, hydrolysis can be even more effective in modifying proteins for use in food systems.

Mild Acid Hydrolysis

It is well-known that, during mild acid hydrolysis, peptide bonds on either side of aspartic acid may be cleaved at a rate 100 times greater than other peptide bonds.[54,55] Possible pathways for the release of aspartic acid from proteins are shown in Fig. 2. Cleavage of the N-peptide bond proceeds via an intermediate containing a six-membered ring, whereas fission of the C-peptide bond proceeds via a five-membered ring.[56] This easy release of aspartic acid under mild acidic conditions provides control of the peptide bond hydrolysis and thus benefits the development of protein functionality. Matsudomi et al.[57] reported that treatment of 2% soy protein solution with 0.05 N HCl at 95°C for 30 min caused significant changes in conformation and improved solubility and emulsifying activity. These workers evaluated protein hydrolysis by using the release of aspartic acid and ammonia as a measure of peptide and amide bond hydrolysis, respectively. Finley[58] treated wheat gluten with 0·5 N HCl at 95°C to increase its solubility in fruit-based acidic beverages. Wu et al.[59] improved the emulsifying and foaming properties of wheat gluten by mild acid hydrolysis.

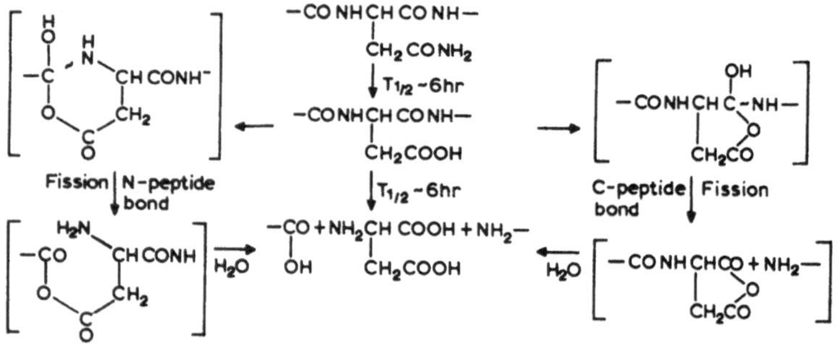

FIG. 2. Possible pathways for the release of aspartic acid from proteins (from Inglis, 1983). Reference 56.

Deamidation

During mild acid treatment of protein, peptide bond hydrolysis seldom occurs without deamidation and vice versa, and the mechanisms for the release of aspartic acid and ammonia are closely related. The conversion products for non-enzymatic deamidation vary, depending on pH, temperature, buffer composition, and sequence and size of amino acids in the deamidated substrate.[60-62] Particularly, pH appears to exert major control over the deamidation. As can be seen in Fig. 3, which shows

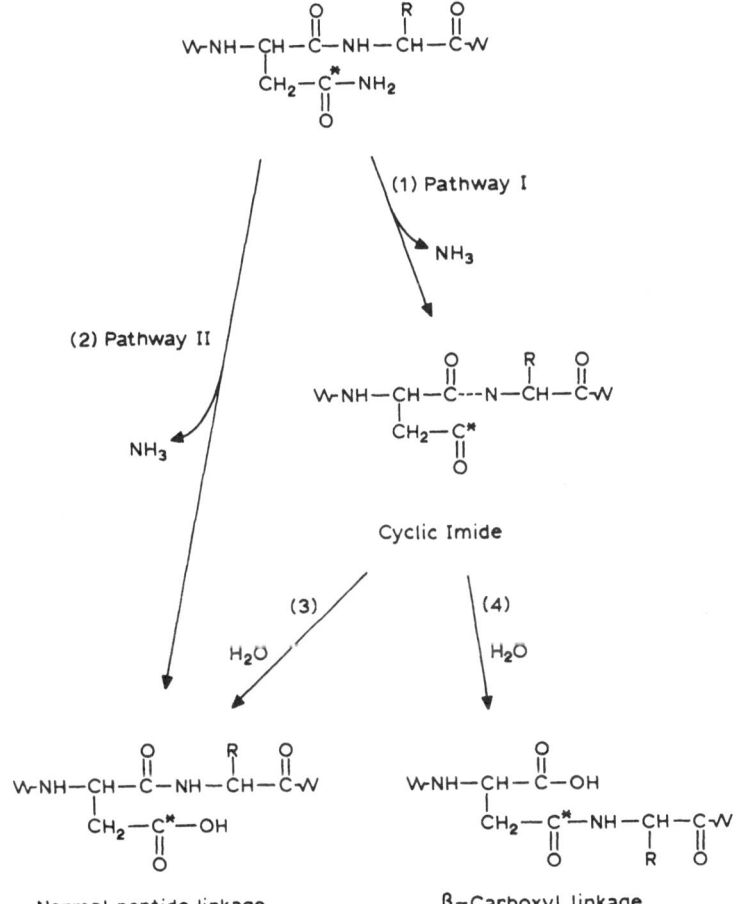

FIG. 3. Mechanism of deamidation: (1) Pathway I, formation of a cyclic imide intermediate and release of ammonia; (2) Pathway II, direct hydrolysis of asparagine (Asn), release of ammonia, and formation of aspartic acid (Asp); (3) hydrolysis of the cyclic imide intermediate to form an α-carboxyl linkage; (4) hydrolysis of the cyclic imide intermediate to form a β-carboxyl linkage. The asterisk marks the position of the side-chain carbonyl carbon. In the formation of a β-carboxyl linkage (iso-Asp), this carbon becomes part of the main chain and increases the chain length by one carbon.

possible pathways of non-enzymatic deamidation, deamidation of asparagine (Asn) residues at alkaline and neutral pH involves the formation of an intramolecular cyclic imide intermediate[63-66] (Fig. 3, Pathway I). The intermediate could be hydrolysed to a normal α-linked aspartate (Asp) residue and a β-linked Asp residue (iso-Asp) with the latter predominating.[67-69] Under acidic conditions, the reaction proceeds via direct hydrolysis of Asn residues to Asp residues without the formation of a cyclic imide intermediate (Fig. 3, Pathway II). This straightforward deamidation under acidic conditions is significant because chemical deamidation of food proteins, normally conducted in mildly acidic solutions, is expected to produce the normal α-linked carboxylic residues instead of the nutritionally undesirable β-linked isomers. The ease of deamidation depends both on the amino acid sequence flanking the carboxamide group[62-70] and on the tertiary structure of the protein.[71] While both asparagine and glutamine can undergo deamidation, asparagine has been shown to deamidate at much higher rates.[67,72]

Treatment of protein under mild acid conditions normally results in significant deamidation (10–20%) but very low and relatively insignificant peptide bond hydrolysis (< 7%).[57] Consequently, most investigators consider mild acid hydrolysis an effective method to achieve deamidation, which is mainly responsible for the ensuing changes in physical and functional properties of the protein.

Methods have been developed in our laboratory in which oilseed proteins are selectively deamidated by non-enzymatic catalysis. Small amounts of long-chain alkyl sulfate or alkyl sulfonate anions enhanced the deamidation of soybean and cottonseed proteins in preference to peptide bond hydrolysis.[73] When a 1% soybean protein suspension in water was stirred in the presence of 0·04 M sodium dodecylsulfate (SDS) (0·05 N HCl, 70°C and 3 h), up to 40% of the amide bonds were hydrolysed, with less than 2% peptide bond hydrolysis. Under comparable conditions without the catalyst, about 20% amide bonds and 4% peptide bonds of the soybean protein were hydrolysed.[73] Aryl sulfonate anions in the form of cation exchange resins were as effective as SDS in catalysing protein deamidation.[74] Because of the easy separation of soluble protein product from the resin catalyst, the use of resin catalysts is particularly attractive to the food industry.

Recently, Shih reported that common anions such as phosphate or bicarbonate could also catalyse the deamidation of soybean protein.[75] To reach the 40% level of deamidation, the common anion catalysts required pH 8 at 100°C for 8 h. This resulted in higher hydrolysis (approx. 4·5%)

of the peptide bonds. Nevertheless, deamidated soybean proteins obtained by all these methods had improved solubility, water binding capacity, foam expansion, emulsion capacity and emulsion viscosity as compared with unmodified soybean proteins.

Alkali Hydrolysis

Alkali also hydrolyses amide and peptide protein bonds, usually at a faster rate than acid. In food processing, high protein plant materials are often treated with alkali to isolate seed proteins[76,77] or to remove toxic constituents.[78] Alkali treatment has been used to solubilize proteins for the preparation of texturized products such as meat analogs. Yet there is a lack of interest in protein modification with alkali because the alkali treatment causes reactions which may be undesirable for food uses.

In addition to peptide and amide bond hydrolysis, alkali treatment of proteins causes elimination reactions involving the side chains of certain amino acids, recemization of amino acid residues, addition of compounds to the proteins, scission of the peptide chain, modification or elimination of non-protein constituents, and the interaction of the protein with alkali-derived products from the environment.[79] As a result, food proteins extensively treated with alkali are not readily digested, the nutritional value is decreased, and toxic effects have been reported. Particularly, β-elimination, one of the most frequent and most important reactions of protein in alkaline solution, could lead to the formation of lysinoalanine.[80] Presumably, amino acids not normally found in nature, including lysinoalanine, are formed via reaction of derivatives from β-elimination with sulfhydryl or amino groups present in the protein. Because lysinoalanine has been reported to cause renal lesions in rats,[81] its presence in food has generated considerable concern about the safety of alkali-treated proteins. Reviews on the effects of alkali treatment on proteins, including possible toxic effects when used in foods, are available in the literature.[79,82–84]

The methods discussed above could complement or be comparable to one another in many uses. Giec et al.[52] reported the preparation of yeast protein by using both succinylation and phosphorylation methods and found the two methods to complement one another. Ma et al.[17] studied the effects of deamidation and succinylation on the bread-making characteristics of gluten. Deamidation led to decreases in bread loaf volume and dough extensibility, whereas succinylation led to a pronounced decrease in dough extensibility but no significant changes in specific loaf volume. Both modifications increased the net charge and surface hydrophobicity of gluten, but deamidation, in addition to hydrolysing the amide groups, caused degradation of gliadin and increase of low molecular weight components. The data indicated that hydrogen bonding by the amide groups of gluten in the breadmaking process may be more important than changes in molecular weight distribution or hydrophobic interaction.

CONCLUDING REMARKS

The discussion in this chapter is by no means a complete review of the non-enzymatic modifications reported in the literature. However, the examples given testify to the importance of non-enzymatic methods in food use. Proteins modified by non-enzymatic methods, particularly those obtained under mild acid hydrolysis conditions, have been popular and continue to attract interest as food ingredients. Further investigations on non-enzymatic modification are needed not only to improve and increase the use of proteins in foods but also to enhance the understanding of how specific forms of protein functionality can be developed.

REFERENCES

1. Feeney, R. E., In *Food Proteins,* ed. R. E. Feeney & J. R. Whitaker. American Chemical Society, Washington, D.C., 1977, p. 3.
2. Feeney, R. E., Yamasaki, B. & Geoghegan, K. F., In *Modification of Proteins. Food, Nutritional, and Pharmacological Aspects,* ed. R. E. Feeney & J. R. Whitaker. American Chemical Society, Washington, D.C., 1982, p. 3.
3. Feeney, R. E. & Whitaker, J. R., In *New Protein Foods,* ed. A. M. Altschul & H. L. Wilcke. Academic Press, New York, 1985, p. 181.
4. Means, G. & Feeney, R., ed. *Chemical Modification of Proteins.* Holden-Day, San Francisco, CA, 1971.
5. Dixon, H. B. F. & Perham, R. N., *Biochem. J.,* **109** (1968) 312.
6. Butler, P. J. G., Harris, J. I., Hartley, B. S. & Leberman, R., *Biochem. J.,* **112** (1969) 679.
7. Hasegawa, K., Fujino, S., Okado, I., Suenobu, K. & Hirose, M., *Agric. Biol. Chem.,* **49** (1985) 2777.
8. Brinegar, A. C. & Kinsella, J. E., *J. Agric. Food Chem.,* **28** (1980) 818.
9. Franzen, K. L. & Kinsella, J. E., *J. Agric. Food Chem.,* **24** (1976) 788.
10. Kim, K. S. & Rhee, J. S., *J. Food Biochem.,* **13** (1989) 187.
11. Kim, S. H. & Kinsella, J. E., *Cereal Chem.,* **63** (1986) 342.
12. Kim, K. S. & Rhee, J. S., *J. Agric. Food Chem.,* **38** (1990) 669.

13. Siu, M. & Thompson, L. U., 1982. *J. Agric. Food Chem.*, **30** (1982) 1179.
14. Girerd, F., Martin, J. F., Mesnier, D. & Lorient, D., *Sci. Aliments*, **4** (1984) 251.
15. Ma, C. Y., *J. Food Sci.*, **49** (1984) 1128.
16. Ponnampalam, R., Goulet, G., Amiot, J., Chamberland, B. & Brisson, G. *J. Food Chem.*, **29** (1988) 109.
17. Ma, C. Y., Oomah, B. D. & Holme, J., *J. Food Sci.*, **51** (1986) 99.
18. Barber, K. J. & Warthesen, J. J., *J. Agric. Food Chem.*, **30** (1982) 930.
19. Rahma, E. H. & Narasinga-Rao, M. S., *J. Agric. Food Chem.*, **31** (1983) 352.
20. Choi, Y. R., Lusas, E. W. & Rhee, K. C., *J. Food Sci.*, **48** (1983) 1275.
21. Ball, H. R. Jr. & Winn, S. E., *Poultry Sci.*, **61** (1982) 1040.
22. Gossett, P. W., Rizvi, S. S. H. & Baker, R. C., *J. Food Sci.*, **48** (1983) 1395.
23. Gossett, P. W. & Baker, R. C., *J. Food Sci.*, **48** (1983) 1391.
24. Thompson, L. U. & Cho, Y. S., *J. Food Sci.*, **49** (1984) 1584.
25. Schwenke, K. D., Raab, B. & Pahtz, W., *J. Food Biochem.*, **13** (1989) 321.
26. Nitecka, E. & Schwenke, K. D., *Die Nahrung*, **30** (1986) 969.
27. Geuguen, J., Bollecker, S., Schwenke, K. D. & Raab, B., *J. Agric. Food Chem.*, **38** (1990) 61.
28. Kabirullah, M. & Wills, R. B. H., *J. Food Technol.*, **17** (1982) 235.
29. Schwenke, K. D., Rauschal, E. J., Zirwer, D. and Linow, K. J., *Int. J. Peptide Protein Res.*, **25** (1985) 347.
30. Schwenke, K. D., Linow, K. J. & Zirwer, D., *Die Nahrung*, **30** (1986) 263.
31. Hasegawa, K., Fujino, Y. & Konami, S., *Nihon Nogei Kagakkai Shi*, **55** (1981) 239.
32. Miller, R. & Groninger, H. S. Jr., *J. Food Sci.*, **41** (1976) 268.
33. Messinger, J. K., Rupnow, J. H., Zeece, M. G. & Anderson, R. L., *J. Food Sci.*, **52** (1987) 1620.
34. Paulson, A. T. & Tung, M. A., *J. Food Sci.*, **52** (1987) 1557.
35. Johnson, E. A. & Brekke, C. J., *J. Food Sci.*, **48** (1983) 722.
36. Yoon, J. U., *Korean J. Food Sci. Technol.*, **12** (1980) 263.
37. Narayana, K. & Narasinga-Rao, M. S., *J. Food Sci.*, **49** (1984) 547.
38. Shyamasundar, R. & Rajagopal-Rao, D., *Lebensm.-Wiss. Technol.*, **15** (1982) 102.
39. Muschiolik, G., Dickinson, E., Murray, B. & Stainsby, G., *Food Hydrocolloids*, **3** (1987) 191.
40. Thompson, L. U., *J. Am. Oil Chem. Soc.*, **64** (1987) 1712.
41. Sui, M. & Thompson, L. U., *J. Agric. Food Chem.*, **30** (1982) 1179.
42. Matheis, G., Penner, M. H., Feeney, R. E. & Whitaker, J. R., *J. Agric. Food Chem.*, **31** (1983) 379.
43. Matheis, G. & Whitaker, J. R., *J. Agric. Food Chem.*, **32** (1984) 699.
44. Frank, A. W., *Phosphorus Sulfur*, **29** (1987) 297.
45. Woo, S. L., Creamer, L. K. & Richardson, T., *J. Agric. Food Chem.*, **30** (1982) 65.
46. Hirotsuka, M., Taniguchi, H., Marita, H. & Kito, M., *Agric. Biol. Chem.*, **48** (1984) 93.
47. Sung, H., Chen, H., Liu, T. & Su, J., *J. Food Sci.*, **48** (1983) 716.
48. Woo, S. L. & Richardson, T., *J. Dairy Sci.*, **66** (1983) 984.
49. Huang, Y. T. & Kinsella, J. E., *Biotechnol. Bioeng.*, **28** (1986) 1690.

50. Huang, Y. T. & Kinsella, J. E., *J. Food Sci.*, 2 (1987) 1684.
51. Sung, H., Chen, H. & Chuan, S., *Proc. Natl. Sci. Counc. ROC(A)*, 7 (1983) 181.
52. Giec, A., Stasinska, B. & Skupin, J., *Food Chem.*, 31 (1989) 279.
53. Ellinger, R. H., In *CRC Handbook of Food Additives*, 2nd edn, Vol. 1, ed. T. E. Furia. CRC, Cleveland, OH, 1972, pp. 640.
54. Schultz, J., Allison, H. & Grice, M., *Biochemistry*, 1 (1962) 694.
55. Han, K. K., Richard, C. & Biserte, G., *Int. J. Biochem.*, 15 (1983) 875.
56. Inglis, A. S., In *Methods in Enzymology*, Vol. 91, ed. C. H. W. Hirs & S. N. Timasheff. Academic Press, New York, 1983, p. 324.
57. Matsudomi, N., Sasaki, T., Kato, A. & Kobayashi, K., *Agric. Biol. Chem.*, 49 (1985) 1251.
58. Finley, J. W., *J. Food Sci.*, 40 (1975) 1283.
59. Wu, C. H., Nakai, S. & Powrie, W. P., *J. Agric. Food Chem.*, 24 (1976) 504.
60. Bhatt, N. P., Patel, K. & Borchardt, R. T., *Pharm. Res.*, 7 (1990) 593.
61. Patel, K. & Borchardt, R. T., *Pharm. Res.*, 7 (1990) 703.
62. Patel, K. & Borchardt, R. T., *Pharm. Res.*, 7 (1990) 787.
63. Aswad, D. W., *J. Biol. Chem.*, 259 (1984) 10714.
64. Murray, E. D. & Clarke, S., *J. Biol. Chem.*, 259 (1984) 10722.
65. Meinwald, Y. C., Stinson, E. R. & Scheraga, A., *Int. J. Peptide Protein Res.*, 28 (1986) 79.
66. Geiger, T. & Clarke, S., *J. Biol. Chem.*, 262 (1987) 785.
67. Sondheimer, E. & Holley, R. W., *J. Am. Chem. Soc.*, 76 (1954) 2467.
68. Battersby, R. & Robinson, J. C., *J. Chem. Soc.*, 1955 (1955) 259.
69. Lura, R. & Schirch, V., *Biochemistry*, 27 (1988) 7671.
70. Harding, J. J., *Adv. Protein Chem.*, 37 (1985) 247.
71. Kossiakoff, A. A., *Science*, 240 (1988) 191.
72. Robinson, A. B., Scotchler, J. W. & McKerrow, J. H., *J. Am. Chem. Soc.*, 95 (1973) 8156.
73. Shih, F. F. & Kalmar, A. D., *J. Agric. Food Chem.*, 35 (1987) 672.
74. Shih, F. F., *J. Food Sci.*, 52 (1987) 1529.
75. Shih, F. F., *J. Food Sci.*, 56 (1991) 452.
76. Edwards, R. H., Saunders, R. M. & Kohler, G. O., *J. Food Sci.*, 45 (1980) 860.
77. Lawhon, J. T., Manak, L. J. & Lusas, E. W., *J. Food Sci.*, 45 (1980) 197.
78. Screenivasamurthy, V., Parpia, H. A. B., Srikanta, S. & Shankar Murti, A., *J. Assoc. Offic. Anal. Chem.*, 50 (1967) 350.
79. Whitaker, J. R. & Feeney, R. E., *CRC Crit. Rev. Food Sci. Nutr.*, 19 (1983) 3.
80. Patchornik, A. & Sokolovsky, M., *J. Am. Chem. Soc.*, 86 (1964) 1860.
81. Woodward, J. C. & Short, C. D., *J. Nutr.*, 103 (1973) 569.
82. Gould, D. H. & Macgregor, J. T., *Adv. Exp. Med. Biol.*, 86B (1977) 29.
83. Dworschak, E., *CRC Crit. Rev. Food Sci. Nutr.*, 13 (1980) 1.
84. Struthers, B. J., *J. Am. Oil Chem. Soc.*, 58 (1981) 501.

Chapter 8

MODIFICATION OF FOOD PROTEINS BY ENZYMATIC METHODS

JAMEL S. HAMADA

*US Department of Agriculture, Southern Regional Research Center,
New Orleans, Louisiana, USA*

INTRODUCTION

Protein structure is modified to improve solubility, emulsification and other functional properties of proteins thereby enhancing their effective use as foods and food ingredients. Modification of protein functionality can also make food products better suited for human nutritional utilization, thus increasing the world's food supply.[1] Therefore, the purpose of modifying the structure of proteins is to create new and unique products that would possess better functional properties in food systems than the unmodified protein. In the past 15 years, protein modification has been the subject of many excellent reviews.[1-9] Existing commercial modification procedures are limited in number and usefulness. There is a need to develop other modification methods. The newly developed method(s) should produce an end product that is safe for human consumption by meeting the food additive standards imposed by the United States Food and Drug Administration (FDA). Most importantly, industrial modification processes must be specfic, reproducible and cost effective.

STRUCTURE AND FUNCTIONALITY OF PROTEINS

Because of their nutritional value, functional properties and versatility in processing, proteins are essential ingredients in the food industry. Wall[10] described some important functional properties of proteins and their applications in foods (Table 1). Gluten is an excellent example of the critical role of proteins in food systems, even though protein is sometimes only a small fraction of the food product. The functional properties of

TABLE 1

FUNCTIONAL PROPERTIES OF PROTEINS IN FOODS AND THEIR APPLICATIONS[a]

Property	Applications
Emulsification	Meats, coffee whiteners, salad dressings
Hydration	Dough, meats
Viscosity	Beverages, doughs
Gelation	Sausages, gel desserts
Foaming	Toppings, meringues, angel food cakes
Cohesion binding	Textured products, dough
Textural properties	Textured foods
Solubility	Beverages

[a]From Wall.[10]

proteins in food systems are dependent on their molecular structure and interactions with the environment.[10-11] *In vivo*, each side-chain group of amino acid residues in proteins, can react with other groups to produce either covalent bonds, i.e. S–S links, which are important in stabilizing chain folding and in forming bonds between chains or non-covalent bonds as illustrated in Fig. 1.[10] Based on the fact that the 21 amino acids commonly found in proteins can combine to form an almost endless number of different primary sequences, the possibilities and versatility of the protein structures are enormous. Outside their cellular environment, protein behavior and thus functionality depends on its new environment. Thus, when proteins are incorporated into foods, they function according to their structures. Although hydrophobic bonds are individually weak, they are able to cause associations of considerable tenacity. These can occur between proteins and substances such as lipids, polysaccharides and low molecular weight organic molecules.[10] Interactions of proteins with other components in the food product ultimately determine the functional properties and applications of the protein.[8,10,11]

Many plant proteins, especially soy protein, have gained importance as food ingredients because of their high protein content, nutritive value and/or functional properties in some foods.[8,12] However, their use in many food systems particularly acidic foods is limited because of lack of solubility and the ability to form emulsions in the pH 3–6 range.[8] Accordingly, these proteins require modification in order to obtain the desirable functional properties for expanded use under the acidic conditions in such food products as coffee whitener and pourable and non-pourable dressings. Amide bond hydrolysis by dilute acids[13-15] or alkaline solutions[4,14]

Bond type	Functional groups involved	Disrupting solvents
Physical		
Electrostatic	Carboxyl	Salt solutions
$-COO^- \cdots {}^+NH_3-$	Amino	High or low pH
	Imidazole	
	Guanido	
Hydrogen bond	Hydroxyl	Urea solutions
$-C{=}O\cdots HO-$	Amide	Guanidine
$\mid$	Phenol	hydrochloride
NH		Dimethylformamide
Hydrophobic bonds	Long aliphatic chains	Detergents
O⌁⌁⌁⌁⌁⌁		Organic solvents
⌁⌁⌁⌁⌁⌁⌁O	Aromatic	
Covalent		
Disulphide bonds	Cystine	Reducing agents
$-S-S-$		Sulfite
		Mercaptoethanol

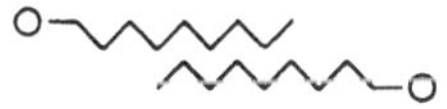

FIG 1. Types of bonds between protein chains (from Wall[10]).

can improve the solubility and other functional properties of food proteins by increasing the number of negative charges in the protein. Conversion of glutamine or asparagine to glutamic acid or aspartic acid in food proteins is expected to have a major effect on proteins, particularly cereal proteins, where up to one-third of the total amino acids is glutamine. Even small levels of deamidation (e.g., 2–5%) could result in a significant improvement of protein functional properties.[13] Phosphorylation of proteins has also been used to increase the numbers of the negative charges on β-lactoglobulin,[16-17] casein,[18] lysozyme.[18] and soy protein.[19] The introduction of negatively charged phosphate groups could improve solubility and emulsification properties of soy protein in mildly acidic foods. Excellent reviews have appeared that discussed chemical phosphorylation of food proteins using phosphorous oxychloride ($POCl_3$) and sodium trimetaphosphate.[9,20]

Glycosylation is another example of a protein modification method, that has already been accomplished by chemical reactions. Waniska and Kinsella[21] covalently attached some sugar groups to both the amino and carboxyl groups of β-lactoglobulin. Glycosylation of β-lactoglobulin resulted in partial unfolding of the protein; increased viscosity; increased exposure of aromatic amino acid residues; increased rate of enzymatic hydrolysis; decreased hydrophobicity; and a decrease in the amount of ordered secondary structure. Surface, foaming and emulsifying properties were also improved. Gloss *et al.*[22] also found that glycosylation of casein improved its surface properties and the modified protein was more flexible, unfolded and hydrated than the unmodified casein.

ENZYMES AS TOOLS FOR PROTEIN MODIFICATION

Protein modification, traditionally carried out by direct chemical means, can also be accomplished using enzyme catalysis. Chemical modification is not very desirable for food applications because of the harsh reaction conditions, non-specific chemical reagents, and the difficulty of removing residual reagents from the final product.[23] Enzymes provide, on the other hand, several advantages including fast reaction rates, mild conditions and, most importantly, high specificity. Under optimum conditions, enzyme-catalysed reactions proceed after 10^8–10^{11} times more rapidly than the corresponding non-enzymatic reactions.[24] This will result in reduction of energy costs and increased processing efficiency. Accordingly, enzymatic methods are more attractive to the food processor and far more compliant to regulatory standards than chemical treatments. Other attractive features of enzyme methodology is that most enzymes can be produced in large quantities, each having appropriate physical, chemical and catalytic properties, the cost of enzyme production is reasonable if one uses microbial fermentation or possibly biotechnological techniques.

Despite the economic, health and safety benefits of using enzymes in protein modification, enzymes must be commercially useful by offering better products than those produced by corresponding chemical modification methods. Because of the relatively low value of food ingredients, which are produced in high bulk, compared for instance, to pharmaceutical materials, processing enzymes must be cheap relative to total costs. Enzymes have the potential of reducing the cost of energy, labor and/or machinery required in the manufacturing process, but this may be off-set by the cost of the enzyme itself relative to the value of the food ingredient. For instance, enzymes sometimes are too specific, so that a number of enzymes may be needed to do the same job as the chemical process can accomplish.

Besides the proteases, which have been investigated extensively and which are the only modifying enzymes currently in use commercially, there are transglutaminase, protein kinase, and peptidoglutaminase. These are the only enzymes that have been reported for use in food protein modification only on a laboratory scale. Whitaker[3] addressed possible approaches applicable to the modification of proteins. He also discussed the impact of these potential modifications on the structure and properties of the proteins in great detail. These chemical processes occur naturally in cells and accordingly the enzymes involved could be isolated from plant, animal or microbial sources and used in protein modification. Furthermore, while some of these enzymes are available commercially for biochemistry and molecular biological research, e.g. glycosylating and deglycosylating enzymes, they may be very expensive. Therefore, isolating the enzymes from their natural sources and subjecting them to partial purification may be a prerequisite for exploring the possibility of any type of new enzymatic modification.

Protease modification

The main commercial use of proteases is in the laundry detergent industry to help remove protein-based stains such as blood and egg from clothing.[25] The literature on these enzymes and their application in the industry is abundant. Hence, the description of their use in protein modification will be very brief relative to the other enzymes covered in this chapter. For greater depth in this subject, extensive reviews, symposia and text books are readily available and may be consulted, e.g. Adler-Nissen.[26] Proteases hydrolyse peptide bonds in proteins to modify their structure. The early use of proteases in the industry has been in the preparation of whipping formulations and milk replacers.[24,27,28] Proteins were hydrolysed to improve dispersibility and nutritional availability without significantly affecting flavor, color, and nutritional value. The characteristics and specificity of the proteases available for industrial use are listed in Table 2. Subtilisin and papain are the most commonly used enzymes.[23] All these enzymes are endoproteases, which attack peptide bonds in the interior of the polypeptide chain producing a range of polypeptides which differ in molecular weight without producing significant amounts of free amino acids as end products. The extent of hydrolysis, expressed as the percentage of the peptide bonds cleaved (%DH)[26] is dependent on the end use, the range of specificity of the protease used, the accessibility of the peptide bonds to enzyme attack (e.g., native vs denatured protein), and reaction conditions such as time, pH and temperature.

TABLE 2

SPECIFICITIES AND pH OPTIMA OF PROTEOLYTIC ENZYMES[a]

Enzyme	pH Optimum	Specificity
Pepsin	1·9	Fairly broad. Hydrolyses proteins and peptides, particularly those with bonds adjacent to aromatic or dicarboxylic L-amino acid residues.
Trypsin	7–9	Hydrolyses peptides, amides, esters etc. at bonds involving the carboxyl groups of L-arginine or L-lysine.
Chymotrypsin	8–9	Hydrolyses peptides, amides, esters etc. especially bonds involving the carboxyl groups of aromatic L-amino acids.
Papain	5·5	Hydrolyses peptides, amides, and esters at bonds involving basic amino acids, leucine or glycine.
Pronase	7–9	Wide.
Subtilisin[b]	6–10	Wide with no clear bond preferences.

[a]From Simmonds & Orth.[24]
[b]From Adler-Nissen.[26]

Kato et al.[29] reported the use of proteases for the deamidation of proteins in a carbonate buffer, pH 10 at 20°C. Protein deamidation using papain, pronase, and chymotrypsin, was as high as 20% but up to 8% of the peptide bonds of some proteins were hydrolysed. Kato et al.[30] later reported the treatment of ovalbumin, lysozyme, 7S globulin, 11S globulin, and gluten with immobilized chymotrypsin on controlled-pore glass at pH 10 at 20°C. Deamidation was 10·0, 8·4, 6·0, 5·0, and 8·0%, respectively. Kato et al.[30] found great improvement of gluten solubility at pH range of 2–12. Both emulsifying and foaming properties of proteins were also improved by treatment with immobilized chymotrypsin. Surprisingly, Shih[33] found no deamidating activities by these proteases at pH 10 and 20°C suggesting that the ammonia might have been released through the non-enzymatic deamidation of the asparagine and glutamine residues in the hydrolysates.

Transglutaminase modification

Folk & Finlayson[32] and more recently Lorand & Conrad[33] discussed the various aspects of transglutaminases (TGases) and focused on their biological roles. TGase, a Ca^{2+}-activated enzyme derived from the soluble fraction of guinea pig liver, catalyses the incorporation of a number of primary amines into proteins and polypeptides through an acyl transfer reaction between the γ-carboxyl groups of glutaminyl residues in proteins and the amines. TGase also catalyses the release of ammonia from proteins in the absence of added amines. Thus TGase action on protein substrates can lead to protein cross-linking, incorporation of small primary amines into the protein substrate or deamidation of a protein-bound glutaminyl side chain.

Cross-linking of proteins by TGase has been investigated by a number of scientists for the preparation of new food proteins with improved functional properties. Ikura et al.,[34,35] Kurth & Rogers,[36] Motoki & Nio,[37] Nio et al.,[38,39] Motoki et al.,[40–42] and Bercovici et al.[43] used guinea pig liver TGase to catalyse the formation of inter- and intramolecular cross-linking in a number of casein, lactoglobulin, gluten and soy protein fractions. TGase can also be used for the covalent attachment of essential amino acids to nutritionally inferior food proteins.[44,45] The modified proteins were targeted for use in such applications as the incorporation of plant proteins into animal proteins, preparing texturized products and producing protein ingredients with improved nutritional and functional properties.

Motoki[46] used guinea pig TGase to catalyse the deamidation of casein. To avoid the formation of ε(γ-glutaminyl) lysyl isopeptide bonds, which can not be broken in digestion, they chemically modified the protein with citraconic anhydride prior to TGase treatment. Although, this citraconylation, aimed at blocking the free amino groups of protein-bound lysyl side chains, were reversible,[47] this pre-treatment poses potential problems for the food industry. Obtaining FDA approval for the product to be used as a food additive; possible reduction in nutritional quality; downstream separation of by-products from the modified protein; and the prohibitive cost of the blocking step itself would render this process not feasible for industrial applications.

Peptidoglutaminase Modification

Bacillus circulans peptidoglutaminase (PGase) catalyses the hydrolysis of the amide of glutamine residues in peptides.[48–50] Gill et al.[51] detected limited deamidating activity of PGase towards casein and whey protein hydrolysates. We also investigated the potential of PGase for food protein modification and observed the PGase readily deamidated glutamine in peptides and protein hydrolysates but its activity toward intact proteins

was small.[52,53] Enzyme activity toward intact protein substrates ranged
from no modification in egg albumin to 6% deamidation in soy protein.[53]
The use of proteolysis prior to PGase treatment substantially enhanced
deamidation.[53,54] In addition to proteolysis, heat treatment and the combination of the heat and proteolysis also significantly enhanced the
limited deamidating ability of PGase towards proteins.[53-55]

Hamada[55] developed new enzymatic methods for the deamidation of
food proteins, in which a range of PGase deamidation can be achieved by
pre-proteolysis alone or a combination of both proteolysis and other means
to disrupt hydrogen and S—S bonds. PGase was used to deamidate soy
protein, casein and gluten after subjecting the proteins to the prior treatments described in Fig. 2. PGase deamidation of proteins increased by
proteolysis as a functioon of percent peptide bond hydrolysis (DH) (A in
Fig. 3). Heat treatment of protein hydrolysates at 100°C for 15 min after
Alcalase hydrolysis increased PGase deamidation of soy protein but not
for gluten or casein (B in Fig. 3). However, heat treatment of proteins at
100°C for 15 min prior to their hydrolysis increased the deamidation of
hydrolysates compared with the proteins of plain proteolysis or post-pro-

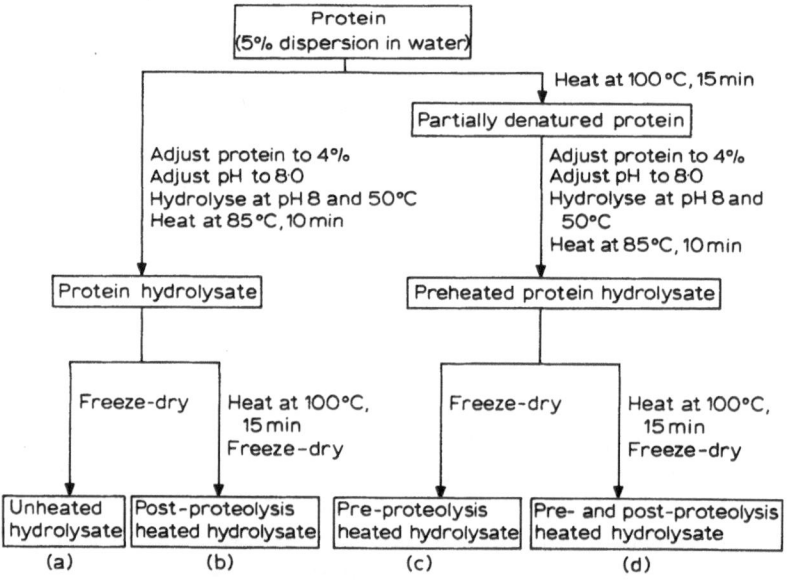

FIG 2. Flow diagram for the preparation of protein hydrolysates with various DH
(degree of hydrolysis) values (from Hamada[55]).

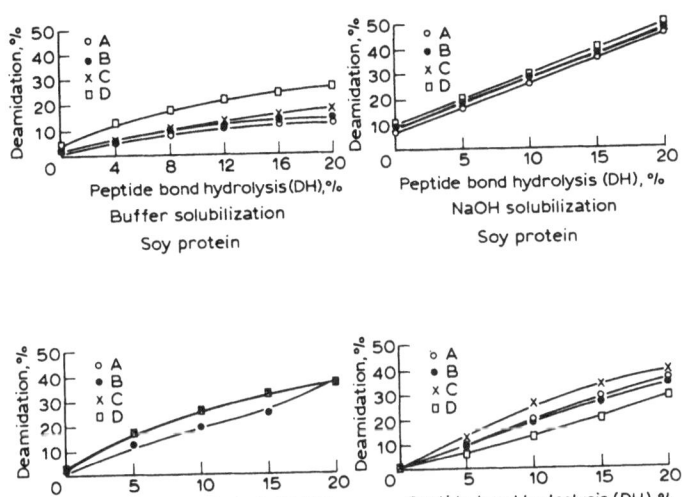

FIG 3. Effect of proteolysis and heat treatment on the peptidoglutaminase deamidation of proteins. Legend for A–D as described in Fig. 1. (from Hamada[55]).

teolysis heat treatment (C in Fig. 3). This was a result of more uniform cleavage of the peptide bonds during protein hydrolysis, which might have increased accessibility of amide groups to PGase attack.

Heating increased deamidation of hydrolysates produced from soy proteins that were pre-heated prior to their hydrolysis (D in Fig. 3). The samples from the first graph (top left) used to study the effect of the pre-treatments on PGase deamidation of soy protein were solubilized in 0·05 M sodium phosphate buffer (pH 7·5) before deamidation.[54] Deamidation increased from 3 to 27% for 20% DH hydrolysates, when proteolysis was combined with heat treatment for heated soy protein (D in Fig. 3). However, when this study was repeated using 0·02 M NaOH followed by the addition of phosphate buffer to solubilize the protein or the protein hydrolysate, deamidation of soy protein increased from 10% to 50% for 20% DH hydrolysates. The extent of PGase deamidation previously observed[54] was nearly half that found using the NaOH solubilization method.[55] The different response of the proteins deamidated by the two methods was attributed to breaking some S—S, hydrogen and other bonds by NaOH.[56] This resulted in maximum amide accessibility and the PGase deamidation was, thus, dependent on the molecular size, i.e. the DH value. Contrary to results for heating soy hydrolysates, heating casein hydrolysates at 100°C for 15 min had no effect on their deamidation by PGase (B and D in Fig. 3). The different responses of the two proteins were due to the colloidal structure of the casein micelles, which are known to be very stable, persisting at 140°C for 20 min.[57] Heat treatment of gluten hydrolysates (B in Fig. 3) or preheated hydrolysates (D in Fig. 3) was also found to have little or no effect on deamidation. During this heat treatment, nearly one-third and one-half of the total amides in the two sets of gluten hydrolysates were hydrolysed non-enzymatically, respectively, decreasing subsequent deamidation by PGase.

Protein Kinase Modification

Protein kinases have been isolated from many plants including soybean,[58–60] wheat, and rice[61] and have been used in the phosphorylation of proteins for physiological and molecular biology studies. Protein phosphorylating enzymes from mammals have also been extensively investigated because of their role in regulating certain cell activities through the regulation of proteins and enzymes.[62–65] Adenosine cyclic 3′,5′-monophosphate-dependent protein kinase (cAMPdPK) from bovine cardiac muscle[66–69] has been the most studied protein kinase, probably because it has been commercially available. Recently, a few reports appeared on the use of protein kinases in the modification of food proteins. Only an update on the advances in the area of food application of protein kinases will be described here.

Seguro et al.[70] and Seguro & Motoki[71] used cAMPdPK for the phosphorylation of several food proteins including soy protein, albumin, and casein. Ross & Bhatnagar[72] isolated the C subunit of cAMPdPK from bovine cardiac muscle to phosphorylate soybean proteins. Ross[73] also used a commercially available protein kinase to phosphorylate a commercial soy protein isolate. Campbell et al.[74] continued the work of Ross[73] using the catalytic subunit of the protein kinase from bovine cardiac muscle to phosphorylate soy proteins. The phosphorylated protein was analysed by SDS-PAGE and autoradiography (Fig. 4). The $^{32}P_i$ initially incorporated into glycinin acidic polypeptides and then to glycinin basic polypeptides. Very little $^{32}P_i$ was associated with the β and α subunits of β-conglycinin. Seguro & Motoki[71] reported that protein kinase phosphorylated only the glycinin acidic polypeptides but not the glycinin basic polypeptides or any of the β-conglycinin subunits. The conflicting results may be explained by assuming that Seguro & Motoki[71] did not allow

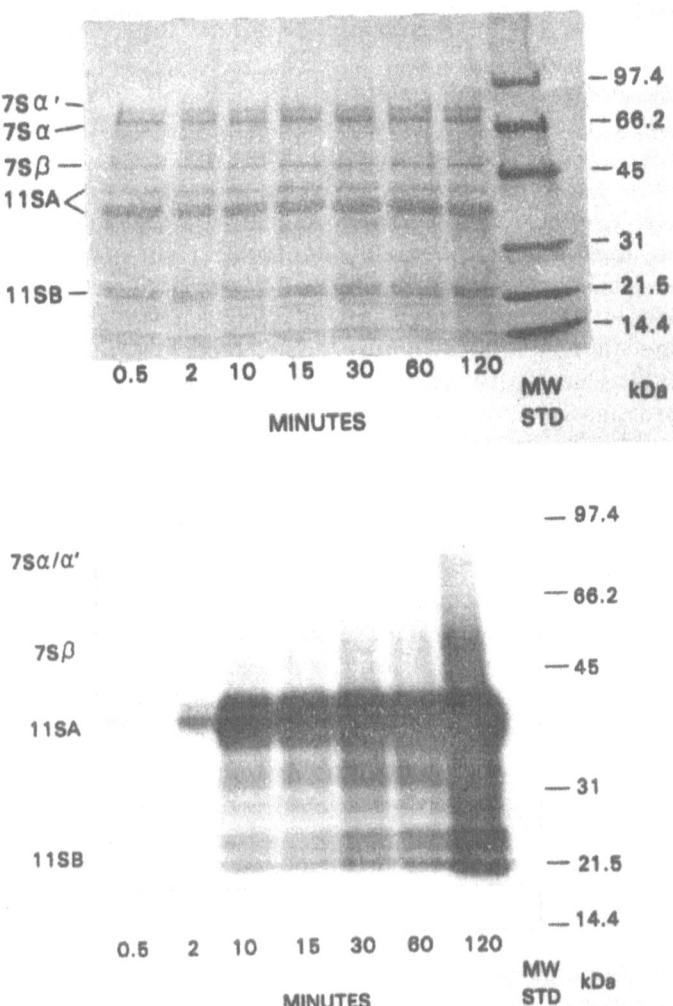

FIG 4. Gradient SDS-PAGE of timed protein kinase assay (A) and the autoradiogram of the SDS-PAGE (from Campbell *et al.*[74]).

sufficient time for reaction to proceed to completion. There, no radiolabeled P_i was incorporated into basic polypeptides of glycinin, β-conglycinin or whey proteins.

ENZYME KINETICS

Kinetics of Transglutaminase

Because of its important physiological function in animal tissues, TGase has been investigated extensively. Therefore, it is not surprising that the literature on TGase kinetics is abundant and still expanding. For instance, Griffin & Wilson[75] described sensitive techniques for the quantitative assessment of the cross-linking of proteins and/or biological materials in human cells. As explained above, TGase can be used to cross-link proteins or to attach amino acids to proteins, Iwami & Yasumoto[45] reported the inhibition and the progress curves of TGase reaction catalysing the attachment of amines to wheat gluten, gliadin, bovine milk casein, soy protein isolate, ovalbumin, zein and gelatin. They found that wheat gluten was the best amine acceptor among those food proteins they tested.

Kinetics of Peptidoglutaminase

PGase activity towards a commercial soy protein hydrolysate was investigated.[53] The reaction velocity in the well-known Michaelis-Menten equation (Eqn 1) is the velocity in the initial part of the reaction when the product concentration $(S_0 - S)$ is very small, i.e. tends to zero. In industrial applications where the reaction usually proceeds to complete conversion, the Michaelis-Menten relationship integrated with respect to time (Eqn 2) is used.

$$v = V_{max}[S]/[S] + K_m \tag{1}$$

where v is the initial rate of enzyme catalysed reaction; and $[S]$ substrate concentration.

$$t = \{S_0 X + K_m \cdot \ln [1/(1-X)]\}/V_{max} \tag{2}$$

where t is reaction time; S_0 is initial substrate concentration; and X is the conversion rate, which is defined as $X = (S_0 - S)/S_0$, where S is substrate concentration at time t.

Contrary to the data from solving Eqn 1 (curve A of Fig. 5), the time course of the reaction predicted by solving Eqn 2 (curve B of Fig. 5), closely matched that measured experimentally (curve C of Fig. 5). As the reaction proceeds, the initial velocity decreases, lowering the substrate concentration, especially with reactions which proceed for long periods of time such as industrial reactions. Accordingly, first- or mixed first- and zero-order kinetics, rather than zero-order kinetics must be used. This

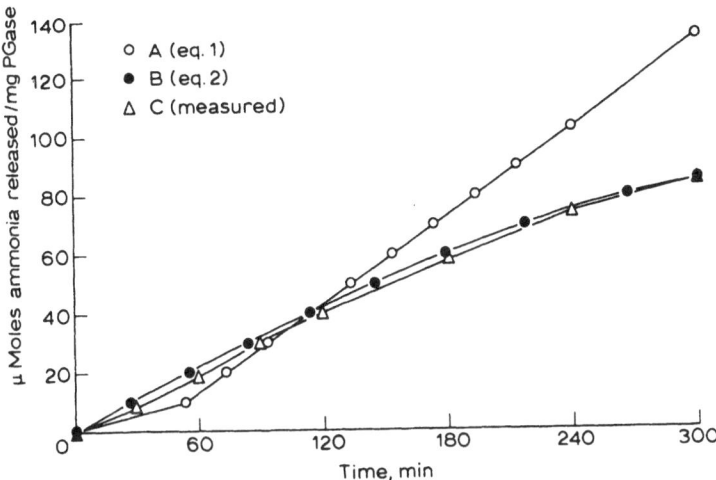

FIG 5. Progress of soy protein hydrolysate deamidation by peptidoglutaminase (from Hamada).

equation provides the basis for modeling and optimizing performance of enzyme reactors to maximize productivity and to achieve complete substrate conversion.

Kinetics of Protein Kinase

Heat treatment of soy protein increased phosphorylation.[71-73] Protein denaturation increased susceptibility to phosphorylation, by causing sequestered serine and/or threonine residues to become more readily available to the kinase. Campbell et al.[74] studied the progress of the protein kinase reaction and found that increasing the number of protein kinase units increased the phosphorylation of heat-treated soy protein. Incorporation of $^{32}P_i$ into soy protein isolate increased with increased incubation time and reached a level of 13 μmol of phosphorus incorporated per gram of protein after 4 h incubation at 37°C (Fig. 6). Campbell et al.[74] also observed sequential and selective phosphorylation of the major soybean globulins and found that the order in which the polypeptides and subunits are phosphorylated corresponds to the number of potential phosphorylation sites in the primary structures known for the proteins.

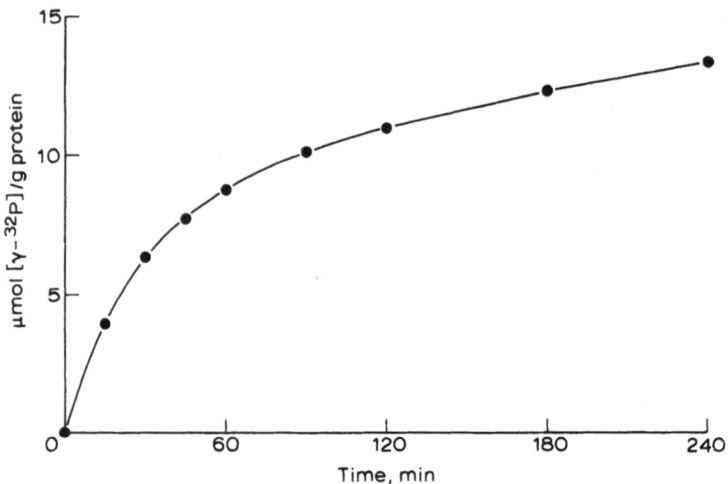

FIG 6. Reaction time course of the phosphorylation of denatured soy protein isolate by protein kinase (from Cambell et al.[74]).

INDUSTRIAL APPLICATIONS OF MODIFYING ENZYMES

Isolation and Purification of Transglutaminases

Rhee and Choi[76] used DEAE-cellulose columns and sequential ammonium sulfate precipitation to purify guinea pig liver TGase 60-fold with 45% recovery. Workers from the same laboratory[77-79] reported the production of transglutaminase from the microbial source *Streptoverticillium* sp. The isolated enzyme had a molecular weight of 44 000, isoelectric point of 8·9 and optimum pH of 6–7. Unlike guinea pig liver TGase, the microbial TGase was not calcium-dependent. Although the importance of transglutaminase in industrial application remains to be seen, this is of great interest to the food processor since the enzyme can be produced in larger quantity for industry by large-scale fermentation methods. The culture filtrate of *Streptoverticillium* strain was purified by ultrafiltration, Amberlite CG-50 (twice) and blue sepharose. This increased the specific activity of the enzyme from 0·13 to 22·6 units with 42% yield. In their studies on the mechanism of action of polyamines, reseachers from different laboratories have recently reported the presence of TGase in plant tissues.[80-83] Animal TGase functions as a catalyst for the post-translational modification of proteins. However, the exact role of TGase in plants is not yet known.[83] Although the TGase from silver beet leaves was not Ca^{2+} dependent, its reaction was simulated by calcium ions.[83]

Membrane Reactor for Peptidoglutaminase

Immobilization of PGase by containment in an ultrafiltration membrane was possible by taking advantage of the relatively high molecular weight and stability of the enzyme. However, proper choice of membrane pore size and membrane material are critical to any practical application.[84] Hamada[85] developed a semi-continuous reactor for the deamidation of soy protein hydrolysate with the PGase retained within a 30 kDa spiral membrane. The reactor was operated at 30°C and 6·5 l/m² per h flux in recycle mode until 60% conversion had been achieved, then in diafiltration mode for 2 h. In order to correct for the influence of ultrafiltration on the progress of the deamidation reaction, an operational effectiveness factor (η_o) was incorporated in the Michaelis-Menten kinetics for mixed zero- and first-order reactions, which was expressed in Eqn (2). The behavior of the reactor and the enzyme dosage required for complete deamidation could be predicted by this equation (now Eqn 3). This method has the potential of being scaled-up for the production of enzyme-free deamidated proteins, which is of great interest to the food processor since inactivation or removal of enzymes used in modification processes is highly desirable.

$$t = \{S_0 X + K_m \ln [1/(1-X)]\}/V_{max}\, \eta_o \qquad (3)$$

Preparative HPLC of Peptidoglutaminase

Preparative HPLC separations are capable of purifying milligram to kilogram amounts of materials per chromatographic run and of giving high resolution and recovery with reduced separation times and elution volumes. Hamada[86] developed large-scale HPLC methodology PGase purification with potential for use in industrial food protein modification. PGase was isolated directly from cell extract of *B. circulans* by gel permeation and ion exchange chromatography with optimal gradient elution programs developed for the latter. The specific activity of cell extract increased more than a thousand-fold by removal of unwanted proteins and all nucleic acids, with 80% recovery.

Regeneration of Protein Kinase Co-substrate

Before protein kinase can be practical for a food application, the problem of the high energy phosphate co-substrate required for the reaction must be tackled. Despite the technical usefulness of the products of enzymatic phosphorylation, finding an effective way to regenerate the ATP, used as a co-substrate, is critical to developing any efficient enzymatic process for industrial protein phosphorylation: the ATP is generally very expensive and its cost is prohibitive. The possibility of the continuous regeneration of ATP was been recently investigated. For instance, glucose 6-phosphate was produced while the ATP was simultaneously regenerated by using immobilized yeast cells[87] or ultrafiltration hollow-fibre reactors.[88,89]

IMPROVEMENT IN FUNCTIONAL PROPERTIES

Proteolysis

Limited hydrolysis of proteins, controlled either by following the rate of proteolysis[26] or by use of highly specific enzymes,[7,15] increased solubility, particularly at the isoelectric pH, with increasing protein hydrolysis.[15,25,90,91] Emulsifying activity[92] increased in the pH range of 2–9[91] and emulsifying capacity increased by limited hydrolysis of soy proteins.[90] However, Casella & Whitaker[15] found that specific limited hydrolysis of zein produced emulsions with less emulsifying activity than the native protein. The emulsion stability of soy protein dispersion decreased upon proteolysis[90,91] but emulsions obtained from heated hydrolysates were more stable than emulsions of unheated hydrolysates.[91] This may indicate that in contrast to emulsifying activity, solubility is not a prerequisite for increased emulsion stability. Casella & Whitaker[15] reported that the emulsion stability of modified zein did not change by proteolysis, with only one exception which was the 1·7% DH hydrolysates. Its emulsion stability increased only at pH values above 8·0. Foamability or foam power[93] increased by limited hydrolysis of proteins.[15,90,91] Limited proteolysis, however, had no effect on foam stability in the case of soy protein hydrolysis[90,91] and increased foam stability for zein hydrolysates.[14] Heat denaturation of soy globulins increased foaming power and foam stability[91] due to increased surface hydrophobicity by the disruption of hydrophobic bonds.[94,95]

Transglutaminase Polymerization:

Motoki et al.[39] found that the emulsifying activity and the unfrozen water content of caseins and soybean globulins polymerized TGase were higher than those of the unmodified proteins. These proteins may be more suitable for the production of new food systems requiring higher hydration ability. TGase was also used to form gels from protein/oil emulsion.[38] Motoki et al.[42] also found that highly concentrated protein solutions could be firmly gelatinized by TGase cross-linking. The casein film had a high tensile strength and strain and was insoluble in water. Choi & Ree[96] found that the sodium caseinate gel formed by guinea pig liver TGase was elastic, water-insoluble and highly digestible. Treatment with acid changed in the structure of the gel to a hard and rubbery texture.

Peptidoglutaminase Deamidation
The PGase modification procedures developed in this laboratory[55] can produce a wide spectrum of modified peptides or proteins with markedly improved solubility and emulsification under mild acidic conditions. Solubility and emulsifying activity of soy protein, deamidated to 6–16% by PGase, increased at pH 4–7 and under alkaline conditions. Deamidation also increased emulsion stability and foaming power.[91]

Protein Kinase Phosphorylation
Seguro & Motoki[71] and Campbell et al.[74] found that the solubility of phosphorylated soy proteins increased at pH 2–7. Seguro & Motoki[97] found that the sensitivity of the protein towards calcium at pH 7 increased after the phosphorylation but Campbell et al.[74] observed that phosphorylation had a negligible effect on the protein sensitivity towards calcium up to 20 mM KCl. The emulsifying activity also increased in the pH range of 2–7[71,74] and the emulsion stability increased at pH 7 and 80°C.[74] However, the water absorption did not change; the foam expansion increased and foam stability decreased after phosphorylation.[74]

Pepides and Flavor Enhancement
For a long period of time, the problem of bitterness of protein hydrolysates has been of major concern to many food processors. One way to avoid the bitterness of hydrolysates was to restrict the DH values to less than 5·0%.[98] However, one should not arbitrarily limit protein hydrolysis to produce hydrolysates with DH less than 5% to avoid bitter peptide bond formation. The taste of protein hydrolysates is a more complicated phenomenon than just whether the bitterness peptides, e.g. proline-rich peptides, are present or absent. Other tastes are contributed by other amino acid residues which include such flavors as sweet, sour, brothy and beefy. Furthermore, bitter peptide formation can be dealt with as a separate issue if and when the problem arises. Several methods have been developed for the debittering of protein hydrolysates.[98] These methods

include the use of immobilized protease,[99] the selective separation of bitterness components, such as adsorption and ultrafiltration, the plastein reaction, the application of exopeptidases (e.g. degrading of proline-rich, bitter peptides) or highly specialised proteases[7] and bitterness masking by a number of organic acids such as glutamic acid and malic acid.

Glutamic acid and its salts have a long history of use in foods to enhance the flavor.[100] Monosodium glutamate (MSG) is by far the most widely used glutamate. However, the potassium, ammonium and calcium salts have been used as salt substitutes. Glutamates may provide a fifth basic taste, in addition to sweet, sour, salty, and bitter. This fifth basic taste is called umami by the Japanese and savoriness by the Americans. MSG is used at a concentration of 0·2–0·8% in a variety of foods such as soups, broth, sauces, gravies, flavoring and spice blends, as well as in a wide variety of canned and frozen meats, poultry, vegetables and combination dishes. Glutamate as part of a protein is not a flavor enhancer,[100] but glutamate bound into a peptide structure may have the flavor-enhancing properties of the free form. Noguchi et al.[101] eliminated the bitterness of protein hydrolysates by incorporating glutamic acid in the hydrolysates through the plastein reaction. The hydrolysate had no bitterness because of the masking effect of glutamic acid residues in the acidic oligopeptide fraction of the plastein hydrolysate. Thus, it is possible to use PGase deamidated protein hydrolysates to obtain oligo- or polypeptide fractions having flavor enhancement capabilities.

FUTURE RESEARCH

Except for the use of proteases in protein modification, all the new enzymatic procedures addressed here for the modification of proteins were limited to basic research. The extent to which any protein modification method can be used by industry will depend on the quantity and the quality of the research devoted to it,[1] as well as demonstrated practicality and economic viability. Innovative advancement in protein modification research will lead to new markets and expand the food use of many proteins in such products as gels, films, protein-fortified beverages, infant formulas, coffee whiteners, emulsifiers and flavor enhancers. Therefore, more research is needed in these areas including: (1) expanding the list of modifying enzymes to include new modification procedures such as glycosylation; (2) exploring less costly sources, preferably from microorganisms, for the existing or newly developed enzymes as a prerequisite for developing a cost-effective process for the enzymatic modification of proteins in the industry. The isolation of microbial or plant transglutaminase is an excellent example of this type of endeavor; and (3) developing enzymatic modification methods for food proteins that are suitable for industrial applications, such as those described for peptidoglutaminase.

REFERENCES

1. Feeney, R. E., & Whitaker, J. R., In *New Protein Foods*, Vol. 5, eds. A. M. Altschul & H. L. Wilcke. Academic Press, New York, 1985, pp. 181–219.
2. Feeney, R. E., In *Food Proteins*, ed. R. E. Feeney & J. R. Whitaker. American Chemical Society, Washington D.C., 1977, pp. 3–36.
3. Whitaker, J. R., In *Food Proteins* ed. R. E. Feeney & J. R. Whitaker. American Chemical Society, Washington D.C., 1977, pp. 95–155.
4. Pearce, R. J., *CSIRO Food Res. Q.*, **41** (1981) 68.
5. Feeney, R. E., Yammasaki, B. & Geoghegan, K. F., In *Modification of Proteins. Food, Nutritional, and Pharmacological Aspects*, ed. R. E. Feeney & J. R. Whitaker. American Chemical Society, Washington, D.C., 1982, pp. 3–55.
6. Fox, P. F., Morrissey, P. A. & Mulvihill, D. A., In: *Development in Food Proteins*, Vol. 1, ed. B. J. F. Hudson. Elsevier Applied Science Publishers, London, 1982, pp. 1–60.
7. Whitaker, J. R. & Puigserver, A. J., (1982). In *Modification of Proteins, Food, Nutritional, and Pharmacological Aspects*, ed. R. E. Feeney & J. R. Whitaker. American Chemical Society, Washington D.C., 1982, pp. 57–87.
8. Kinsella, J. E., Damodaran, S. & German, B., In: *New Protein Foods*, Vol. 5, ed. A. M. Altschul & H. L. Wilcke. Academic Press, New York, 1985, pp. 107–178.
9. Shih, F. F., Hamada, J. S. & Marshall, W. E., In: *Molecular Aspects to Improving Food Quality and Safety*, ed. D. Bhatnagar & T. E. Cleveland. Van Nostraud Reinhold, New York, 1991, pp. 33–69.
10. Wall, J. S., *Cereal Foods World*, **24** (1979) 288.
11. Kinsella, J. E. & Phillips, L. G., In *Food Proteins*, ed. J. E. Kinsella & W. G. Soucie. Amer. Oil Chem. Soc., Champaign, IL, 1989, pp. 52–77.
12. Wolf, W. J., *J. Agric. Food Chem.*, **18** (1970) 969.
13. Matsudomi, N., Sasaki, T., Kato, A. & Kobayashi, K., *Agric. Biol. Chem.*, **49** (1985) 1251.
14. Kossiakoff, A. A., *Science*, **240** (1988) 191.
15. Casella, M. L. A. & Whitaker, J. R., *J. Food Biochem.* **14** (1990) 453.
16. Woo, S. L. & Richardson, T., *J. Dairy Sci.*, **66** (1983) 984.
17. Woo, S. L., Creamer, L. K. & Richardson, T., *J. Agric. Food Chem.*, **30** 65.
18. Matheis, G., Penner, M. H., Feeney, R. E. & Whitaker, J. R., *J. Agric. Food Chem.*, **31** (1983) 379.
19. Sung, H., Chen, H., Liu, T. & Su, J., *J. Food Sci.*, **48** (1983) 716.

20. Matheis, G., *Food Chem.*, **39** (1991) 13.
21. Waniska, R. D. & Kinsella, J. E., In *Food Proteins*, ed. J. E. Kinsella & W. G. Soucie. Amer. Oil Chem. Soc., Champaign, IL, 1989, pp. 100–131.
22. Gloss, B., Courthaudon, J-L. & Lorient, D., *J. Food Sci.*, **55** (1990) 437.
23. Cheetham, P. S. J., In *Handbook of Enzyme Biotechnology*, ed. A. Wiseman. Ellis Horwood Limited Publishers, Chichester and New York, 1986, pp. 274–379.
24. Simmonds, D. H. & Orth, B. A., In *Industrial Uses of Cereals*, ed. Y. Pomeranz. Amer. Assoc. Cereal Chem., St Paul, MN, 1973, pp. 51–120.
25. Anonymous. In *NOVO's Handbook of Practical Biotechnology*, ed. C. O. L. Boycee. Novo industri A/S, Bagsvaerd, Denmark, 1986, pp. 63–69.
26. Adler-Nissen, J., In *Enzymatic Hydrolysis of Food Proteins*. Elsevier Applied Science Publishers, New York, 1986, pp. 9–29.
27. Olsen, H. S. & Alder-Nissen, J., *Process Biochem.*, **14** (1979) 6.
28. Adler-Nissen, J., In *Enzymatic Hydrolysis of Food Proteins*. Elsevier Applied Science Publishers, New York, 1986, pp. 263–313.
29. Kato, A., Tanaka, A., Matsudomi, N. & Kobayashi, K., *J. Agric. Food Chem.*, **35** (1987) 224.
30. Kato, A., Lee, Y. & Kobayashi, K., *J. Food Sci.*, **54** (1989) 1345.
31. Shih, F. F., *J. Food Sci.*, **55** (1990) 127.
32. Folk, J. E. & Finlayson, J. S., (1977) In *Advances in Protein Chem*, Vol. 31, ed. C. B. Anfinsen, J. T. Edsall & F. M. Richards. Academic Press, New York, 1977, pp. 1–140.
33. Lorand, L. & Conrad, S., *Molecular and Cellular Biochem.*, **58** (1984) 9.
34. Ikura, K., Kometani, T., Yoshikawa, M., Sasaki, M. & Chiba, H., *Agric. Biol. Chem.*, **44** (1980) 1567.
35. Ikura, K., Kometani, T., Sasaki, M. & Chiba, H., *Agric. Biol. Chem.*, **44** (1980) 2979.
36. Kurth, L. & Rogers, P. J., *J. Food Sci.*, **49** (1984) 573.
37. Motoki, M. & Nio, N., *J. Food Sci.*, **48** (1983) 561.
38. Nio, N., Motoki, M. & Takinami, K., *Agric. Biol. Chem.*, **50** (1986) 851.
39. Nio, N., Motoki, M. & Takinami, K., *Agric. Biol. Chem.*, **50** (1986) 1409.
40. Motoki, M., Nio, N. & Takinami, K., *Agric. Biol. Chem.*, **84** (1984) 1257.
41. Motoki, M., Nio, N. & Takinami, K., *Agric. Biol. Chem.*, **51** (1987) 239.
42. Motoki, M., Aso, H. & Seguro, K., *Agric. Biol. Chem.*, **51** (1987) 993.
43. Bercovici, D., Gaertner, H. F. & Puigserver, A. J., *J. Agric. Food Chem.*, **35** (1987) 301.
44. Ikura, K., Yoshikawa, M., Sasaki, M. & Chiba, H., *Agric. Biol. Chem.* **45** (1981) 2587.
45. Iwami, K. & Yasumoto, K., *J. Sci. Food Agric.*, **37** (1986) 495.
46. Motoki, M., Seguro, K., Nio, N. & Takinami, K., *Agric. Biol. Chem.*, **50** (1986) 3025.
47. Brinegar, A. C. & Kinsella, E., *J. Agric. Food Chem.*, **28** (1980) 818.
48. Kikuchi, M., Hayashida, H., Nakano, E. & Sakaguchi, K., *Biochemistry*, **10** (1971) 1222.
49. Kikuchi, M. & Sakaguchi, K., *Agr. Biol. Chem.*, **37** (1973) 827.
50. Kikuchi, M. & Sakaguchi, K., *Methods Enzymol.*, **45** (1976) 485.

51. Gill, B. P., O'Shaughnessey, A. J., Henderson, P. & Headon, D. R., *Irish J. Food Sci. Technol.*, **9** (1985) 33.
52. Hamada, J. S., Shih, F. F., Frank, A. W. & Marshall, W. E., *J. Food Sci.*, **53** (1988) 671.
53. Hamada, J. S., *J. Am. Oil Chem. Soc.*, **68** (1991) 459.
54. Hamada, J. S., & Marshall, W. E., *J. Food Sci.*, **53** (1988) 1132.
55. Hamada, J. S., *J. Agric. Food Chem.*, **40**(5) (1992) pp. 719.
56. Ishino, K. & Okamoto, S., *Cereal Chem.*, **52** (1975) 9.
57. Wong, D. W. S., In *Mechanism and Theory in Food Chemistry.*, Van Nostrand Reinhold, New York, 1989, pp. 48–104.
58. Lin, P. P. & Key, J. L., *Plant Physiol.*, **66** (1980) 360.
59. Gowda, S. & Phillay, D. T. N., *Plant Sci. Lett.*, **25** (1982) 49.
60. Putnam-Evans, C. L., Harmon, A. C. & Cormier, M. J., *Biochemistry,* **29** (1990) 2488.
61. Cheng, C-M., Chow, C-K., Hu, N-T. & Kuo, T-T. *Biochem. Biophys. Res. Comm.*, **175** (1991) 467.
62. Bingham, E. W. & Farrell, H. M., Jr., *J. Biol. Chem.*, **249** (1974) 3647.
63. Cohen, P., *Nature,* **296** (1982) 613.
64. Krebs, E. G., In *The Enzymes*. Vol. 17. eds. P. D. Boyer & E. G. Krebs. Academic Press, New York, 1986, pp. 3–20.
65. Pinna, L. A., Meggio, F. & Merchiori, F., In *Peptides and Protein Phosphorylation,* ed. B. C. Kemp. CRC Press, Boca Raton, Florida, 1990, pp. 145–169.
66. Smith, S. B., White J. B., Seigel, J. B. & Krebs, E. G, In *Protein Phosphorylation,* ed. O. R. Rosen & E. G. Krebs. Cold Spring Harbor Laboratory, Cold Spring Harbor, ME, 1981, pp. 55–65.
67. Shoji, S., Ericsson, L. H., Walsh, K. A., Fisher, E. H. & Tetani, K., *Biochem.*, **22** (1983) 3702.
68. Uhler, M. D., Chrivia, J. C. & McKnight, G. S., *J. Biol. Chem.*, **261** (1986) 15360.
69. Okuno, S. & Fujisawa, J., *Biochim. Biophys. Acta,* **1038** (1990) 204.
70. Seguro, K., Nio, S. & Motoki, M., *Japanese Patent* No. 128,843 (March, 1986).
71. Seguro, K. & Motoki, M., *Agric. Biol. Chem.*, **53** (1989) 3263.
72. Ross, L. F. & Bhatnagar, D., *J. Agric. Food Chem.*, **37** (1989) 841.
73. Ross, L. F., *J. Agric. Food Chem.*, **37** (1989) 1257.
74. Campbell, N. F., Shih, F. F. & Marshall, W. E., Enzymatic phosphorylation of soy protein isolate to improve functional properties in foods. *J. Agric. Food Chem.*, **40**(3) (1992) pp. 403.
75. Griffin, M. & Wilson, J., *Mol. Cell. Biochem.*, **58** (1984) 37.
76. Rhee, I. K. & Choi, Y. R., (1990) *51st Annual Mtg. of Inst. Food Technol.*, 16–20 June, Anahiem, CA (Abstr. #200).
77. Nonaka, M., Tanaka, H., Okiyama, A., Motoki, M., Ando, H., Umeda, K. & Matsura, A., *Agric. Biol. Chem.*, **53** (1989) 2619.
78. Ando, H., Adachi, M., Umeda, K., Matsura, A., Nonaka, M., Uchio, Tanaka, H. & Motoki, M., *Agric. Biol. Chem.*, **53** (1989) 2613.

79. Tanaka, H., Nonaka, M. & Motoki, M., *Nippon Suisan Gakkaishi*, **56** (1990) 1341.
80. Icekson, I. G. & Apelbaum, A., *Plant Physiol.*, **84** (1987) 972.
81. Serafini-Fracasani, D., Del Duca & D'Orazi, *Plant Physiol.*, **87** (1988) 757.
82. Margosiak, S. A., Dharma, A., Bruce-Carver, M. R., Gonzales, A. P., Louie, D. & Kuehn, G. D., *Plant Physiol.*, **92** (1990) 88.
83. Signorini, M., Beninati, S. & Bergamini, C. M., *J. Plant Physiol.*, **137** (1991) 547.
84. Hamada, J. S., *J. Food Sci.*, **56** (1991) 1731.
85. Hamada, J. S., *J. Food Sci.*, **56** (1991) 1725.
86. Hamada, J. S., *52nd Annual Mtg. of Inst. Food Technol.*, 1–5 June, Dallas, TX (Abstr. #692). (1991).
87. Asada, M., Kazuhiro, M., Nakanishi, Matsuno, R., Tanaka, A., Kimura, A. & Kamikubo, T., *Agric. Biol. Chem.*, **43** (1979) 1774.
88. Ishikawa, H., Tanaka, T., Takase, S. & Hikita, H., *Biotechnol. Bioengin.*, **34** (1989) 257.
89. Ishikawa, H., Takase, S., Tanaka, T. & Hikita, H., *Biotechnol. Bioengin.*, **34** (1989) 369.
90. Puski, G., *Cereal Chem.*, **52** (1975) 655.
91. Hamada, J. S. & Marshall, W. E., *J. Food Sci.*, **54** (1989) 598.
92. Pearce, K. N. & Kinsella, J. E., (1978) *J. Agric. Food Chem.*, **26** (1978) 716.
93. Wright, D. J. & Hemmant, J. W., *J. Sci. Food Agric.*, **41** (1987) 361.
94. Kato, A., Osaka, Y., Matsudomi, N. & Kobayashi, K., *Agric. Biol. Chem.*, **47** (1983) 33.
95. Nakai, S., *J. Agric. Food Chem.*, **31** (1983) 676.
96. Choi, Y. R., & Rhee, I. K., *52nd Annual Mtg. of Inst. Food Technol.*, 1–5 June, Dallas, TX (Abstr. #301) (1991).
97. Seguro, K. & Motoki, M., *Agric. Biol. Chem.*, **54** (1990) 1271.
98. Adler-Nissen, J., In *Enzymatic Hydrolysis of Food Proteins*, Elsevier Applied Science Publishers, New York, 1986, pp. 57–109.
99. Deter, C. C., Mason, R. D. & Wootall, H. H. *Biotech. Bioeng.*, **17** (1975) 451.
100. Anonymous, *Food Technol.*, **41** (1987) 143.
101. Noguchi, M., Yamashita, M., Arai, S. & Fujimaki, M., *J. Food Sci.* **40** (1975) 367.

Chapter 9

THE PLASTEIN REACTION: FUNDAMENTALS AND APPLICATIONS

MICHIKO WATANABE

Food Science Laboratory, Faculty of Education, Tokyo Gakugei University, Koganei-shi, Tokyo 184, Japan

&

SOICHI ARAI

Department of Agricultural Chemistry, The University of Tokyo, Bunkyo-ku, Tokyo 113, Japan

INTRODUCTION

It has long been known that when a protein that has been hydrolysed by a protease is allowed to stand, it is sometimes transformed into a plastic gel-like product. Surprisingly, this phenomenon was first discovered a century ago.[1] The product was then named 'plastein' and was thought to be formed through a reverse reaction catalysed by the protease.[1,2] It is now known that the substance 'plastein' does exist. Also, the term 'plastein reaction' is commonly used to refer to the protease-catalysed process involved in the formation of a plastein from a protein hydrolysate or an oligopeptide mixture. However, the definition of the plastein varies somewhat. Tauber[3] defined plastein as a protein-like substance occurring in the water-insoluble fraction that results from the centrifugation of an entire plastein reaction product. Wieland[4] defined a plastein as an acetone-insoluble fraction from a plastein reaction product. Determann & Kohler[5] carried out a plastein reaction with a synthetic oligopeptide and obtained a series of polycondensates as the plastein reaction product. In recent studies conducted in order to develop the plastein reaction for various practical purposes, plastein has often been defined as the precipitate formed when an entire plastein-reaction product is treated with a typical protein denaturant such as trichloroacetic acid (TCA).[6-10]

271

It is now universally held that the plastein reaction is a protease-catalysed process which results in the efficient production of a plastein from a protein hydrolysate. The efficient production of plastein entails a reaction that minimizes the further hydrolysis of the protein hydrolysate used as a substrate. Since any protease-catalysed reaction is generally a reversible process in nature, it can be involved in both peptide bond hydrolysis and synthesis. When both happen to proceed at the same rate, it appears that the substrate undergoes no change because of compensation by the two reactions. The synthetic process in such a case, however, is not called a plastein reaction, since it does not contribute to efficient plastein formation.

With respect to the mechanism of plastein formation, several observations were made at an early stage. Horowitz & Haurowitz[11] studied α-chymotrypsin-catalysed synthesis of a plastein and concluded that it proceeded through transpeptidation. Conversely, Determann & Köhler,[5] as mentioned above, investigated pepsin-catalysed synthesis of a plastein from a well-defined oligopeptide and demonstrated the involvement of a condensation reaction. As will be described later in detail, Yamashita et al.[8,9] observed that both transpeptidation and condensation were involved in the efficient formation of a plastein as a 10% TCA-insoluble substance from a soy protein hydrolysate. From the point of view of energy conservation, the transpeptidation hypothesis is the more reasonable in that the condensation process, which is endergonic, requires approximately 0·4 kcal for the formation of one peptide bond.[12] Nevertheless, a peptide–peptide condensation reaction does take place when appropriate reaction conditions are obtained. A variety of techniques have been used in an attempt to synthesize plasteins for food and industrial uses and physiologically active peptides for medical uses.

Reviewing again the history of plastein reaction studies, we find an interesting series of transitions. It seems that during the 1940s the plastein reaction attracted the close attention of biochemists who believed that it might be involved in protein biosynthesis in vivo. Indeed, a report[13] presented at that time bears the title 'Protein Synthesis by Chymotrypsin'. In the 1950s, Fruton and his colleagues[14,15] made extensive investigations on the transpeptidation catalysed by intracellular proteases as well as on protease-catalysed condensation involving N-acylamino acids and amino acid esters. However, they failed to obtain the expected high molecular weight plasteins comparable to the proteins that usually occur in living systems. In the 1960s, when the modern concept of peptide chain elongation in vivo was established, plastein research was directed more toward subjects of enzymological interest.[4,5,16] Interest in plastein intensified in the 1970s when a Tokyo group reinvestigated the plastein reaction in much more detail with the object of applying it to the improvement of food proteins. Several independent groups in many countries then followed with their own studies of the plastein reaction.

This article will be allotted to a discussion of the plastein reaction with reference to much of the recently published data presented by the Tokyo group.

PLASTEIN SYNTHESIS

Required Conditions

The plastein reaction requires at least three conditions different from those required for maximum rates of enzymatic protein hydrolysis. First, the substrate must be a low molecular weight peptide mixture, preferably prepared by enzymatic protein hydrolysis. Virtanen[17] found that for the most efficient plastein formation from a zein hydrolysate, the mixture should contain mostly tetra-, penta-, and hexapeptides. Determann et al.,[16] after investigating plastein formation from synthetic oligopeptides with different chain lengths, concluded that a peptide length of five or six amino acid units is preferable. Tsai et al.,[18] using fractions of a soy protein hydrolysate separated on the basis of molecular size, found that the second and the third fractions, with average molecular weights of 1043 and 635, respectively, produced plasteins (10% TCA-insoluble substances) much more effectively than did the lower or higher molecular weight fractions. Recently, Hofsten & Lalasidis[19] reported that a whey protein hydrolysate fraction with an average molecular weight of less than 1000 can grow to form a plastein product with a molecular weight of 2000–3000. Sciancalepore & Longone also reported that the most favorable size was a molecular weight of 451–1450.[20] Therefore the molecular size of the substrate is one of the most important factors influencing plastein formation.

Secondly, the substrate concentration of the reaction medium is an important factor in peptide bond synthesis and should be around 20–40% (w/v).[21] When a substrate is incubated at a concentration of about 7·5%, no reaction appears to occur, as measured by TCA-solubility/insolubility. A lower substrate concentration is more favorable for the hydrolysis reaction.

Of special interest is pH, the third factor influencing plastein formation. The optima for most proteases to form plastein when measured with a soy globulin hydrolysate were found to lie within a narrow range of pH 4–7 (Table 1). Pepsin, for example, is an acid protease, and one of its characterisics is a high hydrolase activity at low pH. Even at pH 1, it is active and able to hydrolyse proteins. However, as Yamashita et al.[10] demonstrated, no appreciable amounts of plastein are formed by pepsin at pH 1–2, even if all other conditions are satisfied. The optimum for pepsin to produce plastein was around pH 4·5 (Table 1). On the other hand, α-chymotrypsin, a serine proteinase, affords its maximum rate of protein hydrolysis near pH 8. In fact, the optimum lies at pH 7·8 for the hydrolysis of soy globulin by this enzyme (Table 1). However, effective plastein formation from soy globulin hydrolysate using α-chymotrypsin occurs at pH 5·3. In the case of papain, a cysteine proteinase, pH 4–7 is optimal both for protein hydrolysis and for plastein formation. Kim & Lee[22,23] and Winkler et al.[24] confirmed that the pH optima lie in a slightly acidic range.

TABLE 1

OPTIMUM pH VALUES FOR (a) HYDROLYSIS OF SOY PROTEIN WITH PROTEINASES, AND FOR (b) PLASTEIN SYNTHESIS FROM A PEPTIC HYDROLYSATE OF SOY PROTEIN WITH THE SAME PROTEINASES[10]

Enzyme	Source	Optimum pH value	
		(a) Hydrolysis	(b) Synthesis
Aspartic proteinase			
Pepsin	Swine	1·6	4·5
Molsin	Aspergillus saitoi	2·8	5·5
Rapidase	Trametes sanguinea	3·0	4·0
Serine proteinase			
α-chymotrypsin	Bovine	7·8	5·0–6·0
Trypsin	Bovine	8·0	5·5–6·5
Subtilisin BPN	Bacillus subtilis	7·5–8·5	6·0–7·0
Pronase	Streptomyces griseus	8·0–9·0	6·0
Bioprase	Bacillus subtilis	10·0–11·0	6·0
Coronase	Rhizopus	7·0	4·0–5·0
Prozyme	Streptomyces	8·0–9·0	5·5–6·5
Esperase	Bacillus subtilis	9·0–10·0	6·5
Cysteine proteinase			
Papain	Papaya carica	5·0–6·0	5·0–6·0
Bromelain	Ananas sativus	5·0–6·0	5·0–6·0
Ficin	Ficus carica	5·0–6·0	5·0–6·0

In summary, a proper consideration of three conditions, substrate size, substrate concentration, and pH, is required for the effective synthesis of plasteins.

Mode of Peptide Bond Synthesis

A new peptide bond can be formed by either transpeptidation or condensation. It is likely that in plastein reactions both reactions occur to variable extents depending on the experimental conditions and the nature of the peptides available. Below, we discuss some of the features of each of these reactions. Several of the proteases used in plastein formation are known to form transitory acyl-enzyme intermediates during hydrolysis of substrates including proteins.[25,26] The acyl-enzyme intermediate is then broken down through transfer of the acyl group to a nucleophile, generally water in the solution. However, the acyl group can be transferred to other nucleophiles[26] such as the amino groups of amino acids and peptides; this results in the synthesis of peptide bonds. Whether or not there is an increase in molecular size during this synthesis is determined by the relative size of the peptide as it leaves as compared with that of the peptide serving as the nucleophile.

Whether the hydrolytic or the synthetic reaction is more important will depend, among other things, on the substrate (nucleophile) concentration and the pH, as discussed earlier. The higher the nucleophile concentration, the more readily it can compete with the water in the reaction. The addition of organic solvents, such as acetone, will decrease the water concentration as well as speed up the reaction by decreasing the solubility of the synthesized product. The reaction is affected by pH because the acyl-enzyme intermediate generally reaches a higher concentration at lower pH levels (for α-chymotrypsin, see Ref. 26), and the amino group, not protonated, is available as the required active form of the nucleophile.

If plastein formation results in an increase in molecular size, a lower molecular weight product must be formed at the same time; this product can be either an oligopeptide or a free amino acid. In addition to the above workers,[11] Yamashita et al.[9] have observed that free amino acids appear gradually during the plastein reaction, supporting the occurrence of transpeptidation. Although this mechanism clearly explains an increase in molecular weight as a result of the plastein reaction, it does not necessarily lead to a substantial increase in overall size. As measured by gel chromatography, Hofsten & Lalasidis[19] did not usually find an overall increase in size when using a whey protein hydrolysate. However, when a hydrolysate fraction estimated to have a molecular weight of 1000–

20 000 was subjected to the plastein reaction, a product was obtained that had an upper limit of its molecular weight range of approximately 30 000.[19] In this case, a significant amount of a product smaller than the original substrate was formed simultaneously. Tsai *et al.*[18] investigated the molecular weight distribution of a plastein reaction product from a soy protein hydrolysate by means of 7·5% polyacrylamide gel electrophoresis in a phenol–acetic acid–water system. They found that the product contained both high molecular weight and low molecular weight substances; the former, though very small in quantity, appeared to have a molecular weight of approximately 30 000. In a wide range of experimental conditions, Gololobov *et al.*[27] found that transpeptidation was the only chemical reaction causing plastein formation and that the reaction was the result of kinetic rather than thermodynamic factors. Further work is needed, preferably with improved solvent systems, because most plasteins are partially insoluble in the conventional aqueous solvents used for electrophoresis.

In addition to transpeptidation, a certain amount of peptide–peptide condensation, is possibly involved in the plastein reaction. With some of the proteases used for plastein formation, especially α-chymotrypsin,[26,28-30] the acyl-enzyme intermediate can be formed at around pH 5 from the reversal of the hydrolytic reaction. Once the acyl-enzyme intermediate is formed, the acyl group can be transferred to a nucleophile, resulting in peptide bond synthesis.

Virtanen's group,[17,31,32] on the basis of results from terminal amino group analyses, concluded that a condensation reaction occurs during pepsin-catalysed plastein formation from a zein hydrolysate. Wieland *et al.*[4] have isolated two plastein-forming oligopeptides from Witte peptone and demonstrated that these condense when incubated with pepsin. Subsequently, Determann[33] in a model experiment using the synthetic pentapeptide H·Tyr–Leu–Gly–Glu–Phe·OH, obtained a plastein comprising the di-, tri-, tetra-, and pentamers of the pentapeptide as described earlier. Yamashita *et al.*,[34] using a soy globulin hydrolysate in which all the carboxyl groups were labelled with ^{18}O as a substrate, observed that $H_2^{18}O$ was liberated rapidly by α-chymotrypsin at pH 5·3 within the first hour of plastein formation. At the same time there was a corresponding decrease in ninhydrin-reactive groups. N-Acetylation of the soy globulin hydrolysate or reduction of the —COOH groups to —CH$_2$OH significantly prevented plastein formation.[34] Tanimoto *et al.*[35] attempted the production of a plastein with chemically modified α-chymotrypsin. When His-57 of this enzyme was methylated, the plastein-producing ability decreased as the degree of methylation increased. Similarly, the di-isopropylphosphorylation of Ser-195 led to a loss of plastein formation. Although, from the point of view of energy conservation, condensation should be a less favorable contributor to plastein formation than transpeptidation, in fact it can occur as indicated by the data above. Recently, a Japanese group[36] attempted, with fair success, to synthesize biologically active peptides, e.g. angiotensin, secretin, eledoisin, etc., by fragment condensation with various proteases. For example, Val[5]-angiotensin II protected by tert-butyloxycarbonyl (Boc) and ethyl (OEt) groups (Boc–Asp–Arg–Val–Tyr–Val–His–Pro–Phe–OEt) was efficiently synthesized from its three peptide fragments by papain-catalysed condensation between Arg-2 and Val-3 and between Tyr-4 and Val-5. Interestingly, the reaction conditions adopted in this case[36] were almost identical with those required for plastein formation.

Recently another theory has been proposed. Andrews & Alichanidis[37] found that a plastein reaction product was resolubilized and concluded that the plastein synthesis reaction is a purely entropy-driven physical aggregation process. A similar conclusion was reached by Pallavicini & Trentin[38] who pointed out the occurrence of disulfide bond formation during the plastein reaction.

Further investigation is needed of methods for the selective removal of the condensation products from the entire reaction mixture in order to drive the condensation to completion. This might be achieved through changes in solvent composition to minimize the solubility of the product as well as through specific binding of the product by some adsorbents.

Properties of Plasteins

Plasteins are generally characterized by their low solubility in water. If, during the synthesis reaction, a part of the product becomes insoluble, this serves as the driving force mentioned above. Aso et al.[39,40] investigated some physicochemical properties of the water-insoluble fraction of a plastein produced from a soy globulin hydrolysate. This fraction interacted with 1-anilino-8-naphthalene sulfonate (ANS) to give a new and larger ANS emission spectrum at 450 nm (Table 2). The fraction also reacted with sodium dodecylsulfate (SDS) to modify the NMR signal produced by the SDS methylene protons, with a peak broadening to produce an upfield shift of 0·06 ppm. The water-insoluble plastein fraction showed a much higher affinity for a hydrocarbon (n-heptane) than the soy globulin hydrolysate itself. The fluorescence of the plastein fraction originating

TABLE 2

HYDROPHOBIC PROPERTIES OF A PLASTEIN COMPARED WITH ITS SUBSTRATE[40]

Item	Plastein	Substrate[a]
Fluorescence (nm) caused by ANS[b]	450	—
Fluorescence caused by tryptophan	348	356
NMR spectrum (ppm) of the methylene protons of SDS[c]	1·27	1·33
Amount of bound n-heptane (mg/g substrate or plastein)	51·7	12·7
Heat assembly	Clear	No

[a]Peptic hydrolysate of soy globulin.
[b]1-Anilino-8-naphthalene sulfonate.
[c]Sodium dodecylsulfate.

from tryptophan was blue-shifted (by 8 nm) compared with that of the hydrolysate. This probably indicates that the tryptophan residues in the plastein molecules are exposed to a more hydrophobic environment. A temperature effect was distinctly observed for the plastein. When the plastein was suspended in cold water and heated gradually, a larger amount of visible particles appeared at higher temperatures. Plastein formation is sometimes accompanied by gelling. Hofsten & Lalasidis[19] proposed that the gel formation observed in their system was related to some non-covalent rearrangement of peptides, although they did not rule out a rearrangement of peptide sequences in the system. Eriksen & Fagerson[41] speculated that gel formation is an entropy-driven process, the increase in the entropy of water acting as the driving force after an initial concentration of suitable peptides has been formed by either condensation or transpeptidation or both. As mentioned above, Andrews & Alichamidis[37] emphasized the entropy-driven aggregation process. Hartnett & Lincoln[42] found that gel formation from a peptic hydrolysate of soy protein isolate occurred solely by heating, although the presence of an enzyme was necessary for gel formation at 37°C. The gel properties varied in the presence of non-peptide substances.[43,44] At present it is unclear whether this phenomenon is related to the hydrophobic interaction among peptides formed by the plastein reaction. Plastein yield, as determined on the basis of TCA-insolubility, is actually enhanced when both a hydrophilic substrate (i.e., casein hydrolysate) and a hydrophobic substrate (i.e., zein hydrolysate) are used in combination with each other.[45,46] Examples are given in Table 3 which show that plasteins are

TABLE 3

AMINO ACID COMPOSITIONS OF PLASTEINS AND THOSE OF PROTEINS FROM WHICH THE PLASTEINS ORIGINATED[56]

Amino acid	Soybean			S. maxima		R. capsulatus		T. repens L	
	Protein	Plastein		Protein	Plastein	Protein	Plastein	Protein	Plastein
		M[c]	G[b]		LMT[c]		LMT[c]		LMT[c]
	Weight percent			Weight percent		Weight percent		Weight percent	
Lysine	5·28	4·73	3·05	4·59	7·75	5·37	7·39	6·06	8·23
Histidine	2·04	2·20	1·22	1·77	1·91	2·35	2·44	1·94	2·01
Arginine	5·94	5·61	3·82	6·50	6·70	6·27	6·05	3·66	3·84
Aspartic acid[d]	8·70	7·76	11·76	8·60	11·87	8·57	10·21	11·10	10·45
Threonine	2·63	2·11	3·08	4·56	5·42	5·07	4·36	5·60	5·93
Serine	3·53	2·75	4·46	4·20	4·43	3·16	4·63	4·17	4·00
Glutamic acid[e]	15·00	10·20	41·93	12·60	14·68	10·03	9·77	16·00	15·54
Proline	4·32	2·18	2·88	3·90	3·62	4·26	3·45	3·75	3·57
Glycine	4·38	2·55	3·64	4·75	4·76	4·53	5·98	5·01	4·79
Alanine	3·98	2·65	2·51	6·80	5·80	8·74	8·32	6·23	6·05
Valine	3·36	4·29	4·34	4·69	6·00	6·59	6·56	7·45	7·00
Isoleucine	3·00	5·72	2·37	6·03	6·32	4·96	5·30	4·98	4·55
Leucine	5·17	7·26	3·65	8·02	8·98	8·45	8·47	8·86	8·24
Aromatic amino acids	7·03	9·46	3·38	8·92	8·96	8·09	8·16	9·61	9·00
Tyrosine	2·83	3·52	0·88	3·95	3·98	3·21	3·53	4·11	3·88
Phenylalanine	4·20	5·94	2·50	4·97	4·98	4·88	4·63	5·50	5·12
Tryptophan	1·34	1·30	0·70	1·40	2·72	2·05	2·56	1·51	2·73
S-containing amino acids	2·94	9·96	2·80	1·77	8·75	3·73	9·06	1·82	8·13
Methionine	1·18	7·98	1·20	1·37	8·22	2·97	8·29	0·85	7·14
Half-cystine	1·76	1·98	1·60	0·40	0·53	0·76	0·77	0·97	0·99

[a] Methionine-incorporated plastein.
[b] Glutamic acid-incorporated plastein.
[c] Lysine-, methionine-, and tryptophan-incorporated plastein.
[d] Aspartic acid plus asparagine.
[e] Glutamic acid plus glutamine.

richer in hydrophobic amino acid residues than the proteins from which the plasteins are derived.

Plasteins are also characterized by their insolubility in aqueous organic solvents.[4,33,47] It is thus possible to separate plasteins from lower molecular weight products by precipitation with, for example, ethanol or acetone.[47] Either dialysis or membrane filtration is useful for separation.

INCORPORATION OF AMINO ACIDS

Under appropriate reaction conditions, any L-amino acid ester added to a reaction medium containing a protein hydrolysate is usually incorporated in a covalent fashion during the plastein reaction. In this way it is possible to prepare a plastein whose amino acid composition has been altered. Therefore, the plastein reaction, accompanied by new amino acid incorporation, would be expected to be much more valuable than plastein formation carried out in the absence of new amino acids.

In order to obtain basic information concerning amino acid incorporation, Aso et al.[48] investigated a model system in which, instead of a protein hydrolysate, ethyl hippurate, i.e. N-benzoylglycine ethyl ester (Bz–Gly–OEt), was used as the substrate. Ethyl hippurate (50 mM) was first incubated at 37°C with papain (1.7×10^{-2} mM) at pH 6.0. After 15 min, an amino acid ester (50 mM) was added and the incubation was continued under the same conditions. Analysis of the reaction and its products indicated that the following reaction had occurred:

$$Bz–Gly–OEt + AA–OR \rightarrow Bz–Gly–AA–OR + Ethanol$$

where AA–OR refers to the amino acid ester. The amino acid ester reacted as a nucleophile in the aminolysis of the acyl-enzyme intermediate (Bz–Gly–Papain) to result in the formation of a new peptide (Bz–Gly–AA–OR). Various L-amino acid ethyl esters reacted at different velocities depending on the amino acid's side-chain structure; a more hydrophobic side chain was more reactive except in the case of a β-branched chain, as in isoleucine. This observation is similar to that of Alecio et al.[49] Aso et al.[48] also found that even the β-branched amino acid isoleucine was an effective nucleophilic if previously esterified with a hydrophobic alcohol such as n-hexanol. Similarly, n-hexyl esters of glycine and alanine were much more reactive than their ethyl esters.

The data above indicate that more is involved in the plastein reaction than just the amino acid ester serving as a nucleophile in competition with water. Specificity of the enzyme toward the amino acid ester also exists. Perhaps a better understanding of the reaction can be obtained by considering some of the properties of papain. Considerable data are available on the structure of papain, especially on its active site. His-159 is thought to be involved in catalysis as a general base by which Cys-25 becomes acylated in an intermediate step.[50] Schechter & Berger[51,52] postulated from data on the specificity of papain for synthetic substrates that substrate binding occurs by interaction with one or more of the seven sub-sites designated S_1, S_2, S_3, and S_4 toward the N-terminal from the active center of a substrate peptide and S_1', S_2', and S_3' toward the C-terminal from the scissile point. A substrate with amino acid side chains, P_1, P_2, P_3, P_4, P_1', P_2' and P_3', is postulated to fit the corresponding sub-sites S_1, S_2, S_3, S_4, S_1', S_2' and S_3', respectively, and subsequently to undergo hydrolysis. A possible explanation of the data would require that the side chain of the amino acid ester fit into the S_1' sub-site (and possibly the alcohol chain into S_2'), since the amino acid ester specifically serves as a nucleophile to attack the carbonyl carbon of the thioester part of the acyl-papain. Without this specificity of combination at sub-sites S_1' and S_2' there should be no effect of the alcohol constituent, and D-amino acid esters would also function in the reaction, which, however, is not the case.

APPLICATIONS

Improvement of Sensory Properties

In biological systems many proteins interact specifically with various chemical compounds.[53] There is also a less-specific type of interaction by which even denatured proteins can bind miscellaneous non-protein compounds. Such a phenomenon is often observed in the area of food processing. Most food proteins contain non-protein impurities such as odorants, taste substances, coloring materials, lipids and related compounds, etc. These impurities sometimes affect the acceptability of food proteins and their safety. Some of these impurities form such strong complexes with protein molecules that they cannot be completely removed by chemical and physical treatments under conventional conditions. Partial protein hydrolysis is expected to be effective in loosening the interaction and consequently in liberating the protein-bound impurities. Proteases are particularly useful for this purpose, since they usually permit treatments of food proteins under mild conditions.[54,55]

In most cases, the resulting protein hydrolysates, though free from non-protein impurities, are found to have developed another problem, i.e. a bitter taste. As discussed below, the synthesis reaction leading to plastein formation has been found to be very effective in debittering protein hydrolysates.[54,56] As noted later, it is possible to incorporate essential amino acids during the plastein reaction to give a plastein whose amino acid composition has been altered. A process for the improvement of food protein is suggested in which there occurs a combination of enzymatic protein hydrolysis and plastein synthesis (Fig. 1). The details for the

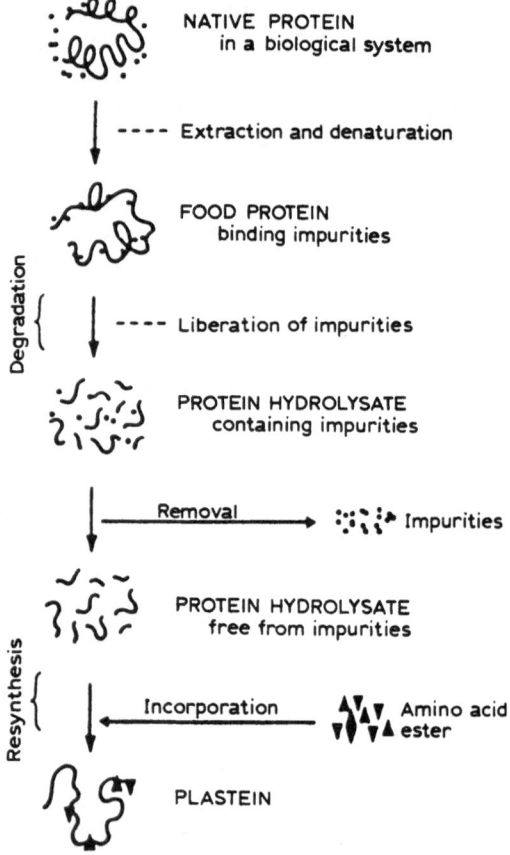

FIG. 1. Schematic representation of a combined process of enzymatic protein hydrolysis and resynthesis for producing a plastein with an improved acceptability and an improved amino acid composition.[56]

removal of impurities are illustrated by means of examples involving the treatment of soy protein and related materials.

Deodorization

The volatile flavor components of soybeans have been investigated in detail.[57,58] Arai *et al.*[59] have studied the interaction of denatured soy protein with *n*-hexanol and *n*-hexanal which are the typical beany flavor compounds of raw and processed soybeans. These protein-bound compounds are liberated by treating the denatured soy protein with pepsin.[59] Noguchi *et al.*[60] observed that not only *n*-hexanol and *n*-hexanal, but also other flavor compounds are effectively liberated and removed from a soy protein isolate during treatment with an acid protease (Molsin). A subsequent study has ascribed this effect to the activity of aspergillopeptidase A, an endopeptidase that has been identified as the main constituent of Molsin.[61] Fujimaki *et al.*[6,55] examined several protease preparations for their usefulness in deodorization and reported that pepsin treatment followed by ether extraction is the most effective for deodorizing some protein preparations of soybean and fish.

Defatting

When tofu, a protein curd made from whole soybeans, is treated with Molsin, volatile as well as non-volatile fatty substances including triglycerides, fatty acids, phosphatidyl choline, phosphatidyl ethanolamine, phosphatidyl inositol, sitosteryl-D-glucoside, genistein, saponins, etc. are liberated.[60]

Generally, compared with storage proteins such as soy globulin, single-cell proteins (SCP) are rather difficult to refine. Fujimaki *et al.*[47] found that a SCP preparation from *n*-paraffin-assimilating yeast (*Candida*) contained significant amounts of lipids and related substances which were not extractable with ether. However, when this SCP preparation was treated with pepsin at 37°C for 24 h, an ether-extractable fraction was obtained in a yield of 0·27%. Analysis showed that this fraction contained various fatty acids including some with odd numbered carbon chains (15:0, 17:0, and 17:1), a series of *n*-paraffin homologues (C_{11}–C_{24}), several polycyclic hydrocarbons (such as anthracene, phenanthrene, and pyrene), and a ubiquinone.

Decolorisation

Much attention is now being given to proteins of photosynthetic origin,[62] especially those from blue-green algae,[63] non-sulfur purple bacteria,[64] and green leaves,[65] as possible food protein sources. These proteins contain large amounts of photosynthetic pigments. Arai et al.[66] investigated the removal of these pigments from Spirulina maxima, a blue-green alga, and from Rhodopseudomonas capsulatus, a non-sulfur purple bacterium. Each was treated with sufficient ethanol to obtain a residue largely free from color. The residue was ground in a mechanical mill and then treated with aqueous alkali to extract the protein. The extract was dialysed against running water to obtain a non-dialysate fraction containing the protein. This was placed on a Sephadex G-15 column and eluted with 10% ethanol. The protein fraction eluted from the column still contained pigments. When the non-dialysate was treated with pepsin and subsequently subjected to similar Sephadex treatment, a colored zone was distinctly separated from the main peptide zone. Completely decolorized protein hydrolysates were thus obtained from both the alga and the bacterium.

Bitter Peptides and Debittering

A Tokyo group[67-70] investigated the bitterness produced during the peptic hydrolysis of soy protein and identified the following bitter peptides: Gly–Leu, Leu–Phe, Phe–Leu, Leu–Lys, Arg–Leu, Arg–Leu–Leu, Tyr–Phe–Leu, Ser–Lys–Gly–Leu, pyrrolidone carboxyl–Gly–Ser–Ala–Ile–Phe–Val–Leu, and a tetracosapeptide bearing –Gln–Tyr–Phe–Leu as the C-terminal structure. Subsequently, a Kyoto group[71,72] identified three bitter peptides in a tryptic hydrolysate of casein. Bitter peptides have also been found in casein treated with a microbial alkaline protease.[73] A number of bitter peptides have been chemically synthesized. A sample of tri- and tetrapeptides produced by chemical synthesis[74] has the following order of bitterness: Phe–Gly–Gly = Gly–Phe–Gly < Gly–Gly–Phe; and Leu–Gly–Gly–Gly = Gly–Leu–Gly–Gly = Gly–Gly–Leu–Gly < Gly–Gly–Gly–Leu. It seems probable that bitterness in general tends to increase when a peptide is enriched with hydrophobic amino acid residues, especially when one of them is located at the C-terminal.

Arai et al.[61] reduced the bitterness of a soy protein hydrolysate by treatment with either carboxypeptidase A or Aspergillus acid carboxypeptidase.[75] This method of debittering is based on the further hydrolysis of bitter peptides by exopeptidases. Debittering can also be achieved by the plastein reaction with endopeptidases. When a peptic hydrolysate of soy protein is concentrated and incubated with α-chymotrypsin at a slightly acidic pH (Table 1), the bitterness gradually decreases as plastein formation proceeds.[8,76] Yamashita et al.[9] reported that during the plastein reaction, the levels of two typical bitter peptides, Gly–Leu and Leu–Phe, decreased in the incubation mixture without being broken down to free amino acids. A sensory test confirmed that the plastein obtained by precipitation from aqueous ethanol is almost completely tasteless.[6]

Improvement of Solubility

Denatured soy proteins are relatively insoluble in aqueous media, which is a disadvantage in their processing. This problem might be solved by incorporating large amounts of certain hydrophilic amino acids into the soy protein through enzymatic hydrolysis followed by plastein synthesis. Yamashita *et al.*[77,78] incorporated glutamic acid into a plastein and obtained a product with greater solubility. They used the α,γ-diethyl ester of glutamic acid which was expected to be more reactive than its α-monoethyl ester. A 2:1 (w/w) mixture of the peptic hydrolysate of a soy protein isolate and the α,γ-diethyl ester of L-glutamic acid was used as the substrate. The plastein reaction was carried out with papain under the following conditions: reaction medium, 20% acetone (pH 5·5) containing 10 mM L-cysteine; substrate concentration, 52·5% (w/v); enzyme-substrate ratio, 2:100 (w/w); temperature, 37°C; and incubation time, 24 h. After incubation the entire reaction mixture was treated with aqueous alkali to hydrolyse the ethyl ester moieties and then dialysed in a cellophane membrane against running water. A plastein was obtained as a non-dialysate in a yield of 9·57 g on a dry-weight basis from 10 g of starting material (soy protein isolate); the yield of the plastein was 6·38 g based on the combined weight of 10 g for a mixture of the hydrolysate and L-glutamic acid α,γ-diethyl ester. The amount of glutamic acid in the plastein was over 40% compared with 15% from the starting protein (Table 3). At room temperatures, this plastein was almost completely soluble in water in the range of pH 1–9. No appreciable amount of any insoluble material appeared when a neutral aqueous solution of the plastein was heated at 100°C for 1 h, whereas similar treatment of a native soy globulin preparation resulted in a decrease in its solubility by about 70%. This glutamic acid-incorporated plastein had an average molecular weight of about 6200 and a broad isoelectric point in the range of pH 1·5–4·0. Circular dichroism indicated that some α-helix structures were formed as a result of the plastein reaction. Although it is unclear whether α-helix formation resulted from the incorporation of large amounts of L-glutamic acid, Noguchi *et al.*[79] have identified the oligomeric L-glutamic acids, Glu–Glu, Glu–Glu–Glu, Glu–Glu–Glu–Glu, and Glu–Glu–Glu–Glu–Glu, in a similar plastein hydrolysed exclusively with Pronase.

A Korean group[80–83] also incorporated glutamic acid diethyl ester into fish protein hydrolysates to increase their water dispersibilities and water-holding capacities.

Improvement of Amino Acid Composition

An important use of the plastein reaction would be to incorporate limiting essential amino acids, thereby improving the nutritional quality of the protein. Yamashita et al.[7] first conducted a study to evaluate the in-vitro and in-vivo digestibilities of a plastein prepared from soy protein. The plastein was readily digested, and its nutritive value, tested in rats, was comparable to that of soy protein itself. Subsequently, in order to increase the sulfur-containing amino acid level of soy protein,[84,85] a soy protein isolate was hydrolysed with pepsin to about 80% splitting of the susceptible bonds. The resulting hydrolysate was neutralized and freeze-dried. L-Methionine ethyl ester (10 g) was added to the dried hydrolysate and the mixture dissolved in a solution of 10 mM L-cysteine containing papain (100 mg). After incubation at 37°C for 24 h, the entire reaction mixture was dialysed against running water to obtain a plastein as a non-dialysate in a yield of about 70% on a dry-matter basis. The methionine content of the plastein was 7·22%; no free methionine was present. Prior to a feeding test in rats, the plastein was diluted with soy protein to a methionine level of 2·74%. This sample gave a protein efficiency ratio (PER) value of 3·38±0·08, compared with a PER value of 2·40±0·05 for casein as a control. According to taste panel evaluation, the methionine-incorporated plastein was almost tasteless and free from any sulfide flavor.[85] Arai et al.[86] have produced a similar plastein from soy protein on an enlarged scale and purified it by precipitation with 70% ethanol. Table 3 shows the amino acid composition of this purified plastein in comparison with that of the soy protein used as the starting material. It has been reported that methionine methyl and ethyl esters can be incorporated into protein hydrolysates by α-chymotrypsin.[87–89]

Although the L-lysine ethyl ester is incorporated less effectively than the L-methionine ethyl ester into protein hydrolysates, it is nevertheless possible to produce a gluten-originating plastein enriched with large amounts of lysine. Fujimaki[90] used commercially available gluten and hydrolysed it in 0·01 N NaOH. A fungal alkaline protease was useful in this case to give 85% maximum gluten hydrolysis. Various amounts (1·0, 2·0, 3·0, 4·0, or 5·0 g) of L-lysine ethyl ester were then added to the gluten hydrolysate (10 g), and the mixtures were incubated with papain under the following conditions: reaction medium, 20% acetone (pH 6·0) containing 10 mM L-cysteine; substrate concentration, 35% (w/v); enzyme-substrate ratio, 1:100 (w/w); temperature, 37°C; and incubation time, 48 h. After incubation, the reaction mixtures were subjected to ultrafiltration to obtain a plastein having a molecular weight greater than 500. The lysine content of the plastein depended on the amount of L-lysine ethyl ester used; there was a sigmoidal relationship between the amount of lysine ethyl ester used and the amount of lysine incorporated. The curve reached a plateau when 5·0 g of the ester was mixed with 10 g of the gluten hydrolysate. The observed covalent lysine content of the plastein was approximately 16% in this case.

Aso et al.[91] incorporated the three essential amino acids, lysine, threonine, and tryptophan, individually into zein, a protein deficient in these amino acids. Commercially available zein was hydrolysed with pepsin to solubilize 90% of the protein. A 10:1 (w/w) mixture of the soluble hydrolysate and the ethyl ester of L-lysine, L-threonine, or L-tryptophan was incubated with papain in 10 mM L-cysteine under the following conditions: substrate concentration, 50% (w/v); enzyme-substrate ratio, 3:100 (w/w); pH 6·0; temperature, 37°C; and incubation time, 48 h. The lysine-, threonine-, and tryptophan-incorporated plasteins were thus obtained as water-insoluble fractions in yields of 30·0, 26·0, and 39·9%, respectively. The lysine, threonine, and tryptophan contents of the respective plasteins were 2·14, 9·23, and 9·71%, whereas those in the zein used as a starting material were 0·20, 2·40, and 0·38%, respectively.

Peptic hydrolysates of proteins extracted from *Spirulina maxima* and *Rhodopseudomonas capsulatus*, after purification on a Sephadex G-15 column to remove pigments,[66] were used as substrates for the plastein reaction. Also, a peptic hydrolysate prepared from a leaf protein extracted from *Trifolium repens* L, a type of white clover, was used as a substrate. Consequently, plasteins in which controlled amounts of lysine, methionine, and tryptophan were simultaneously incorporated, were produced.[66] Table 4 gives the amounts of amino acids used and the yields of the plasteins recovered. In each case the plastein reaction was performed with papain under conditions similar to those used for the production of a plastein from a soy protein hydrolysate.[84] After incubation, the reaction mixtures were ultrafiltered to obtain plastein fractions having molecular weights greater than 500. The yields of the different plasteins are given in Table 4; their amino acid compositions have already been shown in Table 3. The extent of incorporation (on a molar basis) were 22·3–22·5% for L-lysine ethyl ester, 73·1–89·9% for L-methionine ethyl ester, and 86·0–96·9% for tryptophan ethyl ester.

TABLE 4

MATERIALS BALANCE IN PLASTEIN PRODUCTION FROM PHOTOSYNTHETIC PROTEIN RESOURCES[66]

	S. maxima (g)	R. capsulatus (g)	T. repens L (g)
Starting material, dried	10	10	10
Decolorized protein hydrolysate	4·74	3·90	2·00
L-Lysine ethyl ester · 2HCl	1·18	0·65	0·33
L-Methionine ethyl ester · HCl	0·54	0·35	0·25
L-Tryptophan ethyl ester · HCl	0·10	0·04	0·04
Plastein	4·96	4·08	2·03

MODIFIED (ONE-STEP) PLASTEIN REACTION AND ITS APPLICATIONS

As outlined above, the plastein reaction is available as a process for covalently incorporating the ester forms of amino acids. This process, however, may be too expensive for industrial applications, because plastein formation requires two quite different unit processes. The first step is protein hydrolysis with purification of the resulting hydrolysate and the second is plastein synthesis with the covalent incorporation of an expected amino acid. To address this economic problem, Yamashita et al.[92,93] developed a one-step process that omits the first step. This process for a modified plastein reaction has the merit of using a protein directly as a substrate for improvement of its amino acid composition. This section starts with a discussion of basic research on the characteristics of this reaction, followed by a review of its applications to the improvement of the nutritional and functional properties of proteins.

Characteristics of the Reaction

The one-step plastein reaction is a process designed specifically for the direct incorporation of amino acid esters into proteins. It requires a high protein concentration, the use of an alkaline medium of pH 9–10, and the employment of a cysteine proteinase, preferably papain.[92] The mode of the amino acid incorporation is primarily an aminolysis reaction involving a peptidyl enzyme intermediate and an amino acid ester. The aminolysis, being competitive with hydrolysis in this process, proceeds efficiently at high substrate protein concentrations, a condition that results in a fall in the water concentration. Thus, the use of an organic solvent is also useful.[94] Actually, 20% acetone or ethanol in the solvent system does not jeopardize the enzymatic efficiency of amino acid ester incorporation.

Since it is known that aminolysis of the peptidyl enzyme is effected by the non-ionic form of the amino group of any amino acid ester, higher pH values ought to be more effective. However, Yamashita *et al.*[92] found that the most desirable pH level is around 9–10, where no inactivation of the enzyme occurs. Free amino acids cannot be incorporated into the product because of the higher pK values of their amino groups.[92] A change in the amount of amino acid ester added should result in a change in the amino acid composition of the product.[95] The use of cysteine proteinases is justified for two reasons: first, they generally have larger k_{cat}/K_m values than serine proteinases in aminolysis reactions,[96] and second, in the case of papain, subsites S_1' and S_2' favor interaction with amino acid esters.[48] The one-step process is essentially a degradation reaction in which a protein substrate is changed into a relatively low molecular weight species depending on the reaction time, amount of enzyme, temperature, and other factors. However, the process is comparable to the plastein reaction in that both permit the covalent incorporation of the ester form amino acids. In the one-step process, as in the standard plastein reaction, certain amino acid esters are incorporated better than others. The amount of amino acid incorporated into a protein also varies depending on the amount of amino acid ester added to the reaction system. It should be noted that the ratio of the initial concentration of the amino acid ester to the initial protein concentration must be below a certain value. There is a possibility that at ratios above this value the amino acid ester molecules will polymerize with themselves. Arai *et al.*[97] found that in the papain-catalysed incorporation of methionine into soy protein, self-polymerization of the methionine ethyl esters occurred, depending on the initial concentrations of methionine ethyl ester [M] and soy protein [S]. They also defined critical [M]/[S] values, below which self-polymerization proceeded with difficulty. Polymethionine is only slightly digestible in the gastrointestinal tracts of animals[98] and, therefore, it is suggested that the [M]/[S] ratio be suitably adjusted.

Improvement of Nutritional Qualities

Soy protein was first improved by the incorporation of methionine ethyl ester.[92] Methionine is the first limiting amino acid in soy protein and its PER value can be significantly improved by adding free methionine.[92] However, the current practice of adding free methionine causes two problems, i.e. an undesirable flavor and chemical instability. Instead of free methionine, proteins rich in methionine, such as ovalbumin, could be used to fortify soy protein, but changes in the levels of other essential amino acids would then be inevitable. Aso et al.,[95] using the one-step plastein reaction, produced an enzymatically modified protein (EMP) as a peptide mixture whose levels of amino acids other than methionine resembled those of soy protein itself. EMP_{11} (methionine content: 11% on a weight basis) was obtained from L-methionine ethyl ester hydrochloride (20 g) and soy protein (100 g). EMP_{11} is a peptide mixture with a molecular weight range of 1000–8000. From the viewpoint of food quality, EMP has the following three merits: first, the product has no sulfide flavor in spite of the high methionine content; second, the beany flavor originating from soybeans is removed during the one-step process; third, the product retains certain features of functionality intrinsic to soy protein.[95]

Arai et al.[99] also prepared a series of mixtures comprising EMP_{11} and a soy protein isolate and evaluated the nutritive values of the mixtures by feeding tests in rats. By changing the ratio of protein to EMP_{11} from 100:0 to 70:30, the methionine content of the mixture was increased from 1·0 to 4·0%. The dietary protein content was then adjusted to 10%, resulting in dietary methionine levels in the range of 0·1–0·4%. The PER was found to be maximized at dietary methionine level of 0·25%, a level obtained using an 80:20 mixture of protein to EMP_{11}. The mixtures differed only in terms of methionine level. Such methionine-enriched EMP can be used to fortify not only soy protein but also other proteins with poor nutritional qualities.

Improvement of Functional Properties
The study of protein utility is of vital importance since the world-wide demand for functional food proteins has greatly increased in recent years. Hammonds & Call[100] estimated the maximum market potential for protein ingredients at approximately 3·1 billion pounds annually. Kinsella[101] stated that about 80% of these ingredients would have to possess a high degree of functionality. Much is known about protein functionality in terms of water solubility and dispersibility, heat coagulation (or heat setting) properties, water and/or fat adsorbability, oil-emulsifying activity, foamability and whippability, gelling properties, spinning properties, etc. To increase the use of proteins for food, improvements in and maximization of functionality will become more and more important. Recent advances in research in this area have been reviewed by numerous authors.[102–105]

The functionality of food proteins can be modified by enzymes, especially proteases. Several of the enzymatic modification processes could be used practically even on an industrial scale. With the development of the one-step process, the Tokyo group has contributed to the field of nutritional improvement of food proteins through modification by proteases.[92] This enzymatic process may be applied to the modification of protein functionality as well. Provided that a hydrophilic protein as substrate and a highly hydrophobic or lipophilic amino acid ester as nucleophile are used, it is possible to obtain a product with a structure in which the hydrophilic and lipophilic regions of the molecule are separated from each other. It is to be expected that, as a consequence, a proteinaceous surfactant with an adequately amphiphilic function will be produced. The following section deals with the papain-catalysed attachment of lipophilic amino acid alkyl esters to a hydrophilic protein.

In a system containing succinylated α_{s1}-casein as a hydrophile and leucine n-dodecyl ester as a lipophile, the peptide bond between Phe[145] and Tyr[146] of casein was partially hydrolysed to give as a product the succinyl–Arg[1]—Phe[145]–Leu dodecyl ester.[106] This product has improved emulsifying activity as compared with the succinylated α_{s1}-casein starting material, due to the formation of an amphiphilic structure.

Generally, the function of a surfactant depends on its hydrophilicity–lipophilicity balance (HLB). For efficient emulsification of oil, the use of a surfactant with a low HLB value is necessary; on the other hand, whipping characteristics improve at higher HLB values. Watanabe et al.[107] attempted to produce proteinaceous surfactants with different HLB values by enzymatic attachment of amino acid esters with different lipophilicities. For this purpose, L-leucine n-alkyl esters (Leu-OC$_i$) with the alkyl chain length, i, varying from 2 to 12 were used. As hydrophiles, commercially available food proteins such as gelatin, fish protein concentrate (FPC), soy protein isolate (SPI), casein, and ovalbumin were selected. Proteins other than gelatin were succinylated prior to use in order to enhance their hydrophilicity.

Gelatin has long been in use as a food ingredient. Commercially available preparations of gelatin are commonly used as gelling agents, bodying (or binding) agents, emulsion stabilizers, etc.[108] However, gelatin is characterized by its highly hydrophilic nature, which causes poor performance with regard to its whipping and emulsifying functions. It has been proposed that the addition of an appropriate degree of lipophilicity to the gelatin molecule would give rise to an adequately amphiphilic function. The Tokyo group has put this idea into practice. A commercial preparation of gelatin was subjected to an enzymatic reaction where conditions follow primarily those for the one-step process with papain.[107,109] The reaction was terminated by acidification to pH 1 with 1 N HCl, and the acidified mixture was dialysed in running water. Lyophilization of the non-diffusate gave a proteinaceous product in powder form, which was further purified by treatment with acetone to remove any low molecular weight species that might still remain. Each of the purified products was investigated for leucine and alkanol content and mobility in SDS-polyacrylamide gel electrophoresis, and was also tested for three kinds of functionality: whippability, foam stability, and emulsifying activity. To measure whippability, a 1% dispersion (20 ml) of each sample in 0·01 M phosphate buffer (pH 7) was homogenized at 10 000 rev/min for 3 min. The homogenate was transferred immediately into a glass cylinder, and, after 0·5 min, the total and drainage volumes were measured. The whippability was represented by (total volume−drainage volume)/initial volume (20 ml). Subsequently, the whipped sample was allowed to stand at 25°C for another 30 min and the drainage volume was again measured. Using this volume the foam stability could be represented as the ratio of (1−drainage volume/initial volume). To measure the emulsifying activity, 200-mg samples were dispersed in 0·01 M phosphate buffer, pH 7 (20 ml), and homogenized with corn oil (20 ml) at 10 000 rev/min for 3 min. The resulting emulsion was centrifuged at 500 rev/min for 5 min to separate the cream layer. The emulsifying activity was represented by the ratio of cream layer volume to total volume.

Table 5 collates the data obtained, showing first that all of the products have molecular weights in the range of 2000–40 000, with an average of about 7500. Table 5 also shows that for the products other than the gelatin hydrolysate, the observed leucine level increments and alkanol contents are well controlled within a narrow range of 1·4–1·5 and 1·1–1·2 mol/7500 g, respectively. Regardless of these similarities between the products, their functionalities varied greatly. The highest whippability and foam stability resulted from the attachment of L-leucine C_8 and C_6 alkyl esters, respectively, whereas the emulsifying activity tended to increase gradually with the chain length of the alkyl moiety.

Most food protein systems are primarily hydrophobic in nature. Though proteins are often soluble in isolated form, in most cases solubility is due to a native structure in which hydrophilic groups are located on the outside of the molecule. Common treatments that denature proteins often resulted in their buried hydrophobic regions being exposed to the surrounding water. Consequently, denatured proteins tend to aggregate

TABLE 5

CHEMICAL AND FUNCTIONAL PROPERTIES OF PRODUCTS PREPARED FROM GELATIN[a] BY PAPAIN-CATALYSED ATTACHMENT OF L-LEUCINE n-ALKYL ESTERS[107]

Alkyl chain length	Average molecular weight[b]	Leucine increment[c] (mol/7500 g)	Alkanol content[d] (mol/7500 g)	Whippability[e]	Foam stability[f]	Emulsifying activity[g]
0[h]	12 000	0·0	0·0	1·8	0·05	0·00
2	8 300	1·4	1·1	3·4	0·05	0·55
4	8 200	1·5	1·2	4·3	0·40	0·60
5	—	—	—	4·4	0·40	0·62
6	7 500	1·5	1·1	4·5	0·45	0·66
7	—	—	—	4·6	0·35	0·68
8	7 500	1·5	1·2	4·7	0·30	0·69
9	—	—	—	4·7	0·15	0·71
10	7 500	1·5	1·2	2·5	0·10	0·72
11	—	—	—	1·9	0·00	0·73
12	7 500	1·4	1·1	1.3	0·00	0·74

[a] The gelatin sample before the papain treatment gave the following data: whippability, 1·70; foam stability, 0·10; and emulsifying activity, 0·00.

[b] Stained zones were scanned densitometrically. The average molecular weights were roughly estimated from the peak positions. The molecular weight distribution of each product ranged from 2000 to 40 000.

[c] Difference between the leucine content of each sample and that of the gelatin hydrolysate.

[d] Determined by gas chromatography. Columns: 5% polyethylene glycol 20M on Chrom-P (3 mm × 2 mm) for C_2 and C_4 alkanols and 3% silicone OV-17 on Chrom-G (3 mm × 3 m) for C_5–C_{12} alkanols.

[e,f,g] See text.

[h] Referring to the gelatin hydrolysate resulting from the papain treatment in the absence of any L-leucine n-alkyl ester.

with a decrease in their water solubility. Effective solubilization of denatured proteins may call for chemical modifications such as glycosylation, phosphorylation, sulfonation, carboxylation, acylation, etc. One of the conventional processes widely used for this purpose is succinylation, in which protein amino groups are blocked and replaced with carboxyl anions. A number of investigators[104,110-112] have applied succinylation to food proteins, with fair success in improving their water solubilities and/ or dispersibilities. The attachment of anionic succinyl moieties to a protein results in an increase in its molecular surface area. This is apparently due to the contribution of the electric repulsion force occurring between identical charges. The resulting increase in surface area is an important requirement if a protein is to acquire surface activity.

The Tokyo group succinylated FPC, SPI, casein, and ovalbumin, as well as gelatin, with the result that their whippabilities, foam stabilities, and emulsifying activities were improved to a certain extent. Additional studies were conducted to improve further the functionality of succinylated proteins by the covalent attachment of L-leucine n-alkyl esters. Each of the above proteins was succinylated[110] and then subjected to papain treatment under conditions similar to those for the one-step process.[113] The reaction mixtures were purified as described above, and the products were measured for whippability, foam stability, and emulsifying activity. In most cases, the highest whippability resulted from the enzymatic process with the incorporation of C_4–C_8 esters of L-leucine. To improve foam stability, the use of C_4–C_6 esters of leucine was found to be preferable. The use of longer chain alkyl esters of leucine was required to give a very high degree of emulsifying activity to succinylated FPC, casein, and ovalbumin. Somewhat different data were obtained with succinylated SPI, which may be lower in hydrophilicity than the other proteins used. In comparison with the above proteins, the enzymatic process with succinylated gelatin gave distinctly different data; the product resulting from the attachment of C_4 or C_8 alkyl esters of leucine did not make a stable foam, but products with longer-chain alkyl esters of leucine showed high whippability and foam stability. This may reflect the fact that succinylated gelatin is much more hydrophilic than the other succinylated proteins.

In consideration of these results, gelatin–Leu–OC_6 (15 min) and suc–FPC–Leu–OC_4 (30 min) were selected as whipping surfactants, and gelatin–Leu–OC_{12} (15 min) and suc–FPC–Leu–OC_{12} (30 min) as emulsifying surfactants. The abbreviated designation of these surfactants is explained as follows: gelatin–Leu–OC_6 (15 min), for example, is the

TABLE 6
PHYSICOCHEMICAL PROPERTIES OF PROTEINACEOUS SURFACTANTS[56]

Surfactant	Hydrophobicity[a] (recorder response)	T_2(ms)[b]		T_2(ms)[e]		τ[g] ($\times 10^{-9}$ s)
		Before whipping	After whipping	Before emulsification	After emulsification	
Gelatin hydrolysate	28	126	—[c]	—	—[f]	1·6
Gelatin–Leu–OC$_6$ (EMG-6)	60	85	43[d]	81	78	9·3
Gelatin–Leu–OC$_{12}$ (EMG-12)	381	148	—[c]	145	75	10·3
Suc–FPC hydrolysate	131	114	—[c]	74	—[f]	1·7
Suc–FPC–Leu–OC$_4$	146	83	49[d]	74	68	9·4
Suc–FPC–Leu–OC$_{12}$	696	118	—[c]	140	68	9·9

[a] Apparatus: Hitachi model 204 fluorescence spectrometer. Conditions: excitation at 365 nm and measurement at 450 nm.
[b] Apparatus: Plaxis model PR-1005 pulsed NMR analyser. Temperature for measurement: 25°C.
[c] Not measured because no stable foam was formed.
[d] Each of the data obtained with a foam phase separated.
[e] Apparatus: Plaxis model PR-1005 pulsed NMR analyser. Temperature for measurement: 0·5°C.
[f] Not measured because no stable emulsion was formed.
[g] Apparatus: LEOL JES-PE-3X ESR spectrometer. Each ESR spectrum was recorded at 9·2–9·3 GHz by using a modulation frequency of 100 kHz and a 10-mV incident microwave power.

15-min reaction product obtained from gelatin with the attachment of leucine hexyl ester. The hydrolysates of gelatin and suc–FPC were used as controls. All of the samples were tested for their hydrophobicities (lipophilicities) and abilities to interact with water and oil.[109,113] The degree of interaction with water and oil was estimated from pulsed NMR measurement of their spin–spin relaxation times (T_2) and also from ESR measurement of rotational correlation times (τ_0) observed for an added free radical. For one series of samples, the observed hydrophobicities tended to increase in the order of increasing chain length: gelatin hydrolysate, gelatin–Leu–OC_6, and gelatin–Leu–OC_{12} (EMG—12, as mentioned later) (Table 6). A similar relationship was found for the samples suc–FPC hydrolysate, suc–FPC–Leu–OC_4, and suc–FPC–Leu–OC_{12}. This result indicates that the emulsifying surfactant is endowed with a much higher degree of hydrophobicity than the whipping surfactant.

In a stable foam or emulsion system, surfactant molecules are arranged at an interface to interact with air or oil. The interaction takes place in the form of an air–surfactant–water or an oil–surfactant–water complex. In both cases, the rotational freedom of the water molecules is restricted to a greater or lesser extent. The average rotational correlation time of a water molecule is known to correlate well with its T_2 value as measured by pulsed NMR.[114,115] To test this, samples (1·5 g) were dispersed in 10 ml of 0·01 M phosphate buffer (pH 7) and the resultant dispersion was submitted to pulsed NMR measurement. For each of these dispersions of a whipping agent a smaller T_2 was observed than for the protons of pure water (approx. 2 s). For dispersions containing emulsifiable agents, the observed T_2 values were rather large. As Table 6 shows, it was actually observed that, when oil was emulsified with emulsifying agents, T_2 values of water protons decreased. This indicates that it is only in the presence of oil that the emulsifier molecules have any significant ability to prevent water from tumbling. The behavior of a stable free radical added to the medium reflects the average state of molecular rotation of the medium.

Oil-soluble free radicals are commonly used to probe the rotation of oil molecules. The Tokyo group selected an oil-soluble spin probe, 2–(10–carboxydecyl)–2–hexyl–4,4–dimethyl–3–oxazolidinyloxy methyl ester. A τ_0 value of $1·6\times10^{-9}$ s was obtained with this probe in corn oil at 10°C. When the emulsifier dispersions were homogenized in the oil, a distinctly larger τ_0 value was obtained than in the oil alone under the same conditions. Schenouda & Pigott[116] have reported that a one-order increase in τ_0 results when a lipid is bound tightly to a protein. The observation of an increase in τ_0 (Table 6) made by the Tokyo group suggests the occurrence of a similar surfactant–oil interaction. Since the interaction of a surfactant with oil must take place at the side chain of an alkyl moiety, the chain length of the side chain will determine the characteristics of the molecule as a surfactant.

The Tokyo group tested proteinaceous surfactants prepared from gelatin by the papain-catalysed attachment of leucine alkyl esters[117] for their suitability in a series of applications to well-known food products. Using each of the surfactants, the group tried to prepare snow jelly, ice cream, a mayonnaise-like food, and bread. Snow jelly is a typical whipped food. All of the surfactants prepared from gelatin by the covalent attachment of leucine alkyl esters were well suited to the preparation of snow jelly. In particular, the use of gelatin–Leu–OC_6 resulted in a texture comparable to that of snow jelly made from fresh cream. Since ice cream is a colloidal dispersion of foam and oil globules, both the whipping and emulsifying functions of an added surfactant will contribute to the improvement of its structure. From that point of view, gelatin–Leu–OC_4 and gelatin–Leu–OC_{12} were applied in ice cream. Either of these surfactants gave a high degree of overrun by mixing in a short time. Mayonnaise is an oil-in-water emulsion prepared using either whole egg or egg yolk. Instead of egg, gelatin–Leu–OC_{12} was tried as an emulsifier in the preparation of a mayonnaise-like food. Sensory and texturometric data showed that the proteinaceous surfactant gave results between those obtained using whole egg and egg yolk at a concentration range of 0·5–1%. The use of surfactants with lipid-emulsifying activity is essential in bread making. When gelatin–Leu–OC_{12} was used in baking, it was found to be very effective in producing a large loaf volume and maintaining soft crumb during storage.

The Tokyo group also conducted various experiments on the gelatin enzymatically modified by the attachment of leucine dodecyl ester (EMG-12, the same as the gelatin–Leu–OC_{12} mentioned above) in order to evaluate its chemical, physical, and functional properties. Table 7 collates the data comparing this compound with gelatin and its hydrolysate.[118] It can be seen that EMG-12 differs from gelatin and its hydrolysate with respect to almost all of the parameters investigated, and it is clear that the enzymatic modification caused these differences. L-Leucine ethyl ester-incorporated fish proteins have been reported to be good emulsifiers.[80–83]

Recently it was found that EMG-12 added to water is able to retard freezing by supercooling[119]. A similar phenomenon was observed for an oil/water emulsion produced with EMG-12.[120] It was thus expected that

TABLE 7
GENERAL PROPERTIES OF EMG-12 COMPARED TO GELATIN AND ITS HYDROLYSATE[118]

Item	Gelatin	Gelatin hydrolysate	EMG-12
Chemical properties			
Molecular weight	Higher than 30 000	av. 7500	av. 7500
Dodecyl moiety content	—	—	~ 1 mol/7500 g
Effective hydrophobicity	33	11	240
Dispersion properties			
Critical mycelle concentration	No micelle formed	No micelle formed	0·02–0·04%
Decrease in surface tension	Down to 60 dyn/cm	Down to 60 dyn/cm	Down to 35 dyn/cm
Phase characteristic	No phase transition observed	No phase transition observed	Partial formation of liquid crystals
Emulsion properties			
Emulsifying activity	Inferior to Tween-80	Inferior to Tween-80	Superior to Tween-80
Emulsion stability	Inferior to Tween-80	Inferior to Tween-80	Superior to Tween-80
Interfacial molecular area	—	—	48 Å^2
Cryophysical properties			
Supercooling of an o/w emulsion	—	—	Stable at −10°C
Supercooling of an aqueous dispersion	Not stable	Not stable	Stable at −7°C
Interaction with silver iodide crystals	Not observed	Not observed	Adsorption for antinucleation

this proteinaceous surfactant could function as an antifreeze agent as well as an emulsifier. For the crystallization of substances in general, the existence of nuclei is indispensable. The same holds true in the case of freezing water. Without nuclei, water remains liquid even at subzero temperatures, remaining in a supercooled state. The freezing of water starts with ice crystal formation, which is initiated by ice nucleation. This can be classified into two categories: homogeneous nucleation in which water molecules themselves act as nuclei and heterogeneous nucleation which takes place with the aid of exogenous substances termed 'motes'. Plasteins synthesized from egg white and lactalbumin have been reported to act as motes.[121] It is well known that silver iodide crystals can be heterogeneous nuclei with potent nucleating activity. In the presence of even small particles of this compound, it is generally difficult for water to supercool. In such a case, therefore, water freezes at a temperature corresponding to the melting point of ice. The Tokyo group used crystallized silver iodide in order to evaluate how EMG-12 functions as an antifreeze agent despite the presence of this nucleus. Also, in every case, experiments were carried out on a bulk water scale so that the so-called microcompartmentalization effect (leading to an extraordinary decrease in freezing temperature) is excluded.

Antifreeze Emulsions
Since most oil/water emulsions often lose their stability through the formation of ice crystals in the bulk water during freezer storage, it would be of benefit to develop an agent effective in inhibiting ice nucleation. It is preferable in this case to develop a high molecular weight agent which would act in a non-colligative manner, because any colligative decrease in freezing temperature is exclusively accompanied by an osmotic problem.

The Tokyo group produced oil/water emulsions with EMG-12 and with three control agents, gelatin hydrolysate with an average molecular weight of about 7500, polyglycerol stearate (PGS), and Tween-80. Each emulsion was placed in a sample tube fitted for pulsed NMR measurement and cooled gradually to $-10°C$ while the free induction decay (FID) amplitude was recorded automatically. For the emulsions produced with gelatin hydrolysate and Tween-80, the recorder indicated that the bulk water began to freeze when its temperature reached $-4°C$. However, the emulsions produced with either EMG-12 or PGS resisted freezing even at $-10°C$, maintaining a state of supercooling for at least 20 h.[120] Another experiment was conducted to investigate what would happen when ice nucleation was induced by the addition of silver iodide crystals to the emulsions. For this experiment, an EMG-12 dispersion was mixed with linoleic acid and a small amount of silver iodide was added. Subsequently, the mixture was emulsified and the resulting emulsion was allowed to stand at −10°C. Under these conditions, the emulsion with EMG-12 resisted freezing until its temperature reached approximately −9°C.

In order to discover why the supercooled emulsion produced with EMG-12 was stable even in the presence of added silver iodide crystals, its structure was observed by scanning electron microscopy. The results indicated that no crystal pieces existed in the bulk water phase. By magnifying a picture of the oil particle mass that had made up the emulsion, it was found that the crystals were fixed to the oil particle mass at the surface. It is therefore speculated that, as a result of such fixation, the ice-nucleating activity of silver iodide is lost or attenuated to a great extent.

Antifreeze Dispersions

EMG-12 was also found to be effective in retarding the freezing of pure water. In this case, the effect was critically dependent on the amount of EMG-12 used. At a concentration of 0·01% or lower, the water never supercooled. For stable supercooling it was necessary to use EMG-12 at 0·03% or higher. Apparently, a critical zone exists between these concentrations and, interestingly, this accords with the critical micellar concentration of EMG-12 (Table 7).[122] The degree of supercooling observed for an EMG-12 dispersion was around 7°C, significantly greater than that observed for a gelatin hydrolysate dispersion, although polyvinylpyrrolidone (PVP) used as another control resembled EMG-12 in its supercooling effect. As a working hypothesis, the stability of a supercooled emulsion depends on how efficiently the dispersed molecules are adsorbed onto the surface of the heterogeneous nuclei to prevent their nucleating action. To test this hypothesis an experiment, in which the adsorption process was observed by differential thermal analysis, was conducted. By using a thermometer of special design, it was observed that as soon as a silver iodide suspension in water was dropped into an 0·1% EMG-12 dispersion, a distinct exothermic peak appeared. This indicates that an adsorption process did proceed. A similar process took place when PVP was used. However, no clear peak resulted when gelatin hydrolysate was used at 0·1% nor when EMG-12 was used at a level lower than its critical micellar concentration. It is probable that EMG-12 acts as an antifreeze agent only when it is dispersed in water at a sufficiently high concentration.[123]

Extensive studies have been undertaken on the antifreeze glycoproteins (AFGP) existing in the blood of winter polar fish.[124] A great deal of information is available on AFGP.[124] It would be interesting to compare the structure–function relationship of EMG-12 and these glycoproteins. Though both are involved in freezing retardation, this phenomenon occurs in the form of antinucleation in the case of EMG-12 and in the form of ice crystal growth inhibition in the case of AFGP.[124,125] It will be interesting to learn by further experimentation whether a practical application for this modified protein as a cryoprotectant for use in food processing and preservation, can be developed.

It should be noted that EMG-14, an enzymatically modified gelatin product obtained by attachment of alanine myristyl ester instead of leucine dodecyl ester, has been produced on an industrial scale and applied as a proteinaceous surfactant for cosmetics. The product is already on the market.

REFERENCES

1. Danielwski, A., As cited by Henriquez, V. & Gjalda, I. K. Z. Physiol. Chem., **71** (1911) 485.
2. Sawjalow, W. W., Z. Physiol. Chem., **54** (1907) 119.
3. Tauber, H., J. Am. Chem. Soc., **73** (1951) 1288.
4. Wieland, T., Determann, H. & Albrecht, E., Ann. Chem., **633** (1960) 185.
5. Determann, H. & Köhler, R., Ann. Chem., **690** (1966) 197.
6. Fujimaki, M., Yamashita, M., Arai, S. & Kato, H., Agric. Biol. Chem., **34** (1970) 1325.
7. Yamashita, M., Arai, S., Gonda, M., Kato, H. & Fujimaki, M., Agric. Biol. Chem., **34** (1970) 1333.
8. Yamashita, M., Arai, S., Matsuyama, J., Gonda, M., Kato, H. & Fujimaki, M., Agric. Biol. Chem., **34** (1970) 1484.
9. Yamashita, M., Arai, S., Matsuyama, J., Kato, H. & Fujimaki, M., Agric. Biol. Chem., **34** (1970) 1492.
10. Yamashita, M., Tsai, S.-J., Arai, S., Kato, H. & Fujimaki, M., Agric. Biol. Chem., **35** (1971) 86.
11. Horowitz, J. & Haurowitz, F., Biochim. Biophys. Acta, **33** (1959) 231.
12. Borsook, H., Adv. Protein Chem., **8** (1953) 127.
13. Tauber, H., J. Amer. Chem. Soc., **71** (1949) 2952.
14. Greenbaum, L. M. & Fruton, J. S., J. Biol. Chem., **226** (1957) 173.
15. Fujii, S. & Fruton, J. S., J. Biol. Chem., **230** (1958) 1.
16. Determann, H., Bonhard, K., Köhler, R. & Wieland, T., Helv. Chim. Acta, **46** (1963) 2498.
17. Virtanen, A. I., Kerkkonen, H. K. & Laaksonen, T., Acta Chem. Scand., **2** (1948) 933.
18. Tsai, S.-J., Yamashita, M., Arai, S. & Fujimaki, M., Agric. Biol. Chem., **38** (1974) 641.

19. Hofsten, B. V. & Lalasidis, G., *J. Agric. Food Chem.*, **24** (1976) 460.
20. Sciancalepore, V. & Longone, V., *J. Dairy Res.*, **55** (1988) 547.
21. Tsai, S.-J., Yamashita, M., Arai, S. & Fujimaki, M., *Agric. Biol. Chem.*, **36** (1972) 1045.
22. Kim, S. K. & Lee, E. H., *Han'guk Nonghwa Hakhoechi*, **30** (1987) 234.
23. Kim, S. K. & Lee, E. H., *Han'guk Susan Hakhoechi*, **20** (1987) 282.
24. Winkler, H., Noetzold, H. & Ludwig, E., *Nahrung*, **32** (1988) 135.
25. Kézdy, F. J., Jindal, S. P. & Bender, M. L., *J. Biol. Chem.*, **247** (1972) 5746.
26. Bender, M. L. & Kézdy, F. J., *Annu. Rev. Biochem.*, **34** (1965) 49.
27. Gololobov, M. Yu., Antonova, T. V. & Belikov, V. M., *Nahrung*, **30** (1986) 289.
28. Johnson, C. H. & Knowles, J. R., *Biochem. J.*, **101** (1966) 56.
29. Foster, R. J., Shine, H. J. & Niemann, C., *J. Amer. Chem. Soc.*, **77** (1955) 2378.
30. Bender, M. L. & Kemp, K. C., *J. Amer. Chem. Soc.*, **79** (1957) 116.
31. Virtanen, A. I., Kerkkonen, H. K., Laaksonen, T. & Hakala, M., *Acta Chem. Scand.*, **3** (1949) 520.
32. Virtanen, A. I., Laaksonen, T. & Kantola, M., *Acta Chem. Scand.*, **5** (1951) 316.
33. Determann, H., Heuer, J. & Jaworek, D., *Ann. Chem.*, **690** (1965) 189.
34. Yamashita, M., Arai, S., Tanimoto, S. & Fujimaki, M., *Biochim. Biophys. Acta*, **358** (1974) 105.
35. Tanimoto, S., Yamashita, M., Arai, S. & Fujimaki, M., *Agric. Biol. Chem.*, **36** (1972) 1595.
36. Isowa, Y., Ohmori, M., Sato, M. & Mori, K., *Bull. Chem. Soc. Jpn*, **50** (1977) 2766.
37. Andrews, A. T. & Alichanidis, E., *Food Chem.*, **35** (1990) 243.
38. Pallavicini, C. & Trentin, G., *Lebensm.-Wiss. Technol.*, **20** (1987) 74.
39. Aso, K., Yamashita, M., Arai, S. & Fujimaki, M., *Agric. Biol. Chem.*, **37** (1973) 2505.
40. Aso, K., Yamashita, M., Arai, S. & Fujimaki, M., *J. Biochem.*, **76** (1974) 341.
41. Eriksen, S. & Fagerson, I. S., *J. Food Sci.*, **41** (1976) 490.
42. Hartnett, E. K. & Satterlee, L. D., *J. Food Biochem.*, **14** (1990) 1.
43. Hagan, R. C. & Villota, R., *Food Chem.*, **23** (1987) 277.
44. Noetzold, H., Ludwig, E. & Sykora, S., *Nahrung*, **32** (1988) 737.
45. Yamashita, M., Arai, S., Tsai, S.-J. & Fujimaki, M., *Agric. Biol. Chem.*, **34** (1970) 1593.
46. Arai, S., Yamashita, M., Aso, K. & Fujimaki, M., *J. Food Sci.*, **40** (1975) 342.
47. Fujimaki, M., Utaka, K., Yamashita, M. & Arai, S., *Agric. Biol. Chem.*, **37** (1973) 2303.
48. Aso, K., Tanimoto, S., Yamashita, M., Arai, S. & Fujimaki, M., *Agric. Biol. Chem.*, **43** (1979) 1147.
49. Alecio, M. R., Dann, M. L. & Lowe, G., *Biochem. J.*, **141** (1974) 495.
50. Aso, K., Yamashita, M., Arai, S., Suzuki, J. & Fujimaki, M., *J. Agric. Food Chem.*, **25** (1977) 1138.

51. Schechter, I. & Berger, A., *Biochem. Biophys. Res. Commun.*, **27** (1967) 157.
52. Schechter, I. & Berger, A., *Biochem. Biophys. Res. Commun.*, **32** (1968) 898.
53. Lauffer, M. A., *Proteins and Their Reactions*, ed. H. W. Schultz & A. F. Anglemier. AVI, Westport, 1964, p. 87.
54. Fujimaki, M., Kato, H., Arai, S. & Yamashita, M., *J. Appl. Bacteriol.*, **34** (1971) 119.
55. Fujimaki, M., Kato, H., Arai, S. & Tamaki, E., *Food Technol.*, **22** (1968) 889.
56. Fujimaki, M., Arai, S. & Yamashita, M., *Adv. Chem. Ser.*, **160** (1977) 156.
57. Fujimaki, M., Arai, S., Kirigaya, N. & Sakurai, Y., *Agric. Biol. Chem.*, **29** (1965) 855.
58. Arai, S., Koyanagi, O. & Fujimaki, M., *Agric. Biol. Chem.*, **31** (1967) 868.
59. Arai, S., Noguchi, M., Yamashita, M., Kato, H. & Fujimaki, M., *Agric. Biol. Chem.*, **34** (1970) 1569.
60. Noguchi, M., Arai, S., Kato, H. & Fujimaki, M., *J. Food Sci.*, **35** (1970) 211.
61. Arai, S., Noguchi, M., Kurosawa, S., Kato, H. & Fujimaki, M., *J. Food Sci.*, **35** (1970) 392.
62. Gordon, J. F., *Proteins as Human Food*, ed. R. A. Lawrie. AVI, Westport, 1970, p. 328.
63. Clement, G., *Single-Cell Protein*, ed. R. I. Mateles & S. R. Tannenbaum. M.I.T. Press, Cambridge, 1968, p. 306.
64. Kobayashi, M. & Tchan, Y. T., *Water Res.*, **7** (1973) 1219.
65. Gore, S. B., Mungikar, A. M. & Joshi, R. N., *J. Sci. Food Agric.*, **25** (1974) 1149.
66. Arai, S., Yamashita, M. & Fujimaki, M., *J. Nutr. Sci. Vitaminol.*, **22** (1976) 447
67. Fujimaki, M., Yamashita, M., Okazawa, Y. & Arai, S., *Agric. Biol. Chem.*, **32** (1968) 794.
68. Fujimaki, M., Yamashita, M., Okazawa, Y. & Arai, S., *J. Food Sci.*, **35** (1970) 215.
69. Yamashita, M., Arai, S. & Fujimaki, M., *Agric. Biol. Chem.*, **33** (1969) 321.
70. Arai, S., Yamashita, M., Kato, H. & Fujimaki, M., *Agric. Biol. Chem.*, **34** (1970) 729.
71. Matoba, T., Nagayasu, C., Hayashi, R. & Hata, T., *Agric. Biol. Chem.*, **33** (1969) 1662.
72. Matoba, T., Hayashi, R. & Hata, T., *Agric. Biol. Chem.*, **34** (1970) 1235.
73. Minamiura, N., Matsumura, Y., Fukumoto, J. & Yamamoto, T., *Agric. Biol. Chem.*, **36** (1972) 588.
74. Arai, S., *The Analysis and Control of Less Desirable Flavors in Foods and Beverages*, ed. G. Charalambous. Academic Press, New York, 1980, p. 133.
75. Ichishima, E., *Biochim. Biophys. Acta*, **258** (1972) 274.
76. Fujimaki, M., Yamashita, M., Arai, S. & Kato, H., *Agric. Biol. Chem.*, **34** (1970) 483.
77. Yamashita, M., Arai, S., Kokubo, S., Aso, K. & Fujimaki, M., *Agric. Biol. Chem.*, **38** (1974) 1269.

78. Yamashita, M., Arai, S., Kokubo, S., Aso, K. & Fujimaki, M., *J. Agric. Food Chem.*, **23** (1975) 27.
79. Noguchi, M., Yamashita, M., Arai, S. & Fujimaki, M., *J. Food Sci.*, **40** (1975) 367.
80. Kim, S. K. & Lee, E. H., *Han'guk Susan Hakhoechi*, **20** (1987) 481.
81. Kim, S. K. & Lee, E. H., *Han'guk Susan Hakhoechi*, **20** (1987) 582.
82. Kim, S. K., Kwak, D. C., Cho, D. J. & Lee, E. H., *Han'guk Yongyang Siklyong Hakhoechi*, **17** (1988) 242.
83. Kim, S. K., Kwak, D. C., Cho, D. J. & Lee, E. H., *Han'guk Yongyang Siklyong Hakhoechi*, **17** (1988) 312.
84. Yamashita, M., Arai, S., Tsai, S.-J. & Fujimaki, M., *J. Agric. Food Chem.*, **19** (1971) 1151.
85. Arai, S., Yamashita, M. & Fujimaki, M., *Cereal Foods World*, **20** (1975) 107.
86. Arai, S., Aso, K., Yamashita, M. & Fujimaki, M., *Cereal Chem.*, **51** (1974) 143.
87. Hajos, G. & Szarvas, T., *Izotoptechnika*, **29** (1986) 166.
88. Atallah, M. T., *J. Food Sci.*, **52** (1987) 1198.
89. Hajos, G., Halasz, A. & Bekes, F., *Acta Aliment.*, **18** (1989) 325.
90. Fujimaki, M., *Ann. Nutr. Alim.*, **32** (1978) 233.
91. Aso, K., Yamashita, M., Arai, S. & Fujimaki, M., *Agric. Biol. Chem.*, **38** (1974) 679.
92. Yamashita, M., Arai, S., Imaizumi, Y., Amano, Y. & Fujimaki, M., *J. Agric. Food Chem.*, **27** (1979) 52.
93. Yamashita, M., Arai, S., Amano, Y. & Fujimaki, M., *Agric. Biol. Chem.*, **43** (1979) 1065.
94. Arai, S., Yamashita, M. & Fujimaki, M., *Water Activity: Influences on Food Quality*, ed. L. B. Rockland & G. F. Stewart. Academic Press, London, 1981, p. 489.
95. Aso, H., Kimura, H., Watanabe, M. & Arai, S., *Agric. Biol. Chem.*, **49** (1985) 1649.
96. Brubacher, L. J. & Bender, M. L., *J. Amer. Chem. Soc.*, **88** (1966) 5871.
97. Arai, S., Yamashita, M. & Fujimaki, M., *Agric. Biol. Chem.*, **43** (1979) 1069.
98. Arai, S., Watanabe, M., Kimura, H. & Ogiwara, H., *Agric. Biol. Chem.*, **52** (1988) 1873.
99. Arai, S., Aso, H. & Kimura, H., *Agric. Biol. Chem.*, **47** (1983) 2115.
100. Hammonds, T. M. & Call, D. L., *Chem. Technol.*, **2** (1972) 156.
101. Kinsella, J. E., *CRC Crit. Rev. Food Sci. Nutr.*, **7** (1976) 219.
102. Feeney, R. E., Yamasaki, R. B. & Geoghegan, K. F., *Adv. Chem. Ser.*, **198** (1982) 3.
103. Whitaker J. R. & Puigserver, A. J., *Adv. Chem. Ser.*, **198** (1982) 57.
104. Pour-El, A., *Functionality and Protein Structure (ACS Symposium Ser., 92)*, ed. R. F. Gould. American Chemical Society, Washington, D.C., 1979, p. ix.
105. Franzen, K. L. & Kinsella, J. E., *J. Agric. Food Chem.*, **24** (1976) 788.
106. Toiguchi, S., Maeda, S., Watanabe, M. & Arai, S., *Agric. Biol. Chem.*, **46** (1982) 2945.

107. Watanabe, M., Toyokawa, H., Shimada, A. & Arai, S., *J. Food Sci.*, **46** (1981) 1467.
108. Jones, N. R., *The Science and Technology of Gelatin*, ed. A. G. Ward & A. Courts. Academic Press, London, 1977, p. 366.
109. Watanabe, M. & Arai, S., *Adv. Chem. Ser.*, **198** (1982) 199.
110. Hoagland, P. D., *J. Dairy Sci.*, **49** (1966) 783.
111. Beuchat, L. R., *J. Agric. Food Chem.*, **25** (1977) 258.
112. Pearce, K. N. & Kinsella, J. E., *J. Agric. Food Chem.*, **26** (1978) 716.
113. Arai, S., Watanabe, M. & Fujii, N., *Agric. Biol. Chem.*, **48** (1984) 1861.
114. Kuntz, I. D. & Kauzmann, W., *Adv. Protein Chem.*, **28** (1974) 239.
115. Fennema, O., *Food Proteins*, ed. J. R. Whitaker & S. R. Tannenbaum. AVI, Westport, 1977, p. 50.
116. Schenouda, S. Y. K. & Pigott, G. M., *J. Agric. Food Chem.*, **24** (1976) 11.
117. Watanabe, M., Shimada, A., Yazawa, E., Kato, T. & Arai, S., *J. Food Sci.*, **46** (1981) 1738.
118. Arai, S., Watanabe, M. & Hirao, N., *Protein Tailoring for Food and Medical Uses*, ed. R. E. Feeney & J. R. Whitaker. Marcel Dekker, New York, 1986, p. 75.
119. Arai, S., Watanabe, M. & Tsuji, R. F., *Agric. Biol. Chem.*, **48** (1984) 2173.
120. Watanabe, M., Tsuji, R. F., Hirao, N. & Arai, S., *Agric. Biol. Chem.*, **49** (1985) 3291.
121. Honma, K., *Jpn. Kokai Tokkyo Koho*, JP-01-43 (1989) 160.
122. Shimada, A., Yamamoto, I., Sase, H., Yamazaki, Y., Watanabe, M. & Arai, S., *Agric. Biol. Chem.*, **48** (1984) 2681.
123. Watanabe, M., *Nippon Nogeikagaku Kaishi*, **61** (1987) 482.
124. Feeney, R. E. & Yeh, Y., *Adv. Protein Chem.*, **32** (1978) 191.
125. Bush, C. A., Feeney, R. E., Osuga, D. T., Ralapati, S. & Yeh, Y., *Int. J. Peptide Protein Res.*, **17** (1981) 125.

Chapter 10

APPLICATIONS OF ENZYMES IN FOOD

G. M. FROST

21 Spring Walk, Brayton, Selby, North Yorkshire, YO8 9DS, UK

INTRODUCTION

World markets for enzymes amount to several hundred million U.S. dollars annually and usuage is dominated by the food industry, which in the early 1980s accounted for about two-thirds of the total sales value, most of the rest being accounted for by detergents. Reed[1] published a comprehensive account of enzyme use in food up to 1966 and the subject has subsequently been reviewed by Underkofler[2] and by Peppler & Reed.[3]

Reasons for the utility of enzymes in food and food processing are easy to identify. Each step in synthesis and breakdown of the innumerable primary and secondary metabolites which occur in nature is almost universally catalysed by its own particular enzyme, so the choice of enzymes to modify these same natural ingredients of which food is comprised is not surprising. However, most biosynthetic processes are linked to the energy transfer processes of the living cell and involve complex sequences of reactions catalysed by strings of enzymes of which many have relatively poor stability. Much work was done from about 1965 onwards in attempts to develop cell-free multiple enzyme systems to mimic the synthetic capabilities of the cell, but with very little success. The natural processes for breakdown of biomass constituents prior to utilisation are much easier to exploit artificially. The enzymes involved in primary digestion have evolved by adaptation to an extracellular environment and are relatively stable. Furthermore the enzymes catalysing consecutive steps in digestion frequently occur together in nature and can be recovered together for artificial use. Hence to date the principle uses for enzymes in food, as in other areas, involve the hydrolytic enzymes acting on carbohydrates, proteins and fats. These same enzymes can sometimes also be used to catalyse the reverse reactions and following ingenious work on choice and optimisation of reaction conditions they have been exploited to produce new food ingredients such as fats of improved composition[4] and aspartame for sweetening.[5]

Low temperatures of reaction are cited as a major favourable property of enzymes. This is usually true but low temperature stability can be a negative factor in some cases. Food processing often involves heat treatment and application of enzymes would be desirable but they are insufficiently stable under the conditions used. Conversely in the dairy industry hydrolysis of certain components is desirable but pH or thermal treatment would destroy the product. Enzymes such as peptidase or lipase for improving cheese flavour, or lactase for lactose hydrolysis, can be used under mild conditions.

The high specificity exhibited by enzymes can also be of great value in their application. The ability to target a molecule for reaction, often in the presence of other substances of very similar structure, gives enzyme catalysis a unique advantage over other chemical techniques. However, in different circumstances this specificity can become a disadvantage since several different enzymes may need to be supplied to work on a mixture of substrates.

In established applications where efficient methods for production and use have been developed, enzyme technology is extremely cheap. For example, the cost of sufficient enzyme to hydrolyse 1 tonne of starch or lactose in a batch mode is of the order of 1 U.S. dollar. The main factor limiting expansion into new uses is the relatively high cost of developing the manufacturing method for a new product. Manufacturers are reluctant to invest heavily unless the market for the enzyme concerned is reasonably assured and it can be difficult for the potential user to establish the feasibility of the potential application in the absence of a source of supply of quantities of the enzyme. Kilara,[6] in a paper on development of enzyme-modified lipid ingredients, has discussed some of the general problems of introduction of new enzyme technology.

SOURCES OF ENZYMES

Animals, plants, yeasts, moulds and bacteria have all been used for manufacture of enzymes. The oldest uses employed crude enzyme extracts of animal or plant tissues but the majority of processes now use enzymes from specially developed and cultured strains of microorganisms. Initially moulds isolated from eastern food fermentation processes or responsible for benign spoilage of foods were exploited to produce hydrolytic enzymes. Bacteria were introduced later but now predominate among producer organisms. Future new applications will probably use enzymes produced in bacteria for preference, because bacteria are most easily manipulated using in-vitro genetic techniques. It is now possible for genes from any type of organism to be expressed in a chosen host, for example that of calf chymosin (rennin) in *Escherichia coli*,[7] *Bacillus subtilis*[8] or *Saccharomyces cerevisiae*.[9]

Methods for manufacture of enzymes have been reviewed by Aunstrup,[10] Lambert & Meers[11] and Frost.[12] Source species for the more important food enzymes are listed in Table 1.

MODES OF USE FOR ENZYMES

Traditional craft processes have used crude forms of enzymes for centuries or millenia but the essential nature of these processes did not begin to be understood until the late 19th century. Use of malt in brewing and baking, and use of calf stomach (rennet) in cheese manufacture, are enzymatic processes. In the Orient mould fermentations were used to preserve staple foods and enhance flavour.[37] The first commercial enzyme products in the Western hemisphere were extracts of malt and calf stomachs and concentrates of these extracts. In the East the technology of the mould food fermentations was extended to produce mould enzyme extracts for external application, particularly in sweetening of starchy foods or enhancing savoury flavours by proteolysis. This technology was brought to the West when the first patent for an isolated enzyme, 'diastase' (amylase), was applied for in the U.S.[38]

Use of enzymes at present is based largely on the direct addition of a manufactured enzyme preparation to the process stream under ambient conditions which suit the action of the enzyme. Normally the enzyme is not recovered for reuse. In many cases the enzyme activity is destroyed deliberately or gratuitously after it has performed its task but the denatured protein remains in the process stream and in the final product. In some cases the enzyme is stable throughout the process and remains in the product in the active form.

TABLE 1

SOURCES OF THE PRINCIPAL FOOD ENZYMES

Class	Enzyme	Source	Area of use	Reference
Proteolytic	Bacterial neutral	*Bacillus subtilis*	Baking	ter Haseborg[13]
	Thermolysin	*Bacillus stearothermophilus*	Aspartame	Delente[14]
	Fungal acid	*Aspergillus oryzae*	Baking	Peppler & Reed[3]
	Rennet	Animal stomachs	Cheese	Berridge[15]
	Microbial rennin	*Mucor* spp.	Cheese	Sternberg[16]
	Papain	Papaya fruit	Meat, beer	Balls[17]
Amylolytic	Bacterial α-amylase	*Bacillus subtilis*	Starch, brewing	Banks *et al.*[18]
	Bacterial α-amylase	*Bacillus licheniformis*	Starch	Madsen *et al.*[19]
	β-amylase	Barley, soy bean	Starch	Saha & Zeikus[20]
	Fungal α-amylase	*Aspergillus oryzae*	Starch	Saha & Zeikus[20]
	Amyloglucosidase	*Aspergillus niger*	Starch, brewing	Cadmus *et al.*[21]
	Pullulanase	*Klebsiella pneumoniae*	Starch	Fogarty[22] Orskov[23]
	Pullulanase	*Bacillus acidopullulolyticus*	Starch	Jensen & Norman[24]
Other hydrolytic	Pectinases	*Aspergillus niger*	Fruit juices	Pilnik[25]
	Lactase	*Kluyveromyces* spp.	Dairy, whey	Nijpels[26]
	Lactase	*Aspergillus* spp.	Dairy, whey	Gekas & Lopez-Leiva[27]
	Invertase	*Saccharomyces* spp.	Confectionery	Lindley[28]
	β-1–3,1–4-glucanase	*Penicillium emersonii*	Brewing	James *et al.*[29]
	β-1–3,1–4-glucanase	*Bacillus* spp.	Brewing	Huber *et al.*[30]
	Lipase	*Aspergillus niger*	Fats, cheese	Pal *et al.*[31]
	Lipase	*Rhizopus* spp.	Fats, cheese	Iwai & Tsujisawa[32] Laboureur & Labrousse[33]
Other classes	Glucose isomerase	*Streptomyces* spp.	Starch	Bucke[34]
	Glucose isomerase	*Bacillus* sp.	Starch	Oostergaard & Knudsen[35]
	Glucose oxidase	*Aspergillus niger*	Food, drinks	Underkofler[36]

Two perceived disadvantages of this type of enzyme technology, the high cost of non-recoverable enzyme and the contamination of the product with added enzyme, were addressed by devoting a great deal of research to immobilisation methods for enzymes so that continuous or repeated batch use could become possible. This work has been reviewed by Chibata[39] and by Kennedy & Cabral.[40] Work on the applications of immobilised enzymes in food processing has been reviewed in a book edited by Pitcher.[41] It was widely expected that immobilised enzymes would replace soluble forms in existing uses and that immobilised enzyme technology would allow many new uses to develop. One outstanding success has been the development of high fructose corn syrups using immobilised isomerases, but there have been relatively few other applications and in terms of the amount of effort expended in the area enzyme immobilisation has so far had disappointing results, so that it could reasonably be described as an expensive solution looking for a problem. Table 2 lists the factors determining choice between a single use soluble enzyme and an immobilised enzyme system. In the food industry there are two major difficulties in applying immobilised enzymes: firstly, the complex nature of the process streams, which are often semi-solid or contain suspended solids in high amounts: secondly, the substrates on which the enzymes are to act are most commonly biological polymers of high

TABLE 2

FACTORS AFFECTING CHOICE BETWEEN SOLUBLE AND IMMOBILISED
ENZYMES

Soluble favoured	*Immobilised favoured*
Cheap enzyme already available	Enzyme expensive or manufacturing process not developed
Low stability under use conditions	Enzyme stable in use
Enzyme difficult to immobilise	Enzyme susceptible to immobilisation by cheap techniques in high yield
Complex substrate, containing fouling or poisoning agents	Pure substrate
Solid phase substrate	Soluble substrate
High molecular weight substrate	Small molecule substrate
Flexibility in operating parameters important	Operating parameters constant and well defined
Low capital costs desired	Capital resources not limiting
Process susceptible to microbial spoilage	Process resistant to microbial spoilage

molecular weight, which in a number of cases are also insoluble. Free enzymes can penetrate the mixtures whereas satisfactory contact between immobilised forms and their substrates may be extremely difficult to achieve. For processes which used enzymes before immobilisation technology was developed the use of soluble enzymes has universally persisted despite the development of immobilised forms. Amyloglucosidase is probably the most important enzyme applied to foods but despite development of a number of practical immobilised versions Rugh et al.,[42] gave the opinion that they could only be economically used in certain specialised applications, conceding that for the established bulk conversions of starch the soluble enzyme remained the better choice.

The application of glucose isomerase, which is often called xylose isomerase because the enzyme from many source species acts on xylose at lower concentrations than on glucose, developed concurrently with research on immobilisation methods. The high glucose syrups produced from starch by consecutive treatment with α-amylase and amyloglucosidase are converted to sweeter non-crystallising syrups containing a mixture of glucose and fructose. Commercial plants for this process universally use immobilised enzyme reactors. In this case conditions for their use are favourable because the process stream is relatively pure, the substrate is a small molecule, and the enzymes are rather thermostable, allowing use at temperatures which are high enough to minimise the problem of microbial spoilage.[34,43]

Existing uses for enzymes in the food industry largely involve what are essentially digestive processes. These have been well characterised and exploited over a number of years and no large new developments are likely in this area. The present focus in research is on chemical conversions which will find application in the food industry in the areas of additives and ingredients. Here, with purer streams and a probability of low molecular weight soluble reactants, the application of enzymes in immobilised form is likely to be favourable, provided that the development costs are justified by expected market sizes. An example of a relatively new application for immobilised enzymes is in the synthesis of the artificial sweetener aspartame. Here, the condensation of (N-benzyloxycarbonyl)L-aspartate and L-phenylalanine methyl ester to form a peptide bond uses the bacterial protease thermolysin, catalysing its reverse reaction in an immobilised enzyme reactor.[5]

Initial work with immobilised enzymes led to hopes that complex sequences of reactions could be engineered by immobilising the enzymes catalysing the individual steps either together or sequentially in a flow-through reactor. This proved too complex for practical application and attention switched to the use of immobilisation techniques on intact cells. Where the cells remain viable this can be treated as a special case of fermentation technology, which is outside the scope of this review. In other situations the process of immobilisation kills the cell. The resulting catalytically active product should then be classified as an immobilised crude enzyme preparation. Here the biomass acts as an enzyme support and frequently has a stabilising effect.

An extension of the immobilisation principle, which avoids some of its disadvantages, is to design a two-phase aqueous system where enzyme and substrate are partitioned in one phase while the product moves to the other.[44,45] The phases can be separated by settling in a simple liquid–liquid separator and the product can be recovered prior to recycling each phase for re-use. In this system diffusion access of the substrate to the enzyme is not impeded by a solid phase and equilibrium is avoided by continuous removal of the reaction products.

Encapsulation of enzymes in such a way as to allow delayed or gradual release has been used in the application of ripening enzymes to cheese. This technology lends itself to use of enzymes in other foods: the enzyme-containing microcapsules can be mixed into the food during processing or preparation.[46]

Enzymes have traditionally been regarded as creatures of an aqueous environment. However, it is now known that many, such as lipases, are located at lipid–water interfaces in vivo and their physiological functioning requires a two-phase environment. Lipases are used technically in a system where the enzyme is absorbed on moist solid particles and the substrate and products are present in a stream of hexane passing through the solid bed. Under these conditions of restricted water availability the conservation of ester bonds is favoured over hydrolysis and the exchange of fatty acids between the free form and triglycerides can be obtained to achieve a useful improvement in composition of the lipid. For example, increasing the stearyl content of palm oil imparts properties resembling those of cocoa butter.[47] Micellar systems[48] or reverse micellar suspensions[49] can also be used with lipase to induce ester bond synthesis or interesterification.

Enzymes are also active in single-phase organic solvents where lack of conformational stability in the non-aqeous medium can often be ameliorated by using an immobilised form.[50] However, the spontaneous rates of many biological reactions are much faster in water than in organic solvents so that the absolute enzyme catalysed rates in organic solvents are still rather low. Advantages of using organic solvents as media for enzyme reactions have been reviewed by Dordisk.[51] A food industry example of the possible utility of such systems occurs in starch processing where a product is required for soft drinks containing 55% of the sugar content as fructose. The current enzymatic isomerisation technology gives a maximum fructose content of 42% and chromatographic enrichment is used to increase it. Visuri & Klibanov[52] have shown that it is possible to reach 55% directly in an enzymatic reaction by running it in 85–90% ethanol.

Proteolytic enzymes also exhibit esterase activity and can be used in organic solvents for interesterification. Vulfson et al.[53] investigated the interesterification of N-acetylphenylalanine ethyl ester with higher alcohols catalysed by the bacterial protease subtilisin and found that sonication of the mixture enhanced the rate of reaction. The mechanism for this effect could not be assigned with certainty but prevention of formation of water shells on the enzyme or promotion of flexing of the enzyme molecule were suggested. Sonication may find an application in promoting enzyme action in non-aqueous media.

The recent expansion of ideas for methods of use of enzymes will allow scope for introduction of further new applications.

SAFETY AND REGULATIONS CONTROLLING USE OF ENZYMES

Since many fresh foods contain large quantities of active enzymes we normally ingest them as part of our diet. Some food materials are used directly as sources for commerical enzymes, for example, catalase from liver and amylases from barley and malt. The catalytic reactions of enzymes used artificially are well defined and it can readily be established that they do not themselves cause a hazard. Being proteins enzymes do represent a potential cause of allergic reactions but again they are not a special case since any protein-containing food is subject to the same possibility. In deciding whether restrictions on use of enzymes are required, factors favouring their use are that their chemical structure is non-toxic, that the dietary intake is low, that in some cases their source is itself a food and that there is already a substantial history of problem-free use. The main concern is the possibility that harmful or toxic metabolites from the source organism could be present as accidental contaminants of the enzyme preparation, since most commercial products are not highly purified. In selecting source organisms, enzyme producers obviously avoid known pathogenic or toxigenic species. Commercial competition also acts in favour of safety since the most cost-competitive product is the one which has the highest content of active enzyme components and therefore the lowest proportion of contaminants. Nevertheless, the possibility of unwanted contaminants in an enzyme preparation causing harm is a perceptible risk and this has received most attention in considering what safety practices and regulations are required. This has focused attention on the species of organism used as sources.

Commercial enzymes are commonly produced from organisms which were derived originally from food fermentations or have been known as harmless food spoilage organisms (Table 1). In these cases a less rigorous series of tests to verify safety is required than with organisms whose properties are less well known. The Association of Microbial Food Enzyme Producers is a European body founded in 1977 by the producers to coordinate communication with regulatory bodies. It has produced a report[54] listing organisms which are used for the production of food enzymes, and which are recognised as safe sources based on established food use or on toxicological testing. Specifications for chemical and microbiological purity and recommendations for good manufacturing practice are given in the same publication.

Allergic reactions to enzymes have caused actual problems to people involved in their manufacture.[55,56] These cases relate to inhalation of powders of proteolytic enzymes, where allergenicity appears to be enhanced by the catalytic activity of the enzyme. The microbial proteases are now produced in specially developed dust-free forms and producer strains have been mutated to eliminate the more allergenic components.[57] The hazards related to use are much less than those to production workers. A specially commissioned study on protease-containing detergents sponsored by the U.S. authorities concluded that risks from their use which could be attributed to the enzyme content were negligible.[58] However, in handling dried concentrated enzyme preparations, precautions to avoid inhaling dust are necessary.[59]

In most countries the general use of enzymes in foods is not regulated but for individual foods and processes where there are specific regulations these may include stipulations permitting particular enzymes (e.g. the mention of rennet in the U.K. cheese regulations). Other regulations may have the effects of banning enzymes from use, such as the historic and notorious *reinheitsgebot* for beer in Germany which states that only malt, hops, yeast, and water may be used.

In the U.S.A. a list of so-called GRAS (generally recognised as safe) substances is maintained by the Food and Drug Administration (FDA) and this includes a number of enzymes. This list is not exclusive because GRAS status is defined in terms of the opinions of the relevant community of experts and the FDA does not have exclusive power to determine it.[60] However, the FDA, after examining a substance, can order a legally-enforceable ban on its use if not satisfied of its safety. In practice the industry normally petitions the FDA to affirm GRAS status for any newly introduced product. The FDA has also committed itself to a more searching review of long-standing 'GRAS substances and this will include a closer examination of some established enzymes. However, the formal legal situation is that it is permissible to use a product unless the FDA has specifically banned it, subject, of course, to normal civil liability in the event of misjudgement as to its safety. Within the U.S.A. and Canada, use is also subject to the product being made within the officially published guidelines for good manufacturing practice.[61]

In Europe the position varies from country to country at present. The regulatory situation is expected to be harmonised in 1992 but it is not clear at the time of writing how this is likely to be achieved. Use of enzymes in food in the U.K. is not formally regulated but is subject to the general provisions of the 1984 Food Act. The Ministry of Agriculture Fisheries and Food (MAFF) request suppliers and users to keep them informed of new developments on a voluntary basis. A document proposing a permitted list system and reviewing individual enzymes has been published,[62] but the necessity of formal regulation has remained a matter of debate. It was suggested that development of non-specific tests for toxins in enzyme preparations, which did not involve using live animals, would be of value in assuring safety. Some work on such a system of tests has been done but more is required to perfect it and establish its validity. Enzyme use is, of course, subject to the general provision that food may not contain toxic or harmful materials.

In France it is necessary to submit a request to obtain permission to use particular enzymes. In Germany enzymes may be used without special permission but individual decrees may restrict their use in certain specific foods.[64]

Use of recombinant DNA technology ('genetic engineering') on enzyme-producing strains of microbes has introduced a new aspect to safety evaluation. Initial fears of hazards arising from such manipulation in general use have been greatly alleviated by improvements in the precision of the experimental techniques and by experience of problem-free working with what has now become a routine laboratory technology.

There is no logical reason to expect that strain improvement by genetic engineering would introduce new hazards but a cautious approach is clearly justifiable. Future use of recombinant DNA technology has the potential to avoid the possibility of accidental contamination with toxins and to broaden the choice of enzyme sources, since a pure structural gene for any enzyme can, in principle, be taken from any of its source organisms and can then be inserted specifically in a well-characterised and safe receptor strain for use in production. Implications for enzyme technology of introduction of recombinant DNA methods have been discussed in the review by Noordervliet and Toet.[63]

Regulations relating to products from recombinant organisms are in place in some countries. In the U.K. use of such strains must be notified to the Health and Safety Executive. Rules for good laboratory and large-scale practice are specified, relating to measures for containment of the organisms. For laboratory use, only strains which are classified as high risk in microbiological terms are subject to official inspection of the facilities but the production scale plant intended for use with any recombinant microorganism requires initial inspection and approval before use. The MAFF novel food committee has a brief to consider the presence of recombinant organisms in food, there are not yet any specific regulations. Denmark has a system where individual requests for use in enzyme manufacture are considered by a set procedure against predetermined criteria. In Germany, a legal case concerning use of a recombinant strain in pharmaceutical manufacture has been in dispute in the courts for some time and this has had the effect of preventing the use of recombinant strains generally, pending the establishment of a mechanism for regulating them. It is to be hoped that a straightforward and internationally agreed system of regulation can be set up so that the undoubted benefits of the technology can be exploited.

ENZYMES IN FOOD ANALYSIS

Enzymatic methods have been developed for analysis of nearly all types of biological molecules. They have the great advantage of high specificity and convenience in use, and they can usually be applied using simple procedures and detection methods. Frequently an enzymatic assay requires merely the sequential addition of reagents to the sample in the cold and reading in a colorimeter or spectrophotometer. However, enzymes are also adaptable to sophisticated methods of automatic analysis, where immobilised forms may be used. Enzymes are also of central importance in the new technology of biosensors. Bergmeyer[65] has edited a comprehensive treatment of the numerous enzymatic methods used in analysis. Enzyme sensors have been reviewed by Karube[66] who describes some which are specific for particular carbohydrates and amino-acids. The most commonly used enzymatic analyses are those developed for blood samples. These can often usefully be applied to food analysis. Glu-cose oxidase was first shown to be suitable for blood glucose analysis by Keilin and Hartree.[67] It is now used for glucose determination in a number of ways. The most popular form of assay couples the production of hydrogen peroxide in the glucose oxidase reaction to the oxidation of a chromogenic substrate using peroxidase (Eqns 1 and 2). Originally o-dianisidine was the chromogen of choice[68] but this is hazardous and has been superseded largely by 4-aminophenol[69] and 2,2'azino bis(3-ethyl-benzthiazoline sulphonic acid), abbreviated to ABTS.[70] Glucose oxidase can be used in conjunction with polarographic oxygen electrodes in free or immobilised form.[71]

$$\beta\text{-D-glucose} + H_2O + O_2 \xrightarrow[\text{oxidase}]{\text{glucose}} \text{glucono-}\delta\text{-lactone} + H_2O_2 \quad (1)$$

$$H_2O_2 + \text{reduced chromogen} \xrightarrow{\text{peroxidase}} H_2O + \text{oxidised chromogen} \quad (2)$$

Alternatively a different system can be used for glucose with hexokinase and glucose-6-phosphate dehyrdogenase, where the genera-tion of NADPH from NADP is measured in the near uv (Eqns 3 and 4).

$$\beta\text{-D-glucose} + ATP \xrightarrow{\text{hexokinase}} \text{glucose-6-phosphate} + ADP \quad (3)$$

$$\text{glucose-6-phosphate} + NADP \xrightarrow[\text{dehydrogenase}]{\text{glucose-6-phosphate}} \text{6-phosphogluconate}$$

$$+ NADPH \quad (4)$$

Blood triglycerides are measured using lipase to release free glycerol which is determined by an analogous pyridine nucleotide linked system to that described above for glucose, with glycerol kinase and glycerol-1-phosphate dehydrogenase (Eqns 5–7).[72]

$$\text{triglyceride} \xrightarrow{\text{lipase}} 3 \text{ fatty acids} + \text{glycerol} \quad (5)$$

$$\text{glycerol} + \text{ATP} \xrightarrow[\text{kinase}]{\text{glycerol}} \text{glycerol-1-phosphate} + \text{ADP} \qquad (6)$$

$$\text{glycerol-1-phosphate} + \text{NAD} \xrightarrow[\text{dehydrogenase}]{\text{glycerol-1-phosphate}} \text{dihydroxy acetone}$$

$$\text{phosphate} + \text{NADH} \qquad (7)$$

Other analogous reactions using oxidases or dehydrogenases have been exploited in the analysis of a number of metabolites and all can be applied, if required, to foods. The specificity of the analyses can be extended to determine more complex substances by including the appropriate catabolic enzymes in the assay system. For example, the glucose methods can be applied specifically to determine sucrose or maltose by adding invertase or maltase. Free glucose can be distinguished from the disaccharides by taking a measurement before adding the glycosidase. Boehringer Mannheim and other major producers of enzymes for clinical analysis supply a parallel series of test kits specially marketed for food use. A few commercial enzymatic test kits have been developed specifically for use in food where the analytes concerned are important in food but not in clinical analysis. For example gluconate can be determined using a uv coupled assay involving gluconate kinase and 6-phosphogluconate dehydrogenase (Eqns 8 and 9).

$$\text{glucose} + \text{ATP} \xrightarrow[\text{kinase}]{\text{gluconate}} \text{6-phosphogluconate} + \text{ADP} \qquad (8)$$

$$\text{6-phosphogluconate} + \text{NADP} \xrightarrow[\text{dehydrogenase}]{\text{6-phosphogluconate}} \text{Ribulose-5-}$$

$$\text{phosphate} + CO_2 + \text{NADPH} \qquad (9)$$

An enzymatic method has been developed for the determination of starch in foods.[73] The sample can first be extracted with 75% isopropyl alcohol to remove low molecular weight sugars which would otherwise interfere in the assay. Then it is digested with a mixture of α-amylase from *Bacillus licheniformis* and amyloglucosidase from *Aspergillus niger* to convert the starch completely to glucose. This is determined by the glucose oxidase (Eqns 1 and 2) or hexokinase (Eqns 3 and 4) methods. A method for assay of cyanide in cassava using linamarinase extracted from the peel of the plant has been described by Rao and Hahn.[74] The enzyme is used to release bound cyanide which is then determined colorimetrically. The system has been automated to determine 300 samples/h. Smith

& Dacombe[75] have developed a method for determining glucosinolates in rapeseed by utilising the endogenous myrosinase activity. Parallel samples are extracted in acid methanol and water, glucose released by the enzyme from glucosinolates appears only in the aqueous extracts while pre-existing molecular glucose also appears in the methanol extracts. The glucose oxidase method is used to compare glucose levels in the blank and test samples. A specific β-glucanase purified from Novo bacterial amylase 1000S can be used to hydrolyse β-1-3, 1-4 glucans in foods to allow their determination with a colorimetric glucose assay.[76]

The technique of enzyme linked immunoassay (ELISA) can be used to determine the source species of foodstuffs or the degree of mixing or adulteration, for example pork in minced beef can be detected and determined. In ELISA an antibody to the protein to be determined is chemically coupled to an enzyme. The target protein (antigen) can then be determined in a reverse assay by allowing it to react with the enzyme-antibody complex, separating the precipitate, and determining enzyme activity in the supernatant (direct ELISA). Alternatively the chosen antigen can be labelled with enzyme and reacted with the antibody in competition with unlabelled antigen in the sample (indirect ELISA). Ayob et al.[77] have developed direct and indirect ELISA techniques for measuring pork in beef. Using an indirect assay with crude rabbit antiserum to porcine albumin 20% pork in beef can be detected, with a purified anti-porcine albumin the sensitivity can be improved 5% with the indirect method: using the direct assay as little as 1% pork can be found. Yasumoto et al.[78] have perfected an ELISA method to detect soy protein, using antibodies to peptide fragments, which is claimed to be capable of determining 4 g soy/kg protein in either raw or cooked foods.

Assay of endogenous enzymes can be used to determine condition or previous treatment of some foods. α-Amylase activity has been used to verify pasteurisation of eggs since the enzyme is inactivated under proper pasteurisation conditions. Determination of α-amalyase in flour can be used to assess the extent of pre-sprouting damage to the wheat.[79] α-Amylase is absent in the dormant grain but increases rapidly on sprouting.

Presence of specific enzymes in food can also be used as an index of microbial contamination. Laccase is not normally present in must or wine but is produced by the infecting mould Botrytis. Laccase can be determined in wine by a syringaldazine colorimetric test to assess the degree to which the grapes were originally subjected to Botrytis infection.[80]

APPLICATIONS OF ENZYMES IN SPECIFIC FOOD SECTORS

Baking

Endogenous amylases are active during dough mixing, fermentation and baking. Endogenous β-amylase is always sufficient, but it is often useful to add α-amylase to the flour to obtain the best results. α-Amylases in the grain vary according to the variety used and the growth and harvest conditions. In wet climates sprouting of some of the grains produces a large increase in α-amylase but in the dry summer conditions of the North American wheat-growing areas this does not occur and α-amylase is normally added to American flours. The fungal enzyme from *Aspergillus oryzae* is generally used but barley malt is an alternative source.[3] The α-amylase, together with the endogenous cereal β-amalyase, produces maltose from starch to act as a fermentation substrate for the yeast, so a sufficiency of enzymes is needed to ensure the desired loaf volume and texture. However, if the enzyme action persists in the finished loaf an undesirable soft and gummy, open texture is the result. Fungal α-amylase is best suited for flour supplementation because it has relatively low thermal stability, it remains active during the earlier phases of baking as the temperature rises but is destroyed by the time the peak baking temperature is reached, so it can be used at very high levels without adverse effects. Cereal α-amylases are slightly more thermostable but lose most of their activity during baking. However, with flours made from sprouted wheats sufficient activity may persist to cause problems. These can be mitigated by decreasing dough pH and adding more salt. Bacterial α-amylases are too stable to be inactivated under normal baking conditions and they cannot be used in bread. Curves comparing the thermal stabilities of the three types of enzyme are displayed by Underkofler.[2] In the baking of fruit cake, where a sticky texture is desired, bacterial α-amylases can be used and their persistent activity has a beneficial anti-staling effect.

The gluten content of flour determines the structural strength of the bread, which is related to the degree of cross-linking of the gluten molecules by S—S bonds. Flours with a high gluten content are termed 'strong' flours, and those made from North American wheat fall in this category. For these flours it is normal practice to add a fungal protease, again usually obtained from *A. oryzae* to reduce the strength of the cross-linking network by partial degradation of the gluten molecules. This has the beneficial effect of increasing the extensibility of the dough, thus decreasing the mixing time required. It also causes a small increase in loaf

TABLE 3
DAIRY USES OF ENZYMES

Enzyme	Source	Product	Effect
Rennin	Calf stomach	Cheese	Milk coagulation
Pepsin	Bovine, pig stomaches	Cheese	Milk coagulation
Acid protease	*Mucor* spp.	Cheese	Milk coagulation
Esterase	Calf, limb and kid sub-lingual glands	Italian Cheese	Flavour due to lipolysis
Protease, lipase	Various	Cheese	Flavour concentrate
Peptidase	*Brevibacterium linens*	Cheese	Accelerates ripening
Lysozyme	Egg-white, bacteria	Cheese	Prevents spoilage
Lactase	*Kluyveromyces* spp.	Milk	Sweetening, prevents lactose intolerance
Lactase	*Kluyveromyces* spp.	Ice-cream	Prevents lactose crystallisation
Lactase	*Kluyveromyces* spp.	Whey syrups	Solubilisation, sweetening
Lactase	*Aspergillus* spp.	Whey, syrups	Solubilisation, sweetening
Lactase	*Aspergillus* spp.	Whey, feed	Prevents lactose intolerance
Glucose isomerase	*Streptomyces* spp.	Whey syrups	Sweetening

volume and gives the bread a smoother texture. This addition is not normally required with 'weak' European wheat flours. Bacterial neutral proteases are added to cracker biscuit flours to control buckling in the oven which can result in undesirable blackening of the raised edges.

To obtain a paler colour in bread the enzyme lipoxygenase, obtained from defatted soy bean meal, is added to white bread flour. This produces a bleaching effect by oxidising the carotene pigment in a coupled reaction with unsaturated fatty acids and oxygen. The lipoxygenase also promotes the formation of gluten S—S bonds and increases the tolerance of the flour to overmixing.[3]

Rye flours contain a high content of pentosans which make their doughs difficult to mix. Pentosanases are added to some European rye bread flours to reduce dough viscosity. The addition has the further beneficial effect of preventing the separation of the crumb from the crust which sometimes occurs with these breads.

TABLE 4
ENZYMES USED IN ALCOHOLIC DRINKS

Enzyme	Product	Stage	Effect
Fungal α-amylase, Amyloglucosidase	Beer, Spirits	Mashing	Enhance fermentable sugar yield where low activity malts or unmalted grains are used
Bacterial α-amylase,	Beer, Spirits	Mashing	Improves yield from mashes with unmalted grains or high temperature gelling starch
β-1-3,1-4-glucanase, bacterial or fungal	Beer, Spirits	Mashing	Reduces viscosity and improves filterability and yield of wort from high β-glucan mashes
Bacterial neutral protease	Beer, Spirits	Mashing	Increases yield of assimilable nitrogen for yeast nutrition
Amyloglucosidase	Beer, Spirits	Fermentation	Increases fermentability; low calorie beers
Pectinases, fungal	Wine	Extraction	Increases yield of must
β-1-3-glucanases, fungal	Wine	—	Improves filtration and clarity
Glucose oxidase	Beer, Wine	Product	Prevents chill haze
Papain	Beer, Wine	Product	Prevents chill haze

Alcoholic Drinks

Fermentation of alcoholic drinks from grains or other starchy raw materials depends on enzymatic digestion of polysaccharides and proteins present to supply fermentable sugars and nutrients to the yeast. Traditional processes rely on the endogenous enzymes present in the raw materials. The malting process, where grains are allowed to sprout, increases the levels of enzymes important in the subsequent mashing (extraction) stage. The grains usually contain high levels of β-amylase before malting but the other essential enzymes α-amylase, protease and β-glucanase are initially deficient and appear during sprouting. Malted grains are dried and ground then subjected to a temperature profiled process of mashing, during which amino-acids are released by proteases and most of the starch present is converted to fermentable sugars by the amylases. The mash filtrate, called wort, is cooled and inoculated with yeast. Malt contains high levels of enzymes and concentrated extracts from it are sold as commercial enzyme preparations and sometimes used in brewing and in other applications. However, fungal and bacterial enzymes are more commonly used as aids to mashing and for other purposes in brewing, winemaking and distilled spirit production. These enzymes and their uses are listed in Table 4.

The malting stage has always been a nuisance to brewers and the availability of microbial enzymes led to interest in use of unmalted grains with added enzymes to replace those developed during malting. Practical systems for beer production from unmalted barley were described[113] but were not widely adopted because of the increasing public disquiet about alleged deterioration in beer quality resulting from departure from traditional methods. At present the practice in the U.S.A. is to use a mixture of malt and so-called 'adjuncts' such as rice and unmalted barley. The malt used is from barley varieties specially selected for high enzyme levels in their malts, but some microbial enzymes are also needed in mashing. In Germany only malt can legally be used and addition of enzymes is prohibited. Elsewhere in Europe there is some use of added enzymes but not on the same scale as in the U.S.A.

In brewing, the mashing stage is followed by addition of hops and increasing the temperature to boiling. This inactivates the enzymes present. Some unfermentable oligosaccharides ('dextrins') are left from the original starch and these remain throughout the fermentation stage and are present in the final beer where they contribute to its flavour and head retention properties. For low calorie or diabetic beers, amyloglucosidase can be added during the fermentation to ensure conversion to fermentable sugars and ultimately to alcohol. In distilling, the mash is not boiled and the enzymes remain active throughout the fermentation stage.[2,3]

In addition to the requirement in mashing adequately to hydrolyse the starch and protein components, it is necessary to reduce the content of the soluble β-1–3, 1–4-linked glucose polymer which occurs in cereals. This occurs in large quantities in barley and is often called barley β-glucan. It can cause loss of yield of wort due to its high viscosity. The relevant hydrolysing enzyme occurs in malt but sometimes in insufficient amounts, so that an addition is required. This can be from a bacterial source[30,114] or from a mould such as *Penicillium emersonii*.[29]

In extraction of the juice, called must, from grapes for wine fermentation the pectin content is an important factor. Grapes contain endogenous pectolytic enzymes but when fungal pectinase preparations became available it was found to be useful to add them to enhance must yields.[3] The mould *Botrytis cinerea* is normally present on certain types of grapes and this produces a β-1–3-linked glucan polymer which is carried through into the wine and causes problems with filtration and clarification. Addition of a fungal enzyme with β-1-3-glucan hydrolase activity alleviates these problems.[115]

Beer is subject to a phenomenon called chill haze, where a precipitate forms during long-term storage at cool temperatures. The precipitate is a protein–tannin complex and its formation is dependent on the presence of traces of dissolved oxygen. Addition of proteolytic enzymes to beer is widely practised as a means of preventing chill haze. Many types of proteolytic enzyme are effective but papain, obtained from the papaya fruit[17] is the one normally used because it survives pasteurisation in beer and does not adversely affect the flavour or head retention of the product.[116] An alternative approach to the problem is to use a glucose oxidase–catalase preparation to remove dissolved oxygen.[117] The enzyme can be used in an immobilised reactor where the beer is deoxygenated continuously prior to bottling. Immobilised amyloglucosidase can be used, if required, to generate glucose from the dextrins in the beer as a substrate for the oxidase.[118] Glucose oxidase has not found favour commercially in beer but is used to some extent in wines which also suffer from chill hazes and discoloration due to reactions involving dissolved oxygen.[119,120]

Dairy Products
There are several distinct areas of application of enzymes in milk and dairy products involving a number of enzymes. These are listed in Table 3.

Milk coagulation in cheese making is initiated by a proteolytic enzyme reaction. The added rennet is an extract, traditionally from calf stomach, containing the enzyme rennin. Rennin splits the κ-casein fraction of the milk protein at a specific peptide bond, to yield a modified protein with a high capacity for aggregation through the mediation of calcium ions, resulting in the formation from milk of a coagulum. In cheese making this aggregate is cut to release occluded liquid (whey) which contains salts, lactose and a small quantity of soluble protein. The precipitate, or curd, consists of a cross-linked protein network with the fat globules trapped in it. A detailed study of the kinetics of attack of rennin on κ-casein and subsequent coagulation has recently been published.[81–84] The traditional source of the milk-clotting enzyme, rennet, is the fourth stomach (abomasum) of the unweaned calf, or corresponding organ from the lamb or kid in the case of sheep and goat's milk cheeses, and these stomach extracts are still used for a proportion of cheese manufacture. With modern dairy practice the number of young animals slaughtered is insufficient to satisfy the demand for cheese coagulants and other sources have been exploited. All proteolytic enzymes will coagulate milk[15] but most are unsuitable for cheese making, either because their general proteolytic activity is too high relative to milk clotting activity so that excessive amounts of protein are solubilised and yields of cheese are depressed, or because bitter peptides are produced which spoil the flavour of the cheese. Extracts from stomachs of adult mammals contain pepsin, which is a good milk coagulating enzyme, and bovine or pig pepsins, alone or in mixtures with calf rennet, are also used in cheese making.

Shortage of calf rennet also stimulated the search for microbial sources of coagulants. An early patent claims use of bacterial proteinases and specifies *Bacillus* spp. as sources[85] but use of microbial coagulants (microbial rennins) did not establish itself until the 1960s.[86] A mould species *Endothia parasitica*, was screened out as an effective producer,[87] and this was the basis of a commercial product of Chas. Pfizer. It was found to be satisfactory for soft cheeses with short storage lives but was liable to cause flavour defects when used in long-maturing cheeses such as cheddar. Enzymes from *Mucor pusillus*[88,89] and *Mucor miehei*[90-92] were found to be suitable also for long-maturing cheese. The *Mucor meihei* enzyme is now used in a large proportion of cheese production.

During curd separation much of the clotting enzyme passes into the whey, which is increasingly processed further for use as a food ingredient. This disclosed a disadvantage of the *M. meihei* enzyme over calf rennet. Due to its higher thermostability the *M. meihei* enzyme survives the pasteurisation of the whey products which inactivates the calf enzyme. When the whey products are subsequently used in powered milk blends the residual enzyme has been found sometimes to cause them to coagulate on reconstitution. This problem has been circumvented by developing and marketing enzymes which have been chemically modified to reduce their thermostability.[93-96]

As mentioned above, with the establishment of recombinant DNA technology it is possible to produce authentic calf rennin by fermentation in microorganisms. The gene has been expressed in *Escherichia coli*,[7,97] *Bacillus subtilis*[8] and *Saccharomyces cerevisiae*.[9] Commercial microbial calf rennins are starting to become available.

A different approach to assuring adequate supply of rennet is to use it in an immobilised enzyme reactor to increase its productivity. The coagulation which follows attack of rennin on κ-casein occurs only at warm ambient temperatures but rennin action proceeds under refrigerated conditions, so it is possible to pass milk through a low temperature immobilised rennin reactor, then warm up the effluent to promote coagulation. There are numerous reports of success in this approach on the laboratory scale but Carlson,[98] has analysed a number of these and expressed doubt that a practical system can be developed. Many of the results achieved are explicable in terms of release of enzyme into solution in the reactor and the structure of the protein complex in milk would be expected to restrict access to an immobilised enzyme severely.

Italian cheeses have traditionally been made using rennet pastes, which are extracted from offal, which includes the sublingual glands of the animal. These contain esterase enzymes (also called pregastric lipases) which attack milk fat, producing mainly short and medium chain length fatty acids. This partial lipolysis induces characteristic flavour notes in the cheeses. Pregastric esterases are now extracted separately and used in conjunction with liquid rennet extracts of animal or microbial origin in making Italian cheeses such as Romano, Provolone and Mozzarella. The same enzymes are used to prepare lipolysed butter fat or cheese concentrate products for use in flavouring dairy based products.[3] Other lipases and proteases are also used for making a range of dairy flavours of different characteristics.

Long-maturing hard cheeses, of which the predominant type is Cheddar develop flavour as a result of enzymatic processes due principally to the bacteria present but also to some extent to the residual rennet in the curd and to the endogenous enzymes from the milk. The maturing process represents an expense in working capital and in provision of temperature-controlled storage facilities. Addition of proteases and lipases at the cheese making stage can accelerate flavour development with consequent reduction in costs.[99] The smear-ripened cheeses Limburger, Gruyere and Danbo have flavour characteristics in common with Cheddar and important among their ripening flora is *Brevibacterium linens*. This produces peptidases which are involved in generation of amino acids and these are further attacked enzymatically to produce flavour components.[100]

Certain types of cheese such as Gouda, Emmental and Edam are ripened during a storage period of a number of weeks by the action of propionic acid bacteria. If butyric acid bacteria invade the maturing cheese they can proliferate and cause flavour spoilage. In extreme cases the cheeses have been known to explode as a result of excessive gas production. The enzyme lysozyme (*N*-acetyl muramidase) is a natural antibacterial agent which occurs in egg-white. It attacks the butyric acid bacteria, e.g. *Clostridium tyrobutylicum*,[101] but does not significantly affect the ripening bacteria. Egg-white lysozyme can be used in preference to nitrate in cheese making as a prophylactic against spoilage.[102] Human milk also contains a lysozyme but cow's milk contains little or none of this enzyme. The beneficial effect of breast feeding compared with use of milk formulae can partly be attributed to the effect of lysozyme in suppressing undesirable gut bacteria. Egg-white lysozyme has been used to 'humanise' infant formulae based on cow's milk.

Lactase (β-galactosidase) hydrolyses lactose, the major carbohydrate constituent of milk, to the momosaccharides glucose and galactose. These are sweeter and more soluble than lactose and, unlike lactose, they are fermentable by normal commercial yeast strains. Lactose in milk or whey concentrates will normally start to crystallise above about 30% solids. At this level the concentrate is still susceptible to microbial spoilage. Hydrolysis with lactase allows higher solids syrups, which are microbiologically stable and are sweeter, to be produced. Lactase treated whey syrups can be used in bakery products, in confectionery, in soft drinks, and also in animal feeds as a replacement for molasses.[27] Lactase can be used in ice-cream to prevent lactose crystallisation in formulations where high levels of whey powder are used to replace milk powder.[103] Use of lactase can reduce the requirement for sucrose in the ice-cream, although the sweetness from hydrolysed lactose is insufficient to allow sucrose to be eliminated altogether. Using lactase does decrease the calorie content of ice-cream for a given sweetness level to some extent, and this also applies to other sweetened foods containing dairy ingredients.

Human and animal populations sometimes experience difficulty in assimilating lactose from milk products due to a deficiency of intestinal lactase which is required for lactose absorption. This causes proliferation in the lower bowel of bacteria which ferment lactose, resulting in the lactose intolerance symptoms of diarrhoea, flatulence and lower abdominal pain, caused by fermentation of lactose in the bowel by gut bacteria. Where milk is not normally a dietary constituent and is suddenly introduced to such a population, for example in food aid programmes, it is usually found that the adults have a high incidence of lactose intolerance. Young children always have sufficient intestinal lactase except in rare cases of genetic deficiency. Lactase treatment is desirable before drying of surplus milk for storage as an emergency food supply. Small sachets of lactase have been marketed to the public in the U.S.A. under the trade name Lactaid for treatment of individual cartons of milk, for sweetening and to prevent lactose intolerance symptoms.

Two types of lactase are available commercially. The yeast enzymes from *Kluyveromyces* spp. are optimally active at pH 6–7, and can be used directly in milk or skim milk, but have a relatively low thermal stability so that they cannot be used at temperatures above 40–45°C. The mould enzymes have more acid pH optima of 3·5–5·0 and can be used in cheese whey without pH adjustment. They are more thermally stable and generally have longer lifetimes in immobilised enzyme reactors than the yeast enzymes. The *Aspergillus niger* enzyme is slightly more heat-stable than that of *Aspergillus oryzae*. The mould enzymes are used at 50–60°C. A new mould lactase from *Aspergillus fonsecaeus* with greater temperature stability than the *A. oryzae* enzyme has recently been described.[104] No direct comparison with the *A. niger* enzyme was reported but from the data given it seems probable that the *A. fonsecaeus* enzyme will prove to be more tolerant to both temperature and pH conditions than any existing commercial lactases.

Immobilised yeast lactase has been used in commercial plants for lactose hydrolysis in milk.[105,106] The mould enzyme has also been used in a plant-scale reactor.[107] Many types of immobilised lactases and reactors for their use have been described.[27] A recent study[108] showed 80% hydrolysis in skimmed milk in 7 min at 30°C with a spiral flow reactor using the enzyme from *A. oryzae*.

In common with many other glycosidase enzymes, lactases catalyse glycosyl transfer to sugar hydroxyl groups as well as hydrolysis, particularly at high substrate and product concentrations. An accumulation of a variety of oligosaccharides can be detected in the digests, although the predominant products are monosaccharides. Prenosil *et al.*[109] have reviewed literature reports on this subject and have analysed possible models and mechanisms for the reaction. They examined transfer activity of four commercial lactases, two from *Kluyveromyces* spp. and two from *Aspergillus* spp. in experimental studies.[110] The maximum practical yield of monosaccharides under industrial conditions with immobilised enzymes was estimated at 85% and mould enzymes were judged to be preferable for maximal degrees of hydrolysis.

Abril & Stull[111] have described production of whey syrups where hydrolysis of lactose with yeast enzyme is followed by treatment with glucose isomerase from *Streptomyces olivaceus* to convert glucose to fructose, thereby further enhancing sweetness.

Addition of hydrogen peroxide is sometimes used as an alternative to pasteurisation for milk treatment, particularly in milk intended for use in cheese manufacture in the U.S.A. Catalase is subsequently added to the milk to destroy residual peroxide. Liver enzymes or a preparation from

Aspergillus niger can be used. The mould enzyme is the more stable to heat and to hydrogen peroxide inactivation. As a result it gives a higher value for activity in the Baker assay method[112] relative to its titre in the better known initial rate methods, than does the liver enzyme, and is more effective in use per initial rate unit of activity. The Baker assay is unusual in that it is an exhaustion method, where enzyme action is allowed to proceed until it has been completely destroyed by the excess of hydrogen peroxide. Use of hydrogen peroxide in milk is banned in some countries, including the U.K.

FOOD ADDITIVES AND OTHER INGREDIENTS

Enzymatic methods have been developed for several food additives. Examples are listed in Table 5. Most of these processes have not yet been applied commercially but offer promise.

Gaathon *et al.*[49] have described a method using tannase trapped in reverse micelles, which favours the synthesis of the antioxidant propyl gallate from tannic acid and propanol.

Legoy *et al.*[141] used a two-phase system for oxidation of geraniol to the flavour aldehyde geranial using horse liver alcohol dehydrogenase with nicotinamide adenine dinucleotide (NAD) as cofactor. The enzyme and cofactor are located in the aqueous phase while the alcohol and aldehyde are partitioned mainly in the organic phase. The enzyme had a half-life of several hundred hours in systems where the organic solvent used was ethyl acetate or hexane.

Fatty acid esters of monosaccharides have been prepared by lipase transesterification of sugar acetals with triglycerides.[142] The products may offer advantages over the sorbitan esters currently used as emulsifiers. Fayolle *et al.*[146] have synthesised the flavour additive butyl butyrate from the alcohol and acid using *Mucor meihei* lipase in a heptane-aqueous two-phase mixture. Both raw materials were produced by fermentation from enzymatically digested wheat flour to achieve a totally biological process.

The artificial sweetener aspartame (L-aspartyl L-phenylalanine methyl ester) was discovered by G. D. Searle in the 1960s[147] but has only recently reached large scale use. The original manufacturing process was to produce the amino acids by fermentation and use chemical derivatisation and peptide synthesis to complete the process. Various enzymatic steps have now been described. Production of L-aspartate from fumarate using a crude immobilised aspartate ammonia lyase has been in use for many

TABLE 5

ENZYMATIC PRODUCTION OF FOOD ADDITIVES

Product	Use	Enzyme	Reaction	Reference
Propyl gallate	Antioxidant	Tannase	Transesterification of tannin with propanol	Gaathon et al.[49]
Aldehydes	Flavour	Alcohol dehydrogenase	Oxidation of alcohols with NAD cofactor	Legoy et al.[141]
Fatty acid monosaccharides	Emulsifiers	Lipase	Interesterification of sugar acetal with fatty acid triglycerides	Vulfson et al.[142]
Phenylalanine	Aspartame production	Phenylalanine ammonia lyase	trans-Cinnamic acid + NH_3 → L-phenylalanine	Evans et al.[143]
Phenylalanine	Aspartame production	Transaminase	Phenylpyruvate + aspartate → L-phenylalanine + oxaloacetate	Ziehr et al.[144]
Aspartate	Aspartame production	Aspartate ammonia lyase	Fumarate + NH_3 → L-aspartate	Chibata et al.[145]
Aspartame	Sweetener	Thermolysin	N(Benzyloxycarbonyl)L-aspartate + L-phenylalanine methyl ester → Benzyloxycarbonyl aspartame methyl ester	Soda & Yonaha[5]
Butyl butyrate	Flavour	Lipase	Synthesis from butanol and butyric acid	Fayolle et al.[146]

years.[145] More recently Vojtisek et al.[148] (1986) have optimised a pilot scale process using a whole cell preparation from a selected strain of *Alciligenes metaciligenes*. Ziehr et al.[144] reported the use of L-aspartate phenylpyruvate transaminase from *Pseudomonas putida*. Two types of process were compared. In one the isolated enzyme was used in a membrane reactor where the enzyme was retained by the ultrafiltration membrane. In the other a fixed whole cell preparation on chitosan beads was used in a continuous stirred tank reactor. The free enzyme gave a threefold higher productivity than the fixed cell reactor, of 0·072 mol/litre per hour, but the fixed cells displayed higher long-term stability. An alternative route to phenylalanine is to use *trans*-cinnamic acid with the enzyme phenylalanine ammonia lyase. An optimised process using alginate-immobilised cells of *Rhodotorula rubra* has been described by Evans et al.[143] which is claimed to be a significant improvement on previously available methods.

Chemical synthesis of the aspartame peptide from the constitutent amino acids has partly been replaced by enzymatic condensation using immobilised thermolysin.[5]

Flavour concentrates can be prepared from protein sources by the application of proteolytic enzymes. Where dairy materials are used, unwanted intense bitter flavours can be produced, due to certain specific peptides. These peptides are products of the action of some types of proteolytic enzymes but not others, so the choice of enzyme source is important. Controlled low levels of hydrolysis can also be used to enhance such functional properties of proteins as foaming and binding. Use of proteases to modify or generate new food ingredients has been reviewed by Kilara.[149]

Two naturally occurring galactomannan gums, from guar and locust bean, are used as food ingredients. Locust bean gum is the more useful because it forms gels more readily in mixtures with other carbohydrate polymers, but it is less readily produced and is therefore more expensive. In structure both gums consist of a mannan backbone with galactose substituents but the guar polymer has a higher degree of substitution and thus has a higher galactose content (38% against 23% for locust bean gum). Treatment of guar gum with a purified α-galactosidase from the guar seed allows reduction of the galactose content to yield a polymer with similar composition and functional properties to the locust bean gum.[150] The guar α-galactosidase gene has been transferred to yeast where high levels of expression of the enzyme free of interfering β-mannosidase should be possible.

RECOVERY OF BY-PRODUCTS FROM FOOD AND AGRICULTURAL PROCESSES

Enzymatic treatments can be used to recover otherwise intractable residues, or to upgrade by-products, from production processes. Harvesting and processing of crops leaves many millions of tons of unwanted plant biomass constituents, such as sugar cane bagasse, straws from cereal crops and sugar-beet pulp. These are composed largely of complex carbohydrates which are potential sources of sugar for use directly or as fermentation raw materials. Acid hydrolysis processes can be used to hydrolyse these polysaccharides to sugars but are costly in energy and tend to leave unwanted sugar degradation products. Enzyme treatments offer the potential advantages of greater specificity and lower energy utilisation. The principal constituent of many of the biomass residues is cellulose and much attention has been devoted to its enzymatic breakdown. The crystalline nature of cellulose makes it a difficult enzyme substrate and a combination of enzymes working synergistically is required to break it down completely. Initially it was postulated that in addition to glycosidases, which attack the bonds linking the glucose units, other types of enzymes were involved in disrupting the crystalline structure, but it was later realised that the combined actions of glycosidases with different properties were sufficient to account for its degradation.[151] However, natural cellulosic materials are mixtures of cellulose with lignin and with other carbohydrates. Other activities, some of which have not yet been characterised, are involved in the degradation of plant biomass.

An early research objective was to develop practical systems for making fuel ethanol for motor vehicles from the various cellulosic wastes by enzymatic digestion followed by fermentation. A *Trichoderma reesei* enzyme product was developed which gave practical saccharification but was too low in activity for economic use.[152] This was improved by strain manipulation and used in a study of saccharification of a variety of cellulosic residues.[153] Cost of production of ethanol from municipal waste was then estimated at 34 US cents/litre of which 13 cents was the production cost of the enzyme. A mechanical pretreatment was needed for most of the cellulose sources investigated which cost 6 cents/litre of alcohol produced from municipal waste.

Xylan is also a major component of plant biomass. Many cellulase preparations contain adequate xylanase activity to degrade this so that their final products are mixtures of glucose and xylose.

Digestibility of biomass varies with the source of the material. Knapp *et al.*[154] found that digestibility of wheat straws by Trichoderma cellulase varied by up to 22% according to the variety of wheat involved. Hoffman & Wood[155] have selected mutants for effectiveness in digesting barley straw. There are estimated to be several million tonnes of surplus straw produced in Great Britain annually. The residues from mashing and wort separation (brewers' spent grains) have been shown to be susceptible to digestion with a cellulase preparation from *T. reesei* mutant RUT-C30, producing a yield of 42% sugar based on polysaccharide content, without pretreatment.[156]

Sugar beet pulp has a high pectin content but because of its composition the pectin has poor gelling properties compared with those of the pectins from the traditional apple and citrus sources. Matthew *et al.*[157] have reported that treatment of sugar beet pectin with a specially prepared enzyme extract from *A. niger* results in an improvement in gelling properties. This could offer a more productive use for the by-product beet pulp from the sugar industry, which is currently used for animal feed. Three enzymes in the *A. niger* preparation are involved in the effect, a deacetylase, a methylesterase and an arabinohydrolase.

The upgrading of cheese whey for food ingredients and animal feed using the enzyme lactase is discussed above in the section on milk and dairy products.

STARCH PROCESSING

Enzymes Used

The enzymes involved in starch digestion differ in different categories of organisms. For each starch-utilising species two or more enzymes are normally involved in the conversion to sugars. Starch consists of two structurally different glucose polymers. Amylose, present in cereal starches to the extent of 20–30%, is a linear polymer linked by α-1–4 glycosidic bonds. Amylopectin, accounting for the remaining 60–70% of the mass of cereal starch, is a branched polymer with all the subunits linked by α-1–4 bonds but with a number of branch points at each of which a glucose unit is also linked in an α-1–6 bond. Each branch is terminated by a glucose unit which is attached via the 1-carbon to the next unit and has the non-reducing 4-position free. As a result of this structural configuration there is only one reducing sugar group in the whole amylopectin molecule. The so-called exoglycosidases, such as β-amylases and amyloglucosidase, attack amylose and amylopectin by splitting off maltose (disaccharide) or glucose (monosaccharide) units sequentially from the non-reducing chain ends. These enzymes are also described as saccharifying amylases, since their action is to split off sugars. Plant β-amylases, when purified, do not attack the branch points in amylopectin so that a 'limit dextrin' is produced which is the core of the original amylopectin molecule with each branch terminated with a 6-linked glucose residue.

The commercial bacterial α-amylases are endoglycosidases, attacking links deeper within the structure and splitting off larger fragments. It has been said that they attack the chain at random points but this is probably an oversimplification. When they are allowed to catalyse their reactions to completion, a mixture of oligosaccharides is produced which is characteristic of the specific type of α-amylase.[158] With the corresponding enzymes from animals and higher plants the bacterial α-amylases are designated as liquefying amylases, because they have the property of preventing or reversing heat-induced gelatinisation of starch by reducing the polymer size rapidly.

The α and β designations, originally given to the enzymes from malt, relate to the anomeric configurations of the initial hydrolysis products. These depend on the precise reaction mechanism and are deducible from polarimetric data. The glucose residues in starch are bound in the α-configuration and this is retained when the glycosidic bond is broken by an α-amylase but inverted if the attacking enzyme is a β-amylase. Fungal α-amylases are endoamylases but their final products contain a high proportion of maltose, resembling those from β-amylases, and they are sometimes categorised as saccharifying amalyses. Because of their endo-mechanism of attack the fungal α-amylases can by-pass the branch points in amylopectin to produce finally a mixture of small oligosaccharides, mainly maltose and maltotriose.

Fungal amyloglucosidase also releases glucose in the α configuration. The effects of the various sorts of starch hydrolysing enzymes when they act alone on starch are summarised in Table 6. The purpose of Table 6 is to illustrate the reactions of the individual enzymes, which all occur naturally and in commercial preparations in conjunction with other types, so that a complete digestion of starch normally results. For example, malted grains contain a mixture of α and β-amylases, fungi often produce mixtures containing α-amylase and amyloglucosidase, and certain bacteria produce mixtures of β-amylase and a debranching enzyme.[159–161]

Of the enzymes listed in Table 6, only amyloglucosidase attacks α1–6 links, and that only slowly so that the overall hydrolysis rate of amylopectin is limited by occurrence of the branches. Separate enzymes specific for the hydrolysis of the α-1–6 links exist and are called debranching enzymes. The pullulanases attack the regular mixed α-1–4; α-1–6 polymer pullulan to produce maltotriose, and also attack the branch points in amylopectin: isoamylases do not attack pullulan but do attack α-1–6 links in amylopectin and glycogen.[2,162] These enzymes, especially pullulanases, can be used together with β-amylases or fungal α-amylases to produce high maltose syrups from starch, and to increase rates of conversion to glucose by amyloglucosidase. The first pullulanase to be established commercially was obtained from *Klebsiella pneumoniae*. This pullulanase-producing organism has previously been named as *Aerobacter aerogenes*[163] or *Klebsiella aerogenes*.[3] The *Klebsiella* enzyme is well suited to use in the production of maltose syrups in conjunction with cereal β-amylases or fungal α-amylase but it is rather less thermostable than the commercial *A. niger* amyloglucosidase and its pH optimum is higher, so that it is less well suited to dextrose production. A *Bacillus acidopullulyticus* enzyme whose properties more closely match those of amyloglucosidase has been developed[24,164,165,207] and this is sold for use in saccharification to glucose. Suzuki *et al.*[166] have described a newly discovered pullulanase with a very high thermal stability, from the thermophile *Bacillus flavocaldarius*. It is said to be active in 30% starch at 105°C and has an optimum pH of 6·2.

TABLE 6

PRODUCTS FROM STARCH HYDROLYSING ENZYMES

Enzyme	Products
Liquefying: (α-amylases)	
Cereal and animal	Oligosaccharides
Bacillus subtilis	Oligosaccharides[a]
Bacillus licheniformis (high temperature)	Oligosaccharides[b]
Saccharifying:	
Fungal α-amylase (*Aspergillus oryzae*)	Largely maltose
Cereal β-amylase	Limit dextrin, maltose[c]
Bacterial β-amylase	Limit dextrin, maltose[c]
Amyloglucosidase (*Aspergillus niger*)	Glucose[d]

[a]Predominantly 6-membered (maltohexaose) all smaller oligomers also present.
[b]Mainly 5-membered (maltopentaose) with maltose and maltotriose.
[c]Amylose gives solely maltose, amylopectin also gives limit dextrins.
[d]Completed slowly due to low rate on α-1,6 links.

Glucose has relatively low sweetness, whereas fructose is very sweet. Isomerisation of starch-derived glucose to fructose can be achieved enzymatically using an isomerase. A large proportion of the glucose syrup produced is now further processed to fructose-containing sweet syrups. A great deal of work has been devoted to enzymes for this application and numerous species have been investigated as sources. Most of these enzymes act *in vivo* as xylose isomerases but are also capable of acting on glucose.[34,43] Established commercial products include those from *Bacillus coagulans*,[167,168] *Actinoplanes missouriensis*[169] and *Streptomyces* spp.[170–172] These enzymes all have neutral or slightly alkaline pH optima for activity and stability. They are inhibited by calcium ions, and activated by certain other divalent metal ions such as magnesium. A glucose syrup obtained directly from the conventional process contains calcium, carried over from the liquefaction step where it is required to stabilise α-amylase, and has a pH of about 4·5, the optimum for amyloglucosidase. It is therefore necessary to remove the calcium by ion exchange, to add a different ion for activation, and to bring the pH up to a minimum of about 7·0 before applying the isomerisation step. Furthermore, the isomerisation is endothermic in the fructose production direction so that the equilibrium fructose concentration increases with temperature. At the process temperature of 60°C which is normally used the equilibrium mixture contains 55% of the sugars present as fructose and practical considerations restrict the product to 42% fructose. It would be an advantage to operate at higher temperatures to achieve higher concentrations of fructose but the commercially available enzymes are insufficiently stable to permit this. The ideal commercial glucose isomerase would be activated by calcium, would have a lower pH optimum for activity and stability, and a higher temperature tolerance. Enzymes from *Streptomyces acidodurans*,[173] *Streptomyces thermviolaceus*,[174] and from the extreme thermophile *Thermus aquaticus*[175] have been characterised and reported to have higher acid and thermal stabilities than other sources, although direct comparisons were not described. The *T. aquaticus* enzyme appears promising in its capability to work at lower pH values. The complete X-ray structure and action mechanism has been elucidated for an *Arthrobacter* enzyme and work on structure modification for improved properties by protein engineering has been initiated.[176–178]

An increasingly important group of products from starch are cyclodextrins. These are cyclic glucose oligomers, linked through 1 and 4 positions and therefore devoid of free reducing sugar groups, containing 6–8 glucose residues. They are able to form inclusion complexes with a variety of organic molecules and can be used as carriers for flavour or other additives and components of food.[179] Cyclodextrins are produced in high yield from starch by the action of cyclodextrin glycosyl transferase enzymes, which can be obtained from a variety of bacterial sources. Most produce mainly cyclohepta amylose (β-cyclodextrin), with minor amounts of cyclohexa amylose and cycloocta amylose (α and γ-cyclodextrins), but changes in the conditions and selection of the appropriate enzyme source can be used to vary the proportion of the products.[180,181]

Products obtained

The degree of hydrolysis of starch products is usually expressed as dextrose equivalent (DE). This is the proportion of the total saccharides present with free reducing groups expressed relative to molecular glucose as 100. The DE alone does not define the total composition of the mixture, for example a β-amylase digest of DE15 would contain about 30% maltose and the remainder as large molecules but a bacterial α-amylase digest of the same DE value would comprise mainly intermediate oligosaccharides up to about 12 glucose units. Thus the food ingredient properties of starch products such as viscosity and sweetness depend on the production process used as well as on the DE.

The acid hydrolysis of starch to sweet syrups was discovered by Kirschoff in 1811[18] and a commercial plant was in use by 1814.[2] Unlike enzymatic digestion, acid hydrolysis is random so that only one composition exists for a product of a given DE value. Acid hydrolysis of starch has largely been replaced by enzyme hydrolysis but was recently still in regular use to produce syrups up to DE42. Above this degree of hydrolysis with acid the products are unusable due to objectionable bitter taste and brown colour. The first use of an enzyme was treatment of DE42 acid syrups with fungal α-amylase to give a DE62-65 product of acceptable flavour and greater sweetness, containing a high proportion of maltose.[182] With the availability of commercial amyloglucosidase it became possible to produce high DE syrups in which the starch had been converted almost completely to glucose, and to use these to produce solid crystalline glucose. When amyloglucosidase was first prepared around 1950 it contained a transglucosidase impurity which depressed glucose yields and prevented commercial development at first. In the following decade this problem was solved by developing methods for removing transglucosidase[2] and by developing a mutant transglucosidase-free strain.[183] By the 1970s several million tonnes of starch were being converted annually to glucose.[158] By the early 1980s there was an additional production of about 5 million tonnes of fructose syrups, principally in the U.S.A. These use as feedstock a glucose syrup made with amyloglucosidase.

The major types of hydrolysis and isomerisation products of enzymatic treatment of starch are listed in Table 7. The soluble starches and maltodextrins are normally sold as dried solid products. They have a low level of sweetness and are used as thickeners, gums and bulking agents. The higher DE and isomerised syrups are used primarily as sweeteners in the beverage, baking, canning and confectionery industries. Of the sugars present in these products maltose is the least sweet, with about one-third the sweetness of sucrose on a weight basis, glucose has approximately three quarters of the sweetness of sucrose, while fructose is considerably sweeter than sucrose.[184] Lower DE syrups are stable with respect to crystallisation but if the glucose content exceeds about 44% in a commercial high solids syrup crystallisation occurs at ambient temperatures. The dextrose syrups need to be stored at a temperature of 60°C or above to avoid crystallisation.

Products of specific composition are made for particular applications using appropriate combinations and proportions of the various commercially available enzymes. Acid hydrolysis is still used to some extent for the initial liquefaction stage, mainly for the maltose and other intermediate DE syrups. For crystalline glucose or isomerised syrups where the maximum possible glucose yield is needed at the saccharification stage, bacterial α-amylase liquefaction is the method of choice.

Fructose syrup (42%) has the same sweetness on a solids basis as sucrose and is used for direct replacement of sucrose, particularly in soft drinks. The 42% fructose effluent from the enzyme reactors can be separated by chromatography into a fructose-rich fraction containing up to 90% fructose which is blended back to give a 55% fructose product. This has the same level of sweetness as invert sucrose and can be used to replace it in use. The glucose fraction from the chromatography stage can be recycled via the isomerisation column.

Process conditions

The processes used to obtain the various types of starch products are summarised in Fig. 1.

Starch has the property of gelatinising when it is heated as a suspension in water, this occurs at varying temperatures, characteristic of the source species, between 55°C and 75°C. Prior to gelatinisation, starch is less readily attacked by hydrolytic enzymes so an initial heating step is always used. At high concentrations gelatinised starch suspensions become

TABLE 7
PRODUCTS FROM ENZYMATIC STARCH PROCESSING

Name	DE	Enzymes used[a]	GLU	MAL	HSS	FRU
			Typical composition[c]			
Soluble starches	3–10	—	0	3	97	—
Maltodextrins	10–20	—	2	6	91	—
Maltose syrups	40–48	Fungal α-amylase[b]	6	47	47	—
Extra high maltose syrups	45–55	Fungal α-amylase[b] + pullulanase	2	75	23	—
High conversion syrups	65–70	Fungal α-amylase + amyloglucosidase	40	40	20	—
Dextrose syrups	94–96	Amyloglucosidase	94	3	3	—
Crystalline glucose	99	Amyloglucosidase	99	0	1	—
Isomerised syrup (42%)	—	Amyloglucosidase + glucose isomerase	52	2	4	42
Isomerised syrup (55%)	—	Amyloglucosidase + glucose isomerase	39	2	4	55

[a]Bacterial α-amylases always used in enzymatic liquefaction.
[b]Or cereal β-amylase.
[c]As % of total carbohydrates. GLU = glucose; MAL = maltose; HSS = higher oligosaccharides; FRU = fructose.

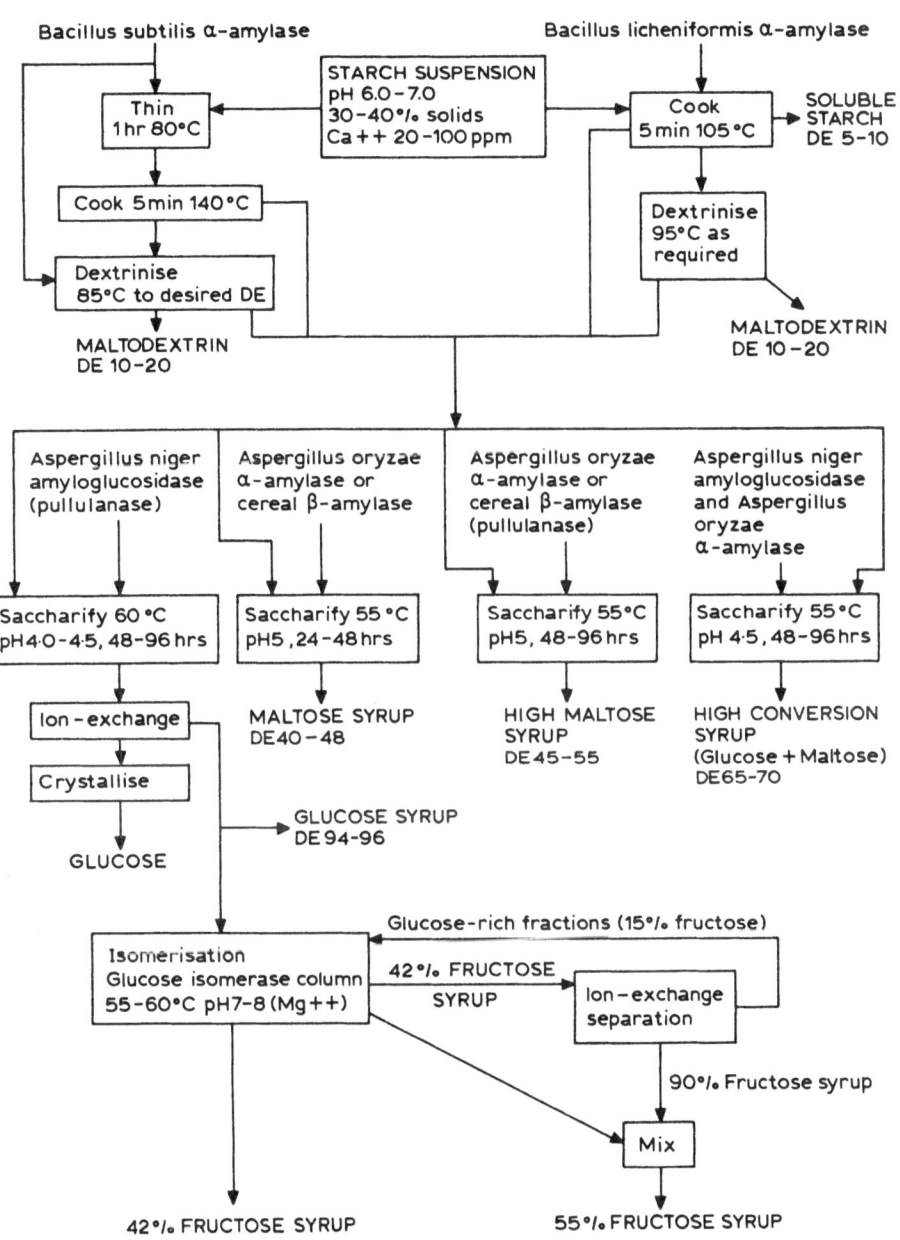

FIG 1.

extremely viscous and impossible to work so that some hydrolysis is essential to maintain a fluid consistency. Bacterial α-amylases, have a sufficiently high thermal stability to act above starch gelatinisation temperatures, and are used for the liquefaction stage. Cereal enzymes are marginally stable at the temperatures required and fungal α-amylases are ineffective because of inactivation before gelatinisation temperatures are reached. The first bacterial α-amylases used in the liquefaction process were made with *B. subtilis* or *B. stearothermophilus* and the normal process for their use involves a holding period of 1 h at around 80°C followed by briefly heating to 140°C in a pressure vessel. The mix is cooled to 85°C and more enzyme is added. The introduction of a more stable α-amylase from *B. licheniformis* permits a single addition to be used with an initial heating period of a few minutes at 105°C in a jet cooker, followed by a further incubation at 95°C.[19,185] The bacterial α-amylases are stabilised by calcium ions in solution and Ca^{2+} is added at 20–100 ppm to the starch feedstock. Incubation period following the high temperature treatment at 140°C or 105°C can be varied to obtain products of different DE values. For soluble starches the DE is low. Products of lower viscosity can be made by allowing enzyme action to proceed near to completion at around DE20. If the product from the liquefaction stage is to be used in further processing to obtain dextrose isomerised syrups it is best to restrict the DE of the liquefied product to the lowest possible value.

The saccharifying stages in starch processing involve a protracted incubation of up to 4 days, usually at 60°C for production of glucose with amyloglucosidase or at 50–55°C where maltogenic enzymes are involved. The pH of the liquefied starch is first adjusted to 4–4·5 for amyloglucosidase or 5–5·5 for the other enzymes. Total conversion to glucose is never quite achieved, because of the transglycosidase activity of amyloglucosidase. A final mixture with a DE one or two points short of 100 is obtained containing small amounts of oligosaccharides, of which the major component is the α-1–6-linked disaccharide isomaltose. Effectively complete hydrolysis to glucose is possible at low starch concentrations, where relatively few potential acceptor molecules are available, but requirements for high plant throughput dictate the use of 30–40% starch suspensions in industrial practice. For crystalline dextrose production or use for isomerised syrup manufacture, where maximum glucose levels are a paramount consideration, the starch concentration used in the saccharification stage is restricted to 30%.[2]

Maltose syrups of DE 40–48 are produced using a single enzyme incubation with either cereal β-amylase or fungal α-amylase. These products contain low levels of glucose and up to 50% by weight of maltose. 'Extra' high maltose syrups can be made by using a debranching enzyme with the maltogenic enzyme and prolonging the incubation to 2–4 days. Maltose contents in the product of more than 80% on a dry solids basis are possible.[20]

High conversion syrups are prepared using mixtures of fungal α-amylase or cereal β-amylase with amyloglucosidase and stopping the reaction by heating to prevent too much of the maltose from being converted to glucose. The products contain a mixture of glucose and maltose, have a very high fermentability, and are stable to crystallisation at lower temperatures than dextrose syrups made with amyloglucosidase alone.

Feedstocks for isomerisation are treated by ion-exchange to remove calcium ions and evaporated to 40–50% dry solids. An activating divalent cation for glucose isomerase is added, for example 10 ppm Mg^{2+} for Sweetzyme from *B. coagulans*.[35] The reaction is conducted using an immobilised enzyme in a packed column. Enzymes from a number of different sources using a number of immobilisation techniques have been exploited and these have slightly different properties.[34] All use temperatures of 60–65°C and require some adjustment to the ions in the feed. The enzymes used have relatively high pH optima and this is a disadvantage because a double pH adjustment, before and after treatment, is required and because the sugar mixtures tend to deteriorate and form coloured products if subjected to alkaline conditions. Initially, some enzymes were used at pH 8·5 and 65°C and under those conditions a maximum exposure of a few hours was specified. pH values of 7–7·5 are now more commonly used but an even lower pH would be desirable. Lifetimes in use for the enzyme columns are measured in months but a gradual deterioration in performance is seen. This can be circumvented by reducing the feed rate so that the % conversion to fructose at the outlet remains the same but this means that a slow decline in throughput occurs. This can be balanced by using a number of columns at different stages in their life cycle so that total output remains approximately constant.[35]

A process for the production of β-cyclodextrin (cycloheptaamylose) using cyclodextrin glycosyl transferase (CGOT) from an alkalophilic *Bacillus* sp. has been described by Matsuzawa *et al.*[186] Fifteen percent potato starch is liquefied using *B. subtilis* α-amylase and incubated with CGOT at 60°C for 20 h and pH 8·5 in the presence of 10 mM Ca^{2+}. After a heating step to stop transferase action linear oligosaccharides are hydrolysed using α-amylase and amyloglucosidase and the cyclodextrin is concentrated and crystallised with a final yield after recrystallisation of

24% β-cyclodextrin. A continuous laboratory process with an immobilised form of the same enzyme has also been described.[187]

Bender[188] has characterised a CGOT from *Klebsiella pneumoniae*. The capability to vary the proportions of the differently-sized cyclodextrins using this enzyme has been investigated.[181] Under standard conditions the predominant product is the 7-membered β-cyclodextrin but if sodium acetate and bromobenzene, which forms an inclusion complex with cyclodextrins, are present throughout the process a predominance of α-cyclodextrin is formed. If sodium acetate is used, and the addition of bromobenzene delayed until several hours into the reaction, the ultimate product contains up to 18% γ-cyclodextrin, the 8-membered analogue, which is valuable because of its greater solubility in water and its ability to complex larger molecules. Although this digest still contained 34% of the β form, the possibility of economic production of the γ-cyclodextrin was claimed. The other conditions for incubation are: unthinned starch suspension at 15%; 5 mM Ca^{2+}; pH 6·9; temperature 40°C; duration 24 h. Yagi *et al.*[180] describe a process using an enzyme from *Micrococcus* spp. with 10% soluble starch containing 12 mM Ca^{2+} at pH 6 with incubation at 60°C for 48 h, yielding 60% of mixed cyclodextrins.

Possible Developments
In industrial starch processing a gelatinisation stage is always necessary but it would be an advantage if it were possible to run the digestion without prior cooking. 'Raw' or native starch is attacked only slowly by amylolytic enzymes, and action by the saccharifying enzymes is very restricted. Commercial amyloglucosidases contain two forms of the enzyme, the larger of which has a peptide molecular weight of 71 000 and has some action on native starch while the smaller 62 000 Da form has no action at all on raw starch, probably because it lacks a starch binding site. In at least two different commercial strains of organism the two forms of the enzyme are derived from a single gene and their synthetic processes diverge at the mRNA stage.[189–193] With the existing amyloglucosidase products the rates of attack on raw starch are too low for commercial application. Fairbairn *et al.*[194] screened for raw starch digestion and selected a strain of *Streptomyces limosus* producing an enzyme which is capable of converting raw starch granules largely to maltose. However, in their opinion the temperature stability of this enzyme was too low for commercial exploitation. Fujii *et al.*[195] report that amyloglucosidase and bacterial α-amylase exhibit synergism in attacking starch granules and attribute this to the action of the amyloglucosidase in 'peeling' starch

molecules from the granule surface. In current industrial practice α-amylase and amyloglucosidase are always used consecutively. The joint action of the enzymes on raw starch should be investigated further.

Bacterial β-amylases have long been known[196,197] but have not been used industrially because they offer no advantage over the fungal α-amylases in production of maltose syrups. Outtrup & Norman[198] describe a maltogenic α-amylase from *Bacillus stearothermophilus* which has higher acid and thermal stability than either bacterial β-amylase or fungal α-amylase. Production of 70% maltose syrup using a 30% liquefied starch suspension at pH 5·5 and 60°C was described. With a pullulanase addition the maltose content could be enhanced to 80%. The enzyme was produced in low yield by the parent strain but gene transfer into *B. subtilis* was undertaken to enhance the yield. Saha and Zeikus[199] have described a process for producing maltose syrups using an even more thermostable enzyme from *Clostridium thermosulphurogenes* at 75°C and pH 5·0 which gave 65% maltose from 35% dry solids, DE10 maltodextrin when used alone and up to 80% maltose when *B. acidopullulyticus* pullulanase was also used. Only slightly lower maltose levels were obtained using raw starch suspensions.

Starch is a poor substrate for application of immobilised enzymes because of its high molecular weight and complex structure. However, thinning treatment with α-amylases leaves a soluble product which should be susceptible to application of immobilised saccharifying enzymes. Introduction of glucose isomerase made immobilised saccharifying enzymes more attractive because an integrated continuous process became a possibility. In practice a large amount of work on immobilised amyloglucosidase has so far failed to produce a commercially successful system. Products of high glucose content can be made but these prove to be a little lower in DE than conventionally made syrups. There are two major reasons for this. At DE values below 15–20 the thinned starches contain some particulate and colloidal matter which induces severe problems in operation of columns but not in stirred tanks used for soluble enzymes. To obtain maximum glucose yield from saccharification it is desirable to restrict the thinning treatment to avoid by-product formation so that if higher DE feedstock is used to avoid operating problems with columns the glucose yield is slightly depressed. Frequently the immobilised systems depend more strongly on diffusion than stirred tank reactors and this seems to be associated with a higher level of transfer activity, causing marginal increases in isomaltose and panose in the product at the expense of glucose. The resultant reduction of 1·5–2% in final yield has prevented use in major applications.[42] Further attention to immobilisation technique and design of reactors can mitigate this. It has been shown that, while physically trapped enzyme gave a 5% lower DE than soluble enzyme, an enzyme covalently linked to carrier particle surfaces gave a DE only 1% lower than the soluble version.[200] Use of surface-immobilised enzyme in plug-flow reactors using high enzyme concentrations and short path lengths at high shear rates might allow competitive performance with the soluble enzyme.

Novel systems for use in starch processing with enzymes have been described. Larsson et al.[44] worked with a dextran–polyethylene glycol two-phase aqueous mixture, used in a phase separation-recirculation system incorporating membrane separation of the products. The starch feed did not require to be gelatinised and B. licheniformis α-amylase and A. niger amyloglucosidase were used together. Glucose purity in the ultrafilter permeate was reported to be higher than that achieved with conventional processes. A study has also been made of an ultrafiltration reactor for use in the hydrolysis of extruded cassava starch by amyloglucosidase.[201] Glucose of high purity was obtained but the concentrations of starch used, at 1–8%, were low relative to commercial practice.

SUGAR

In processing sugar-beet, amounts of raffinose sufficient to inhibit sugar crystallisation may be present, and this reduces the yield of sugar. Enzymatic hydrolysis of raffinose to sucrose plus galactose by α-galactosidase gives a potential improvement in these cases. Yamane[202] described a process where third-stage molasses was treated at pH 5·2, 50°C, with a washed mycelium of Mortierella vinacea to reduce raffinose content. An improvement in yield and productivity was obtained after adoption of the process at the Hokkaido Sugar Company plant in Japan. Kobayashi & Suzuki[203] optimised fermentation of the same organism to obtain mould pellets which could be used directly in a column reactor, but under these conditions leakage of enzyme from the pellets and resultant loss in the product stream restricted column life severely. Delente et al.[137] isolated a new strain of B. stearothermophilus which produced a thermostable neutral α-galactosidase. An advantage over the mould enzymes was anticipated because it was said that the slightly acid pH values required by them could cause sucrose inversion and protein precipitation in the process. Raffinose content can reach as high as 10% in mother liquors. Treatment of final molasses with *M. vinacea* α-galactosidase can yield a further crop of sugar crystals.

Recovery of sugar from cane offers different problems which are also susceptible to solution using enzymes. Infection of cane juice with *Leuconostoc mesenteroides* leads to production of dextran polymers from sucrose. This causes high viscosity and problems with handling the juice in the refinery. A method for removing the dextran enzymatically using a dextranase from *Penicillium funiculosum* has been described.[204] Sugar-cane contains significant quantities of starch which can also interfere with the sugar refining process. Use of amylase treatment to alleviate this problem has been known for over 30 years.[202] More recently the problem has been reinvestigated.[205] It was confirmed that α-amylases from *B. licheniformis* or *B. subtilis* were useful in this application. An *A. oryzae* preparation was found to contain invertase and was therefore unsuitable for use in sugar refining, but it could be used in cane juice intended for fermentation to alchohol.

Invertase from yeast can be used for industrial scale production of invert syrups from sugar. It has the advantage over acid hydrolysis that the salt content of the syrup is not increased as a result of the process. Use of invertase has been reviewed by Woodward and Wiseman,[206] who recommend use of a *Candida utilis* enzyme with a higher thermal stability than the better-known *Saccharomyces cerevisiae* product. Various immobilisation techniques are listed and critically reviewed by the same authors.

EGGS

Egg-white contains about 0·4% glucose and when bulk whole or separated egg is dried this can participate in Maillard spoilage reactions with the egg proteins. Fermentation methods have been used to remove glucose ('desugar') prior to drying but current practice is to use the same glucose oxidase–catalase preparations which are applied to drinks to convert the glucose to gluconic acid in an enzymatic reaction. Bulk whole egg is sometimes treated before drying but the main use of the enzyme is in separated bulk egg-white (albumen) which is an important ingredient in the baking and confectionery industries. There is insufficient dissolved oxygen in egg to react with all the glucose present so an additional source is required. Air sparging is possible but the usual method is to add hydrogen peroxide continuously or intermittently while gently stirring the batch. The peroxide is decomposed by the catalase present in the enzyme preparation to maintain a high oxygen content. Treatment usually takes

place over a period of several hours at low temperature.[121] Numerous
methods for the immobilisation of glucose oxidase, alone or together with
catalase, have been described but their use in egg treatment is problema-
tic because of the viscous nature of the system and the requirement for
dissolved oxygen. Work on an immobilised process is nevertheless continuing. Sankaran et al.[122] have described a system where the enzymes are
immobilised on cotton cloth and used in a batch reactor to desugar whole
egg.

Dried egg albumen must have a high foaming capability and the foam
needs to be stable. These properties are sometimes inhibited by fat carried over from the egg-yolk during separation and there is also a small
amount of fat which originates in egg-white. Much of this fat is coagulated
by the foaming induced by use of hydrogen peroxide in desugaring and
can be removed mechanically but some remains and pancreatic or fungal
lipase is sometimes added to ensure a very low fat content.[2]

FATS

The characteristic flavours of cheese and some other dairy products are
owed in part to the presence of shorter-chain fatty acids released from the
milk fat during maturation. Lipases, especially those also known as pregastric esterases, which are obtained from calf, lamb or kid sublingual
glands, have been used to prepare flavour concentrates for adding back to
cheeses or to other foods where a 'cheesy' flavour is desired.[2] A *Mucor
meihei* enzyme can also be used.[6]

Generation of objectionable rancid flavours in fats is initiated by oxidative cleavage of unsaturated fatty acids. The oxygen scavenging action of
glucose oxidase has been found to help prevent deterioration in storage of
animal lard[123] and butter.[124] This finding does not seem to have been
exploited commercially.

The use of lipases in reduced water systems to exchange fatty acids in
triglycerides, mentioned above, has been described by Macrae.[4,47] Palm
oil interesterified with stearic acid using *Rhizopus delemar* lipase gives a
product containing high levels of glycerol 1 (3)-palmitoyl-3 (1)-stearoyl-2-
monooleate and 1,3-distearoyl-2-monooleate, which are the main constituents of cocoa butter. In addition, using interesterification of an oil
with a free fatty acid or with a different oil can produce lipids with previously unavailable composiitons having new properties and possibilities
for new users. Goderis et al.[125] have studied the kinetics of interesterification of triolein with palmitic acid by immobilised *Rhizopus arrhizus* lipase
at different temperatures, water contents, and substrate concentrations.

They concluded that control of the water content was critical in determining the composition of the products. Wisdom *et al.*[126] described a pilot-scale packed bed reactor where interesterifications were achieved between shea oleine and stearic acid or shea oil and myristic acid using *R. arrhizus* lipase.

In reaction mixtures where glycerol is the primary solvent component, lipases can be used to synthesise glycerides from free fatty acids. Tahoun and Ali[127] found that up to 50% of the oleic acid in a mixture could be converted to mixed glycerides, mainly monoolein and diolein, by *R. delemar* lipase.

VEGETABLE MILKS

Legumes and certain root vegetables contain members of a homologous series of non-reducing oligosaccharides composed of a sucrose residue substituted on the 6-position of glucose with galactose or a short chain of galactose residues linked α-1–6. The trisaccharide raffinose and the tetrasaccharide stachyose are common but smaller amounts of the higher homologues occur with them. Because the lower gut in mammals possesses no enzymes capable of hydrolysing these sugars they are not absorbed and as a result they ferment in the bowel, causing flatulence and discomfort. Among the plant organs which contain raffinose oligosaccharides is the soy bean. Soy bean 'milk' can be fed as a substitute to children who are allergic to cow's milk. It has a nutritional value equal to cow's milk but needs to be treated to remove the raffinose sugars. This can be achieved using an enzyme treatment with α-galactosidase. Use of column reactors containing the immobilised enzymes from *B. stearothermophilus*,[128,129] has been demonstrated using pure sugars as model substrates. Smiley *et al.*[130] have used the enzyme from *Aspergillus awamori* in a hollow-fibre reactor fed with soy bean milk. An endogenous α-galactosidase occurs in soy beans: its purification has been described by Porter *et al.*[131] Alternative sources of the enzyme are guar seedlings,[132] and others listed by Lindley.[28]

FRUIT JUICES AND SOFT DRINKS

Pectolytic enzymes are used in the extraction of juices from fruits. Pectin is a polymer composed mainly of galacturonic acid residues containing substituent methyl ester groups. Pectin may also contain some rhamnose and also certain other sugars. Two mechanisms exist for enzymatic depolymerisation of pectin. In one the methyl groups are first removed by a pectinmethylesterase, then the chain is broken down by polygalacturonases, which are usually unable to act on pectin until most of the substituent methyl groups have been removed. A third type of enzyme, pectin transeliminase (also called pectin lyase), breaks the glycoside bonds to depolymerise the molecule by an elimination mechanism, it is able to act directly on pectin, without prior removal of methyl substituents. All three enzymes are present in most commercial pectinase preparations, which usually come from A. niger. Pectinases are used for two purposes: to reduce the viscosity of pressed fruit to allow an easier and higher-yielding juice separation, and to remove suspended pectin and related substances from extracted juices where the requirement is for the juice to be sparkling clear. In the latter case enzymes to hydrolyse other potential haze-forming polymers may be added separately to the juice, or may be added to the pectinase preparation by the enzyme manufacturer. For example, haze in apple juice is often due to starch and here an α-amylase is required. The commercial pectinases contain a variety of other endogenous side activities, some of which may be involved in producing the optimum effect. These side reactions are still not completely understood and the manufacture and uses of pectinases have been likened to a 'black art'. Methods for pectinase use in juice preparation have been reviewed by Charley[133] and more recently by Whitaker.[134] An immobilised pectinase has been made by Lozano et al.[135] and tested in treatment of fresh apricot juice.

Glucose oxidase–catalase products have been used to protect fruit juices and soft drinks from oxidative deterioration of flavour and colour.[208] Only certain specific types of drinks are affected and the enzymes are not in general use. The most extensive application is in citrus drinks, containing a proportion of natural juice. These are displayed in bottles in the open in Latin America and are susceptible to flavour deterioration under the conditions of strong light and high ambient temperature which prevail.[2]

MEAT

Meat toughens *post mortem* due to the locking of the actin and myosin filaments to form actomyosin. In storage the endogenous proteases gradually reduce toughness by dissociating the actin filaments from their anchorages on the Z discs and by attacking collagen. This process may take up to 2 weeks in some species and when it reaches maximum tenderness some meat is still tougher than is desirable.[136] Plant proteases are applied to meat to tenderise it.[138] Ficin, from figs, bromelain, from pineapple stems, and papain, from papaya fruits, are all used. Most of the market is served by papain. In the normal method of use most of the tenderising occurs during cooking, between the times when the collagen in the meat melts and the enzyme is inactivated by heat at about 80°C

Lean meat wastes, for example that left on heads or on bones, can be recovered for use in edible products using proteolytic digestion to solubilise them. The recovered products can be used for enriching the flavour of soups and canned meat products.[139] O'Meara and Munro[140] compared the performance of thirteeen commercial proteases from diverse sources and selected 'Alcalase' from *B. licheniformis* as the most suitable for this use.

REFERENCES

1. Reed, G., *Enzymes in Food Processing*. Academic Press, New York, 1966.
2. Underkofler, L. A., Enzymes. In *CRC Handbook of Food Additives*, Vol. 2, 2nd Edn, ed. T. E. Furia. CRC Press, Boca Raton, Florida, 1980, pp. 57–124.
3. Peppler, H. J. & Reed, G., Enzymes in food and feed processing. In *Biotechnology*, Vol 7a, ed. H.-J. Rehm & G. Reed. Verlag, Weinheim, 1987, pp. 547–603.
4. Macrae, A. R., Enzyme-catalysed modification of oils and fats. *Phil. Trans. R. Soc. Lond. Ser. B.*, **310** (1985) 227–233.
5. Soda, K. & Yonaha, K., Application of free enzymes in pharmaceutical and chemical industries. In *Biotechnology*, Vol 7a, ed. H.-J. Rehm & G. Reed, Verlag, Weinheim, 1987, pp. 605–652.
6. Kilara, A., Enzyme modified lipid food ingredients. *Process Biochem.*, **20** (1985) 35–45.
7. Carey, N. H., Harris, T. J. R., Lowe, P. A., Doel, M. T. & Emtage, J. S., A process for the production of a polypeptide. U.K. Patent 2100737 (1983).
8. Beppu, T., Uozumi, T., Tsuchiya, M. & Sekine, S. Expression plasmids useful in *B. subtilis*. European Patent 154351 (1985).
9. Mellor, J., Dobson, M. J., Roberts, N. A., Tuite, M. F., Emtage, J. S., White, S., Lowe, P. A., Patel, T., Kingsman, A. J. & Kingsman, S. M., Efficient synthesis of enzymatically active calf chymosin in *Saccharomyces cerevisiae. Gene*, **24** (1983) 1–14.
10. Aunstrup, K., Production, isolation and economics of extracellular enzymes. In *Applied Biochemistry and Bioengineering*, Vol. 2, eds. L. Wingard, E. Katchalaski-Katzir & L. Goldstein. Academic Press, New York, 1979, pp. 27–69.
11. Lambert, P. W. & Meers, J. L., The production of industrial enzymes. *Phil Trans. R. Soc. Lond. Ser. B.*, **300** (1983) 263–282.
12. Frost, G. M., Commercial production of enzymes. In *Developments in Food Proteins 4*, ed. B. J. F. Hudson. Elsevier Applied Science Publishers, London, 1986, pp. 57–134.

13. ter Haseborg, E., Enzymes in flour and baking applications especially waffle batters. *Process Biochem.*, **16** (1981) 16–19.
14. Delente, J., Process for the Preparation of Heat Resistant Neutral Protease Enzyme. U.S. Patent 3796635 (1974).
15. Berridge, N. J., Manufacture, purification and properties of rennin. In Production and Applications of Enzyme Preparations in Food Manufacture, (SCI Monograph No 11). Society of Chemical Industry, London, 1961, pp. 64–70.
16. Sternberg, M., Microbial rennets. In *Adv. Appl. Microbiol.*, Vol. 20, ed. D. Perlman, 1976, pp. 135–157.
17. Balls, A. K., U.S. Dept. Agriculture, Circular No. 631. Protein-Digesting Enzymes from Papaya and Pineapple, 1941.
18. Banks, G. T., Binns, F. & Cutcliffe, R. L., Recent developments in the production and industrial applications of amylolytic enzymes derived from filamentous fungi. In *Progress in Industrial Microbiology*, Vol. 6, ed. D. J. D. Hockenhull, Iliffe, London, 1967, pp. 95–139.
19. Madsen, G. B., Norman, B. E. & Slott, S., A new heat-stable bacterial amylase and its use in high temperature liquefaction. *Staerke*, **25** (1973) 304–308.
20. Saha, B. C. & Zeikus, J. G., Biotechnology of maltose syrup production. *Process Biochem.*, **22** (1987) 78–82.
21. Cadmus, M. C., Jayko, L. G., Hensley, D. E., Gasdorf, H. & Smiley, K. L., Enzymatic production of glucose syrup from grains and its use in fermentation. *Cereal Chem.*, **43** (1966) 658–668.
22. Fogarty, W. M., Microbial amylases. In *Microbial Enzymes and Biotechnology*, ed. W. M. Fogarty. Applied Science Publishers, London, 1983, pp. 1–92.
23. Orskov, I., Klebsiella. In *Bergey's Manual of Systematic Bacteriology*, Vol. 1, ed. N. R. Krieg & J. G. Hill. Williams & Wilkins, Baltimore, 1984, p. 321.
24. Jensen, B. F., & Norman, B. E., *Bacillus acido pullulyticus* pullulanase. Application and regulatory aspects for the food industry. *Process Biochem.*, **19** (1984) 129–134.
25. Pilnik, W., Enzymes in the beverage industry, (fruit juices, nectar, wine, spirits and beer). In *Util. Enzymes Technol. Aliment. Symp. Int.*, ed. P. Dupuy. Tech. Doc. Lavoisier, Paris, 1982, pp. 425–450.
26. Nijpels, H. H., Lactases and their applications. In *Enzymes and Food Processing*, ed. G. G. Birch, N. Blakeborough & K. J. Parrer. Applied Science Publishers, Barking, 1981, pp. 89–104.
27. Gekas, V. & Lopez-Leiva, M., Hydrolysis of lactose—a literature review. *Process Biochem.*, **20** (1985) 2–12.
28. Lindley, M. G., Cellobiase, melibiase and other disaccharidases. In *Developments in Food Carbohydrates*, Vol. 3, ed. C. K. Lee & M. G. Lindley. Elsevier Applied Science Publishers, London, 1982, pp. 141–165.
29. James, A. E., Fare, G., Sagar, B. F., Lucas, F. & Mitchell, I. de G., Improvements in or Relating to Enzymes. U.K. Patent 1421127, (1976).
30. Huber, J., Mueller, H., Dickscheit, R., Haefner, B., Herfort, B. & Riedel, K., Process for the Production of Glucane Decomposing Enzymes. U.K. Patent 1222396 (1971).

31. Pal, N., Das, S. & Kundu, A. K., Influence of culture and nutritional conditions on the production of lipase by submerged culture of *Aspergillus niger*. *J. Ferment. Technol.*, **56** (1978) 593–596.

32. Iwai, M. & Tsujisaka, Y., Fungal lipase. In *Lipases*, ed. B. Borgstrom & H.L. Brockman. Elsevier, Amsterdam, 1984, pp. 443–469.

33. Laboureur, P. & Labrousse, M., Lipase de *Rhizopus arrhizus*. Obtention purification et proprietees de la lipase de *Rhizopus arrhizus* var delemar. *Bull. Soc. Chim. Biol.*, **48** (1966) 747–770.

34. Bucke, C., Enzymes in fructose manufacture. In *Enzymes and Food Processing*, ed. G. G. Birch, N. Blakeborough & K. J. Parker. Applied Science Publishers, London, 1981, pp. 51–72.

35. Oestergaard, J. & Knudsen, S. L., Use of Sweetzyme in industrial continuous isomerisation. Various process alternatives and product types. *Staerke*, **28** (1976) 350–356.

36. Underkofler, L. A., Properties and applications of the fungal enzyme glucose oxidase. *Proc. Intern. Symposium Enzyme Chem. Tokyo and Kyoto, 1957*, Vol. 2. (1958) pp. 486–490.

37. Hesseltine, H. W. & Wang, H. L., Fermented foods. *Chemistry and Industry*. (1979) 393–399.

38. Takamine, J. Process of Making Diastatic Enzyme. U.S. Patent 525823 (1894).

39. Chibata, I., *Immobilised Enzymes, Research and Development*. Wiley, New York, 1978.

40. Kennedy, J. F. & Cabral, J. M. S., Enzyme immobolisation. In *Biotechnology*, Vol. 7a. ed. H.-J. Rehm & G.Reed. Verlag, Weinheim, 1987, pp. 347–404.

41. Pitcher, W. H., *Immobilised Enzymes in Food Processing*. C.R.C. Press, Boca Raton, Florida, 1980.

42. Rugh, S., Nielsen, T. & Poulsen, P. B., Application possibilities of a novel immobilised glucoamylase. *Staerke*, **31** (1979) 333–337.

43. Bucke, C., Industrial glucose isomerase. In *Topics in Enzyme and Fermentation Technology*, Vol. 1, ed. A. Wiseman. Ellis Horwood, Chichester, 1977, pp. 147–171.

44. Larsson, M., Arasaratnam, V. & Mattiason, B., Integration of bioconversion and downstream processing: starch hydrolysis in an aqueous two-phase system. *Biotechnol. Bioengng.*, **33** (1989) 758–766.

45. Andersson, E. & Hahn-Haegerdahl, B., Bioconversions in aqueous two-phase systems. *Enzyme microb. Technol.*, **12** (1990) 242–254.

46. Kirby, C., Delivery systems for enzymes.*Chem. Britain*, Sept. (1990) 847–850.

47. Macrae, A. R., Lipase-catalysed interesterification of oils and fats. *J. Am. Chem. Soc.*, **60** (1983) 291–294.

48. Fletcher, P. D. I., Freedman, R. B., Robinson, B. H., Rees, G. D. & Schomaecker, R., Lipase-catalysed ester synthesis in oil-continuous microemulsions. *Biochim. Biophys. Acta*, **912** (1987) 278–282.

49. Gaathon, A., Gross, Z. & Rozhanski, M., Propyl gallate: enzymatic synthesis in a reverse micelle system. *Enzyme Microb. Technol.*, **11** (1989) 604–609.

50. Fukui, S. & Tanaka, A., Enzymatic reactions in organic solvents. *Endeavour, New Series*, **9** (1985) 10–17.
51. Dordisk, J. S., Enzymatic catalysis in monophasic organic solvents. *Enzyme Microb. Technol.*, **11** (1989) 194–211.
52. Visuri, K. & Klibanov, A. M., Enzymatic production of high fructose corn syrup (HFCS) containing 55% fructose in aqueous ethanol. *Biotechnol. Bioengng.*, **30** (1987) 917–920.
53. Vulfson, E. N., Sarney, D. B. & Law, B. A., Enhancement of subtilisin-catalysed interesterification in organic solvents by ultrasound irradiation. *Enzyme Microb. Technol.*, **13** (1991) 123–126.
54. Anonymous, *Regulatory Aspects of Food Enzymes*, Third edition. The Association of Microbial Food Enzyme Producers, Brussels, 1988.
55. Wuthrich, B. & Ott, F., Berufsasthma durch proteasen in der Waschmit-telindustrie. *Schweiz. Med. Wochenschr.*, **99** (1969) 1584–1586.
56. Flindt, M. L. H., Respiratory hazards from papain. *The Lancet* (8061) (1978) 430–432.
57. Tang, P., Nielsen, G. C., Gibson, K., Aunstrup, K. & Schiff, H. E., Protease Product of reduced Allergenicity. U.K. Patent 2024830, (1979).
58. Anonymous, Report PB204118. Ad hoc Committee on Enzyme Detergents, US National Academy of Sciences, National Research Council, 1971.
59. Flindt, M. L. H., Health and safety aspects of working with enzymes. *Process Biochem.*, **13** (1978) 3–7.
60. Roland, J. F., Regulation of food enzymes. *Enzyme Microb. Technol.*, **3** (1981) 105–110.
61. Anonymous, Code of Federal Regulations. 21CFR, Sect.110.1. Office of the Federal Register, National Archives and Records Service, General Services Administration. Washington, D.C., 1981.
62. Anonymous, Ministry of Agriculture Fisheries and Food. Food Additives and Contaminants Committee Review of Remaining Classes of Food Additives Used as Ingredients in Food. Report on the Review of Enzyme Preparations. HMSO London, 1982.
63. Noordliviet, P. F. & Toet, D. A., Safety in enzyme technology. In *Biotechnology*, Vol. 7a, ed. H.-J. Rehm & G. Reed. Verlag, Weinheim, 1987, pp. 711–741.
64. Grampp, E. G., Modification of certain foodstuffs by enzymes. *Process Biochem.*, **17** (1982) 3–6, 12.
65. Bergmeyer, H.-U., ed. *Methods of Enzymatic Analysis*, 3rd edition, 10 vols. Verlag Chemie, Weinheim, 1983.
66. Karube, I., Analytical applications of enzymes: enzyme sensors for clinical, process and environmental analysis. In *Biotechnology*, ed. H.-J. Rehm & G. Reed. Verlag, Weinheim, 1987, Vol. 7a, pp. 685–708.
67. Keilin, D. & Hartree, E. F., The use of glucose oxidase (notatin) for the determination of glucose in biological materials and for the study of glucose-producing systems by manometric methods. *Biochem. J.*, **42** (1948) 230–238.
68. Teller, J. D., Direct quantitative colorimetric determination of serum or plasma glucose. *Abstrs. 130th Meeting Am. Chem. Soc.* (1956) 69C.

69. Trinder, P., Determination of glucose in blood using glucose oxidase with an alternative oxygen acceptor. *Ann. Clin. Biochem.,* **6** (1969) 24–27.
70. Werner, W., Rey, H.-G. & Wielinger, H., Uber die eigenschaften eines neuen chromogens fur die blutzuckerbestimmung nach der GOD/POD methode. *Fresenius Z. Anal. Chem.,* **252** (1970) 224–228.
71. Updike, S. J. & Hicks, G. P., Reagentless substrate analysis with immobilised enzymes. *Science,* **158** (1967) 270–272.
72. Gore, H. G., An automated enzymic method for the determination of glycerol. *Anal. Biochem.,* **75** (1976) 604–610.
73. Karkalas, J., An improved enzymic method for the determination of native and modified starch. *J. Sci. Food Agric.,* **36** (1985) 1019–1027.
74. Rao, P. V. & Hahn, S. K., An automated enzyme assay for determining the cyanide content of cassava (*Manihot esculenta* Crantz) and cassava products. *J. Sci. Food Agric.,* **35** (1984) 426–436.
75. Smith, C. A. & Dacombe, C., Rapid method of determining total glucosinolates in rapeseed by measurement of enzymatically released glucose. *J. Sci. Food Agric.,* **38** (1987) 141–150.
76. Henry, R. J., Evaluation of a general method for measurement of (1–3), (1–4)–β-glucans. *J. Sci. Food Agric.,* **44** (1988) 75–87.
77. Ayob, M. K., Rayab, A. A., Allen, J. C., Farag, R. S. & Smith, C. J., An improved rapid ELISA technique for detection of pork in meat products. *J. Sci. Food Agric.* **49** (1989) 103–116.
78. Yasumoto, K., Sudo, M. & Suzuki, T., Quantitation of soy protein by enzyme linked immunosorbent assay of its characteristic peptide. *J. Sci. Food Agric.,* **50** (1990) 377–389.
79. Henry, R. J., Rapid α-amylase assays for assessment of presprouting damage in wheat. *J. Sci Food Agric.,* **49** (1989) 15–23.
80. Grassin, C. & Dubourdieu, D., Quantitative determination of Botrytis laccase in musts and wines by the syringaldazine test. *J. Sci. Food Agric.,* **48** (1989) 369–376.
81. Carlson, A., Hill C. G. & Olson, N. F., Kinetics of milk coagulation I. Kinetics of kappa casein hydrolysis in the presence of enzyme deactivation. *Biotechnol. Bioengng.,* **29** (1987) 582–589.
82. Carlson, A., Hill C. G. & Olson, N. F., Kinetics of milk coagulation II. Kinetics of the secondary phase, micelle flocculation. *Biotechnol. Bioengng.,* **29** (1987b) 590–600.
83. Carlson, A., Hill, C. G. & Olson, N. F., Kinetics of milk coagulation III. Mathematical modelling of the kinetics of curd formation following enzymatic hydrolysis of kappa casein — parameter estimation. *Biotechnol. Bioengng.,* **29** (1987) 601–611.
84. Carlson, A., Hill, C. G. & Olson, N. F., Kinetics of milk coagulation IV. The kinetics of the gel-forming process. *Biotechnol. Bioengng.,* **29** (1987) 612–624.
85. Shimwell, J. L. & Evans, E. E., Improvements in the Manufacture of Cheese and Other Substances Depending on the Clotting of Milk. U.K. Patent 565788 (1944).
86. Sardinas, J. L., New sources of rennet. *Process Biochem.,* **4** (1969) 13–16, 21.

87. Sardinas, J. L., Rennin enzyme of *Endothia parasitica*. *Appl. Microbiol.*, **16** (1968) 248–255.
88. Arima, K., Iwasaki, S. & Tamura, G., Milk clotting enzyme from microorganisms. Part I. Screening test and identification of the potent fungus. *Agric. Biol. Chem.*, **31** (1967) 540–545.
89. Iwasaki, S., Tamura, G. & Arima, K., Milk clotting enzyme from microorganisms. Part II. The enzyme production and the properties of the enzyme. *Agric. Biol. Chem.*, **31** (1967) 546–551.
90. Aunstrup, K., Improvements in or Relating to a Milk-Coagulating Enzyme. U.K. Patent 1108287 (1968).
91. Charles, R. L., Gertzman, D. P. & Melachouris, N., Milk-clotting Enzyme Product and Process Therefor. U.S. Patent 3549390 (1970).
92. Feldman, L. I., Verfahren zur Herstellung von Mikrobiellen Labferment. German Patent Application No. 1767183.
93. Cornelius, D. A., Process for Decreasing the Thermal Stability of Microbial Rennet. U.K. Patent 2024828. (1980).
94. Branner-Jorgensen, S., Thermal Destabilisation of Microbial Rennet by Acylation Thereof. U.K. Patent 2038339 (1980).
95. Branner-Jorgensen, S., Schneider, P. & Eigtved, P., A Method of Modifying the Thermal Destabilisation of Microbial Rennet, Rennet so Modified and a Method of Cheese Making Using Rennet so Modified. U.K. Patent 2045773 (1980)
96. Cornelius, D. A. Process for Decreasing the Thermal Stability of Microbial Rennet. U.K. Patent 2058082 (1981).
97. Nishimori, K., Kawaguchi, Y., Hidoka, M., Uozumi, T. & Beppu, T., Expression of cloned calf prochymosin gene sequence in *Escherichia coli*. *Gene,* **9** (1982) 337–344.
98. Carlson, A. Coagulation of milk with immobilised enzymes. *Enzyme Microb. Technol.,* **6** (1984) 46–47.
99. Law, B. A., Flavour development in cheeses. In *Advances in the Microbiology and Biochemistry of Cheese and Fermented Milk,* ed. F. L. Davies & B. A. Law. Elsevier Applied Science Publishers, London, 1984, pp. 187–208.
100. Hayashi, K. & Law, B. A., Purification and characterisation of two aminopeptidases produced by *Brevibacterium linens*. J. Gen. Microbiol., **135** (1989) 2027–2034.
101. Wasserfall, F. & Teuber, M., Action of egg-white lysozyme on *Clostridium tyrobutylicum*. Appl. Environmental Microbiol., **38** (1979) 197–199.
102. Teuber, M., Lysozyme as a substitute for nitrate in cheese making. *Milchwirtsch. Ber. Bundesanst. Wolfpassing Rotholz*, **63** (1980) 129–130.
103. Gregory, K. W. The application of yeast lactase to the production of ice cream. In *Util. Enzymes Technol. Aliment. Symp. Int.*, ed. P. Dupuy. Tech. Doc. Lavoisier, Paris, 1982, pp. 249–252.
104. Gonzalez, R. R. & Monsan, P., Purification and characterisation of β-galactosidase from *Aspergillus fonsacaeus*. *Enzyme Microb. Technol.*, **13** (1991) 349–352.
105. Dinelli, D., Fibre-entrapped enzymes. *Process Biochem.*, **7** (1972) 9–12.
106. Marconi, W., Bartoli, F., Morisi, F. & Marani, A., Improved whey treatment by immobilised lactase. *Enzyme Engineering 5, Proc. Int. Conf., Henniker, New Hampshire, (1979)*. Plenum Press, 1980, pp. 269–278.

107. Dohan, L. A., Baret, J. L. Pain, S. & Delalande, P., Lactose hydrolysis by lactase, semi-industrial experience. In *Enzyme Engineering 5, Proc. Int. Conf., Henniker, New Hampshire, 1979*. Plenum Press, 1980, pp. 279–293.

108. Bakken, A. P., Hill, C. G. & Amundson, C. H., Hydrolysis of lactose in skim milk by immobilised β-galactosidase in a spiral flow reactor. *Biotechnol Bioengng.* **33** (1989) 1249–1257.

109. Prenosil, J. E., Stuker, E. & Bourne, J. R., Formation of oligosaccharides during enzymatic lactose: part I. State of the art. *Biotechnol. Bioengng.*, **30** (1987) 1019–1025.

110. Prenosil, J. E., Stuker, E. & Bourne, J. R., Formation of oligosaccharides during enzymatic lactose hydrolysis and their importance in a whey hydrolysis process: part II. Experimental. *Biotechnol. Bioengng.*, **30** (1987) 1028–1031.

111. Abril, J. R. & Stull, J. W., Lactose hydrolysis in acid whey with subsequent glucose isomerisation. *J. Sci. Food Agric.*, **48** (1989) 511–514.

112. Scott, D. & Hammer, F. E., Assay of catalase for industrial use. *Enzymologia*, **22** (1960) 194–198.

113. Wieg, A. J., Technology of barley brewing. *Process Biochem.*, **5** (1970) 46–48.

114. Tomkins, A. L. & Aunstrup, K., Improvements in or relating to preparation of an enzyme product. U.K. Patent 1380451 (1975).

115. Villetatz, J.-C., Steiner, D. & Trogus, H., The use of a beta glucanase as an enzyme in wine clarification and filtration. *Am. J. Enol. Vitic.*, **35** (1984) 253–256.

116. Brocklehurst, K., Baines, B. S. & Kierstan, M. J. P., (1981) Papain and other constituents of *Carica papaya*. In *Topics in Enzyme and Fermentation Biotechnology*, Vol. 5, ed. A. Wiseman. Wiley, New York, 1981, pp. 262–335.

117. Ohlemeyer, D. W., Use of glucose oxidase to stabilise beer. *Food Technol.*, **11** (1957) 503–507.

118. Hartmeier, W. & Willox, I. C., Immobilised glucose oxidase and its use for oxygen removal from beer. *Techn. Q., Master Brew. Assoc. Am.*, **18** (1981) 145–149.

119. McLeod, R. & Ough, C. S., Some recent studies with glucose oxidase in wine. *Am. J. Enol. Vitic.*, **21** (1970) 54–61.

120. Ough, C. S., Further investigations with glucose oxidase—catalase enzyme systems for use in wine. *Am. J. Enol. Vitic.*, **26** (1975) 30–36.

121. Scott, D., Applications of glucose oxidase. In *Enzymes in Food Processing*, ed. G. Reed, 2nd edition. Academic Press, New York, 1975, pp. 519–547.

122. Sankaran, K., Godbole, S. S. & D'Souza, S. F., Preparation of spray-dried sugar-free egg powder using glucose oxidase and catalase co-immobilised on cotton cloth. *Enzyme Microb. Technol.*, **11** (1989) 617–619.

123. Osadchaya, I. F., The use of glucose oxidase and catalase in animal fat production. *Mol. Biol.* (Kiev), **6** (1971) 115–116.

124. Dedek, M., Hanus, J. & Vedlich, M., Method for the Production of Long-life Butter. Czech. Patent 132372, (1968).

125. Goderis, H. L., Ampe, G., Feyten, M. P., Fouwe, B. L., Guffens, W. M., van Cauwenbergh, S. M. & Tobback, P.P., Lipase-catalysed ester exchange

reactions in organic media with controlled humidity. *Biotechnol. Bioengng.*, **30** (1987) 258–266.

126. Wisdom, R. A., Dunnill, P. & Lilly, M. D., Enzymic interesterification of fats: laboratory and pilot-scale studies with immobilised lipase from *Rhizopus arrhizus*. *Biotechnol. Bioengng.*, **29**, (1987) 1029–1085.

127. Tahoun, M. K. & Ali, H. A., Specificity and glyceride synthesis by mycelial lipases of *Rhizopus delemar*. *Enzyme Microb. Technol.*, **8** (1986) 429–432.

128. Reynolds, J. H., An immobilised α-galactosidase continuous flow reactor. *Biotechnol. Bioengng.*, **16** (1974) 135–147.

129. Korus, R. A. & Olson, A. C., The use of α-galactosidase and invertase in hollow fiber reactors. *Biotechnol. Bioengng.*, **19** (1977) 1–8.

130. Smiley, K. L., Hensley, D. E. & Gasdorf, H. J., Alpha-galactosidase production and use in a hollow-fibre reactor. *Appl. Environ. Microbiol.*, **31** (1976) 615–617.

131. Porter, J. E., Ladisch, M. R. & Hermann, K. M., Ion-exchange and affinity chromatography in the scale-up of the purification of α-galactosidase from soybean seeds. *Biotechnol. Bioengng.*, **37** (1991) 356–363.

132. Shivanna, B. D., Ramakrishna, M. & Ramados, C.S., Enzymic hydrolysis of raffinose and stachyose in soybean milk by α-galactosidase from germinating guar (*Cyamopsis tetragonolobus*). *Process Biochem.*, **24** (1989) 197–199.

133. Charley, V. L. S., Use of enzymes in the processing and storage of juices and other fruit products. In *Production and Applications of Enzyme Preparations in Food Manufacture*. (SCI Monograph No 11). Society of Chemical Industry, London, 1961, pp. 107–120.

134. Whitaker, J. R., Pectic substances pectin formation and haze formation in fruit juices. *Enzyme Microb. Technol.*, **6** (1984) 341–349.

135. Lozano, P., Manjon, A., Romojano, F. & Iborra, J. L., Properties of pectolytic enzymes covalently bound to nylon for apricot juice clarification. *Process Biochem.*, **23** (1988) 75–78.

136. Dransfield, E. & Etherington, D., Enzymes in the tenderisation of meat. In *Enzymes and Food Processing*, ed. G. G. Birch, N. Blakeborough & K. J. Parker. Applied Science Publishers, London, 1981, pp. 177–194.

137. Delente, J., Johnson, J. H., Kuo, M. J., O'Connor, R. J. & Weeks, L. E., Production of a new thermostable neutral α-galactosidase from a strain of *Bacillus stearothermophilus*. *Biotechnol. Bioengng.*, **16** (1974) 1227–1243.

138. Caygill, J. C., Sulphydryl plant proteases. *Enzyme Microb. Technol.*, **1** (1979) 233–242.

139. Petersen, B. R., Recovery and use of proteins. In *Enzymes and Food Processing*, ed. G. G. Birch, N. Blakeborough & K. J. Parker. Applied Science Publishers, Barking, 1981, pp. 149–175.

140. O'Meara, G. M. & Munro, P. A., Selection of a proteolytic enzyme to solubilise lean beef tissue. *Enzyme Microb. Technol.*, **6** (1984) 181–185.

141. Legoy, M. D., Kim, H. S. & Thomas, D., Use of alcohol dehydrogenase for flavour alcohol production. *Process Biochem.*, **20** (1985) 145–148.

142. Vulfson, E. N., Pickersgill, R. W. & Law, B. A., Lipases and phospholipases: media optimisation and protein engineering. *Proc. Int. Conf. Ind. Appl. Enzymes*, Pisa, 1990.

143. Evans, C. T., Choma, C., Peterson, W. & Misawa, M., Bioconversion of

trans-cinnamic acid to L-phenylalanine in an immobilised whole cell reactor. *Biotechnol. Bioengng., 30* (1987) 1067–1072.

144. Ziehr, H., Kula, M.-R., Schmidt, E., Wandrey, C. & Klein, J., Continuous productionof L-phenylalanine by transamination. *Biotech. Bioengng., 29* (1987) 482–487.

145. Chibata, I., Tosa, T. & Sato, T., Applications of immobilised biocatalysts in pharmaceutical and chemical industries. In *Biotechnology,* Vol. 7a, eds. H.-J. Rehm & G. Reed, Verlag, Weinheim, 1987, pp. 653–684.

146. Fayolle, F., Marchal, R., Monot, F., Blanchet, D. & Ballerini, D., An example of production of natural esters: synthesis of butyl butyrate from wheat flour. *Enzyme Microb. Technol., 3* (1991) 215–219.

147. Mazur, R. H., Schlatter, J. M. & Goldkamp, A. H., Structure-taste relationships of some dipeptides. *J. Am. Chem. Soc.,* **91** (1969) 2684–2691.

148. Vojtisek, V., Guttman, T., Barta, M. & Netrval, J., Preparation of L-aspartic acid by means of immobilised *Alcaligenes metalcilagenes* cells. *Biotechnol. Bioengng., 28* (1986) 1072–1079.

149. Kilara, A., Enzyme modified protein food ingredients. *Process Biochem.,* **20** (1985) 149–157.

150. Bulpin, P. V., Gidley, M. J., Jeffcoat, R. & Underwood, D. R., Development of a biotechnological process for the modification of galactomannan polymers with plant α-galactosidase. Carbohydr. Polymers, 12 (1990) 155–168.

151. Wood, T. M. & McCrae, S. I., Synergism between the enzymes involved in the solubilisation of native cellulose. In *Adv. Chem. Ser. 181, 1978. Hydrolysis of Cellulose. Mechanisms of Enzymatic and Acid Catalysis.* American Chem. Soc., Washington D.C., 1979, pp. 189–209.

152. Ryu, D. D. Y. & Mandels, M., Cellulases: Biosynthesis and applications. *Enzyme Microb. Technol., 2* (1980) 91–102.

153. Allen, A. A., Enzymic hydrolysis of cellulose to fermentable sugars. In *Liquid Fuel Developments,* ed. D. L. Wise. CRC Press, Boca Raton, Florida, 1983, pp. 49–64.

154. Knapp, J. S., Parton, J. H. & Walton, N. I., Enzymic saccharification of wheat straw derived from different cultures of winter wheat. *J. Sci. Food Agric.,* **34** (1983) 433–439.

155. Hoffman, R. M. & Wood, T. M., Isolation and partial characterisation of a mutant of *Pencillium funiculosum* for the saccharification of straw. *Biotechnol. Bioengng.,* 27 (1985) 81–85.

156. Khan, A. W., Lamb, K. A. & Schneider, H., Recovery of fermentable sugars from the brewers' spent grains by the use of fungal enzymes. *Process Biochem.,* 23 (1988) 172–175.

157. Matthew, J. A., Howson, S. J., Keenan, M. H. J. & Belton, P. S., Improvement of gelation properties of sugarbeet pectin following treatment with an enzyme preparation derived from *Aspergillus niger*. Comparison with a chemical modification. *Carbohydr. Polymers,* **12,** (1990) 295–306.

158. Norman, B. E., New developments in starch syrup technology. In *Enzymes and Food processing,* ed. G. G. Birch, N. Blakeborough & K. J. Parker. Applied Science Publishers, Barking, 1981, pp. 15–50.

159. Harada, T., Yokobayashi, K. & Misaki, A., Formation of isoamylase by Pseudomonas. *Appl. Microbiol.*, **16** (1968) 1439–1444.

160. Takasaki, Y., Studies on amylases from Bacillus effective for production of maltose. Part I. Production and utilisation of β-amylase and pullulanase from *Bacillus cereus* var. mycoides. *Agric. Biol. Chem.*, **40** (1976) 1515–1522.

161. Takasaki, Y., Studies on amylases from Bacillus effective for production of maltose. Part II. Purification and enzymic properties of β-amylase and pullulanase from *Bacillus cereus* var. Mycoides. *Agric. Biol. Chem.*, **40** (1976) 1523–1530.

162. Manners, D. J., Specificity of debranching enzymes. *Nature*, **234**, (1971) 150–151.

163. Bender, H. & Wallenfels, K., Untersuchungen an pullulan II. Spezifische abbau ein bacterielles enzym. *Biochem. Z.*, **334** (1961) 79–95.

164. Nielsen, G. C., Diers, I. V., Outtrup, H. & Norman, B. E., Debranching Enzyme Product. Preparation and Use Thereof. U.K. Patent 2097405. (1982).

165. Norman, B. E., A novel Bacillus pullulanase, its properties and application in the glucose syrups industry. *J. Jpn. Soc. Starch Sci.*, **30** (1983) 200–211.

166. Suzuki, Y., Hatagaki, K. & Oda, H., A hyperthermostable pullulanase produced by an extreme thermophile, *Bacillus flavocaldarius* KP1228, and the evidence for the proline theory of increasing enzyme thermostability. *Appl. Microbiol. Biotechnol.*, **34** (1991) 707–714.

167. Yoshimura, S., Danno, G. & Natake, M., Studies on D-glucose isomerising activity of D-xylose grown cells from *Bacillus coagulans* strain HN68. Part I. Description of the strain and conditions for formation of the activity. *Agric. Biol. Chem.*, **30** (1966) 1015–1023.

168. Zittan, L., Poulsen, P. B. & Hemmingsen, St. H., Sweetzyme—a new immobilised glucose isomerase. *Staerke*, **27** 236–241.

169. Anheuser-Busch Inc., Method of Making Glucose Isomerase and Using Same to Convert Glucose to Fructose. UK Patent 1399408.

170. Takasaki, Y., Kosugi, Y. & Kanbayashi, A., Streptomyces glucose isomerase. In *Fermentation Advances*, ed. D. Perlman. Academic Press, London, 1969, pp. 561–589.

171. Iizuka, H., Ayukawa, Y., Suekane, S. & Kanno, M., Production of Extracellular Glucose Isomerase by Streptomyces. U.S. Patent 3622463 (1971).

172. Dworschack, R. G., Chen, J. C., Larnon, W. R. & Davies, L. G., Process for Producing Glucose Isomerase, U.K. Patent 1284218, (1972).

173. Bok, S. H., Jackson, L. G., Schroedel, C. J. & Seidman, M., Carbohydrases from Thermophilic Streptomyces. Canad. Patent 1131143 (1982).

174. Hafner, E. W., Constitutive Mutant of a Thermostable Glucose Isomerase. U.S. Patent 4551430 (1985).

175. Lechmacher, A. & Bisswanger, H., Isolation and characterisation of an extremely thermostable D-xylose isomerase from *Thermus aquaticus* HB8. *J. Gen. Microbiol.*, **136** (1990) 679–686.

176. Henrick, K., Collyer, C. A. & Blow, D. M., Structures of D-xylose isomerase from Arthrobacter strain B3728 containing the inhibitors xylitol

and D-sorbitol at 2·5Å and 2·3Å resolution respectively *J. Mol. Biol.*, **208** (1989) 129–157.

177. Collyer, C. A., Henrick, K. & Blow, D. M., The mechanism for aldose ketose interconversion of D-xylose isomerase involving ring opening followed by a [1,2] hydride shift. *J. Mol. Biol.*, **212** (1990) 211–235.

178. Collyer, C. A. & Blow, D. M., Observations of reaction intermediates and the mechanism of aldose ketose interconversion by D-xylose isomerase. *Proc. Natl. Acad. Sci.*, **87** (1990) 1362–1366.

179. Horikoshi, K., Production and industrial applications of β-cyclodextrin. *Process Biochem.*, **14** (1979) 26–28, 30.

180. Yagi, Y., Kuono, K. & Inui, T., Process for Purifying Cyclodextrins. U.S. Patent 4317881, (1982).

181. Bender, H., An improved method for the preparation of cyclooctaamylose, using starches and the cyclodextrin glycosyltransferase of *Klebsiella pneumoniae* M5 al. *Carbohydr. Res.*, **124** (1983) 225–233.

182. Dale, J. K. & Langlois, D. P., Syrup and Method of Making Same. U.S. Patent 2201609 (1940).

183. Aunstrup, K., Preparation of Amyloglucosidase. U.K. Patent 1092775 (1967).

184. Biester, A., Wood, M. W. & Wahlin, C. S., Carbohydrate studies I, the relative sweetness of pure sugar. *Am. J. Physiol.*, **73** (1925) 387–400.

185. Aschengreen, N. H., Helwieg-Nielsen, B., Rosendal, P. & Oestergaard, J. Liquefaction, saccharification and isomerisation of starches from sources other than maize. *Staerke*, **31** (1979) 64–66.

186. Matzuzawa, M., Kawano, M., Nakamura, N. & Horikoshi, K., An improved method for the preparation of Schardinger β-dextrin on an industrial scale by cyclodextrin glycosyl transferase of an alkalophilic *Bacillus* sp. (ATCC21783) *Staerke*, **27** (1975) 410–413.

187. Kato, T. & Horikoshi, K., Immobilised cyclodextrin glucanotransferase of an alkalophilic *Bacillus* sp. no 38.2. *Biotechnol. Bioengng.*, **26** (1984) 595–598.

188. Bender, H., Cyclodextrin glucantransferase von *Klebsiella pneumoniae*. Synthese, reinigung und eigenschaften des enzymes von *K. pneumoniae* M5 al. *Arch. Microbiol.*, **111** (1977) 271–282.

189. Svensson, B., Pedersen, T. G., Svendsen, I., Sakai, T. & Ottesen, M., Characterisation of two forms of glucoamylase from *Aspergillus niger*, *Carlsberg Res. Commun.*, **47** (1982) 55–69.

190. Svensson, B., Larsen, K., Svendsen, I. & Boel, E., The complete aminoacid sequence of the glycoprotein glucoamylase G1 from *Aspergillus niger*. *Carlsberg Res. Commun.*, **48** (1983) 529–544.

191. Boel, E., Hjort, I., Svensson, B., Norris, F., Norris, K. E. & Fiil, N. P., Glucoamylases G1 and G2 from *Aspergillus niger* are synthesised from two different but closely related mRNAs. *EMBO J.*, **3** (1984) 1097–1102.

192. Boel, E., Hansen, M. T., Hjort, I., Hoegh, I. & Fiil, N. P., Two different types of intervening sequences in the glucoamylases gene from *Aspergillus niger*. *EMBO J.*, **3** (1984) 1581–1585.

193. Nunberg, J. H., Meade, J. H., Cole, G., Lawyer, F. C., McCabe, P.,

Schweickart, V., Tal, R., Wittman, V. P., Flatgaard, J. E. & Innis, M. A., Molecular cloning and characterisation of the glucoamylase gene of *Aspergillus awamori*. *Mol. cell. Biol.*, **4** (1984) 2306–2315.

194. Fairbairn, D. A., Priest, F. G. & Stark, J. R., Extracellular amylase synthesis by *Streptomyces limosus*. *Enzyme Microb. Technol.*, (1986) 89–92.

195. Fujii, M., Homma, T. & Taniguchi, M., Synergism of α-amylase and glucoamylase on hydrolysis of native starch granules. *Biotechnol. Bioengng.*, **32** (1988) 910–915.

196. Robyt, J. & French, D., Purification and action pattern of an amylase from *Bacillus polymyxa*. *Arch. Biochem. Biophys.*, **104** (1964) 338–345.

197. Fogarty, W. M. & Griffin, P. J. Purification and properties of a β-amylase produced by *Bacillus polymyxa*. *J. Appl. Chem. Biotechnol.*, **25**, (1975) 229–238.

198. Outtrup, H. & Norman, B. E., Properties and application of a thermostable maltogenic amylase produced by a strain of Bacillus modified by recombinant DNA techniques. *Staerke*, **36** (1984) 405–411.

199. Saha, B. C. & Zeikus, J. G., Improved method of preparing high maltose conversion syrup. *Biotechnol. Bioengng.*, **34** (1989) 299–403.

200. Kennedy, J. F., Cabral, J. M. S. & Kalogerakis, B., Comparison of action patterns of gelatin-entrapped and surface-bound glucoamylase on an α-amylase degraded starch substrate: a critical examination of reversion products. *Enzyme Microb. Technol.*, **7** (1985) 22–28.

201. Darnoko, D., Cheryan, M. & Artz, W. E., Saccharification of cassava starch in an ultrafiltration reactor. *Enzyme Microb. Technol.*, **11** (1989) 154–159.

202. Yamane, T., The decomposition of raffinose by α-galactosidase. *Sucr. Belg.*, **90** (1971) 345–348.

203. Kobayashi, H. & Suzuki, H., Studies on the decomposition of raffinose by α-galactosidase of mold. II Formation of the mold pellet and its enzyme activity. *J. Ferment. Technol.*, **50** (1972) 625–632.

204. Tilbury, R. H., Improvements in the production of sucrose. U.K. Patent 1290694 (1972).

205. Park, Y. K., Martens, I. S. H. & Sato, H. H., Enzymatic removal of starch from sugarcane juice during sugarcane processing. *Process Biochem.*, **20** (1985) 57–59.

206. Woodward, J. & Wiseman, A., Invertase. In *Developments in Food Carbohydrates, 3*, ed. C. K. Lee & M. G. Lindley. Elsevier Applied Science Publishers, London, 1982, pp. 1–21.

207. Norman, B. E., A novel debranching enzyme for application in the glucose syrup industry. *Staerke*, **34** (1982) 340–346.

208. Barton, R. R., Rennert, S.S. & Underkofler, L. A., Glucose oxidase in the protection of foods and beverages. *Food Technol.*, **11** (1957) 683–686.

Chapter 11

PROTEINS AS A SOURCE OF FLAVOUR

G. S. D. WEIR

Brooke Bond Foods Ltd., Leon House, High Street, Croydon CR9 1JQ, UK

INTRODUCTION

Proteins are of course one of the essential components in our nutrition. The earliest development of proteins as a contribution to flavour in our diet is subject to much speculation. Proteins may themselves be seen as part of the evolutionary process that had its earliest origins in the synthesis of laevo amino acids and thus through to the evolution of the simplest form of life capable of reproduction.

This development has reached the stage when today there is considerable choice in the style of living, particularly in the developed countries where there is a substantial movement towards enhancement of the quality of life, part of which is an increased desire for improved flavour in food. Flavour for the purposes of this review is defined as the organoleptic perception of the combination of taste and aroma.

One of the earliest developments in appreciation of food flavour derived from proteins is the anecdotal evidence attributed to the Chinese who allegedly recognised that the accidental roasting of captured early pig varieties yielded a highly desirable flavour. This is said to be the origin of the appreciation of the flavour of roast pork, an accident which was to be repeated many times.

Proteins in their natural state do not contribute much to flavour directly, but do to some extent influence the taste perception of other components in food through binding reactions. The products of the hydrolysis of proteins, peptides and amino acids, are much more reactive, as are other breakdown products. Peptides through their reactions with taste sites may give flavours which may be bitter or sweet. Amino acids also have distinctive taste profiles in some instances, but more import-

363

antly they take part in a variety of reactions which give rise to an extensive range of flavours. The most studied of these is the Maillard reaction, which is reviewed elsewhere in this volume. Other important flavour reactions of amino acids have also been similarly extensively studied, but not necessarily in such a well-documented manner.

The raw materials of proteinaceous foods include the familiar natural or man-altered agricultural products. The food industry has developed these natural products to give convenience, long-life and efficiency for the food supply, particularly for the developed nations. The acceptability of protein foods where scarcity or price constraints are minimal is dependent to a large extent on the acceptability of the flavour.

PROTEINS

Proteins for the purposes of this review are taken to be the large molecules (molecular weight in excess of 10 000) whose synthesis in the natural state is designed essentially to confer some sort of functional property to the living system. A useful classification of proteins may be found in a review.[1]

It is therefore not to be expected that such large molecules will themselves directly contribute flavour. Through size alone, no perceptible organoleptic response seems possible. Any reaction products which produce flavours will be preceded by a fragmentation of the molecule into more response-invoking moieties. However, binding by proteins, particularly of aroma compounds, will alter the perception of flavour in some cases.

Where flavours from individual proteins within food types is mentioned in the literature, almost invariably such flavoured products are derived from fragmentation of the protein, generally by hydrolysis reactions, but also by other breakdown processes, especially thermal breakdown.

Dairy Proteins—Milk and Cheese

It is recognised that the origin of bitter components in dairy products is complex and diverse, as bitter off-flavours may arise from a variety of chemical mechanisms.[2] Breakdown of milk (and cheese to a lesser extent) protein material is one of the causes of off-flavour development in these products after irradiation.[3]

Breakdown of α-casein in UHT milk has been found to be the cause of astringent off-flavours.[4] In milk, proteolysis often leads to the formation of bitter peptides leading to 'unclean' off-flavours. Further degradation of these peptides may remove the bitterness but this may then be replaced by other undesirable flavours.[5] However, it is possible to improve the quality of a Domiati cheese by the addition of extra whey protein and CMC without adversely affecting flavour.[6]

It is claimed that the antimicrobial action of the LP system for preserving raw Buffalo milk prevented off-flavours developing as a result of proteolytic action of enzymes in milk.[7] Milk stored at 8°C for 2 days developed substantial proteolytic activity compared with the same milk stored at 3°C for the same period. Both can be stabilised by a high temperature sterilisation process and then have a shelf-life in excess of 100 days at 30°C. A sample inoculated with a *Pseudomonas fluorescens* strain, however, showed recovery of proteolytic activity after the same sterilisation treatment.[8]

The role of amino acids and peptides in the flavour of hard or semi-hard cheese released by proteolytic enzymes is not clearly understood. Proteolysis is only one mechanism for flavour development in cheese which can give rise to both favourable flavour development and off-flavours (often described as unclean). In addition amino acid catabolism caused by surface flora is said to be important for developing some of the distinctive flavour and aroma of Camembert cheeses.[9] Proteolysis is probably the most important biochemical event in cheese ripening. The formation of amino acids and peptides during ripening may, however, also cause bitterness.[10]

Different peptides from those found in milk may be responsible for bitterness in cheese, as any peptides in milk tend to pass into the whey portion during cheese making.[11] Proteases added to cheese to accelerate the development of flavour also give rise to bitterness, and conditions which allow peptides to be formed are likely to lead to bitter flavours, especially those containing the non-polar whey concentrate, which when added to whole milk, produced Cheddar cheese with flavour defects described as 'unclean'.[12] It has been found, however, that denatured whey protein could be incorporated into Cheddar cheese with only a relatively small off-flavour development discernible to some panellists.[13] A combination of heat and acid treatment will allow incorporation of relatively high levels of whey protein into Cheddar cheese manufacture with virtual elimination of the bitter off-flavours. The quality of the Cheddar cheese was, however, impaired.[14]

Modification of the casein protein in cheese was seen to be the key to controlling cheese flavour in an enzyme investigation.[15] A rapid production of a cheese flavour may be obtained by mixing cheese curd with a protease and a lipase and warming to between 75°C and 95°C.[16]

Proteins from Meat and Fish Sources

Of all the proteins available, the least used are meat and blood plasma.[17] The utilisation of protein from blood is more attractive if undesirable off-flavours can be removed. Blood may be heated to obtain a coagulate suitable for incorporation into meat products such as luncheon meat.[18]

Soluble proteins from meat tissue or their hydrolysates may be used to impart a meaty flavour to meat mixes for sausages, burgers etc. reducing or eliminating monosodium glutamate.[19] Meat flavour development is thought to be primarily dependent on combinations of amino acids and nucleotides with other breakdown products.[17]

During the heating of meat, many reactions occur, resulting in a very complex mixture of flavour compounds arising through breakdown of the components of meat and combinations of various breakdown components. The most important precursors of the brothy flavour are the low molecular weight and water-soluble dialysable components of protein and carbohydrate breakdown.[20] Differences have been found between the cooked peptides and meat which has been cooked, stored and reheated. The reheated meat contained predominately hydrophobic peptides, which are usually associated with bitter, and often sour, tastes. The once-cooked meat had approximately equal amounts of hydrophilic peptides (often associated with sweet taste) and hydrophobic (bitter) peptides.[21]

The 'warmed-over flavour' (WOF) in meat has previously been considered to be an off-flavour which develops in cooked and uncooked meat due principally to lipid oxidation. A role for proteins in this process has been proposed.[22] Good hygiene resulting in freedom from bacterial contamination is one obviously important factor in preventing WOF.[23] The WOF may be reduced significantly by the addition of antioxidants to the precooked meat product. Various antioxidants are effective in reducing WOF development, including Maillard reaction products.[24] The infusion of calcium chloride and sodium ascorbate into freshly slaughtered lambs accelerated post-mortem tenderisation and retarded WOF.[25] The amine content, particularly tyramine, of fresh beef in vacuum packs is an indication of deterioration due to bacterial action by lactobacilli, some of the amines formed having highly undesirable sensory properties.[26] The use of meat extract for providing meaty flavours is somewhat in decline as the flavours derived from meat extract may not be as good as alternative flavourings.[27]

Irradiation of meat for preservation produces off-flavours in meat which make the technique somewhat limiting.[3]

The flavour of meat can, however, be improved after post-mortem conditioning. This is noticeable in pork and chicken (though not beef) and has been attributed to the role of free amino acids and oligopeptides in improving the brothy taste intensity.[28] Using technologies derived from the pharmaceutical industry for gland extraction, meat flavours have been developed from meat tissue extraction, with claimed advantages of high 'saltiness' without a high salt content.[29] Protein hydrolysis leading to flavour precursors was accelerated in salted beef by a combination of pH control with electrical and mechanical treatment.[30] The sodium content of salted pre-cooked, recombined beef chuck roasts may be reduced by substituting potassium chloride without any measurable loss of flavour precursors.[31]

Collagen is thought to contribute to the flavour of meat. Older animals have a stronger, more desirable meaty flavour. This is possibly related to the increase in glycosylation which occurs with age in the animal.[32] One of the sources of objectionable flavours in gelatin was thought to be due to the production of collagen breakdown products due in part to thermal degradation of the protein.[33]

The utilisation of food sources is partly dependent on the removal of unpleasant off-flavours. For example, treatment of a hydrolysate of a fish protein concentrate by crystalline wheat carboxypeptidase reduced the bitterness and produced a corresponding increase in the free amino acid content.[34] The fermentation of a fish sauce product from capelin was aided by addition of enzymes from squid. Some of the desired flavour characteristic arose from hydrolysis to form the larger molecules (up to 300 Da), as removal of the larger molecules lowered the acceptability.[35]

During storage of canned mackerel over 36 months, denaturation and hydrolysis of protein occurred. This was considered to be the primary cause of ripening changes, the most active changes in the protein components found in the broth. This liquid portion plays an important role in the sensory quality of the product.[36] The taste components in sun-dried Ray were thought to be derived from protein hydrolysis, whereas flavour components were derived from other causes in addition to protein sources.[37] Careful use of modern manufacturing techniques can, however, prevent deterioration of fish protein in producing dehydrated fish powder with little or no loss of taste.[38]

Proteins from Vegetable and Microbial Sources

The utilisation of proteins from vegetable sources is expected to grow from the approximately 50% currently used by developed nations towards the present 83% usage by developing countries as world population grows. Such growth is likely to come from increased use of proteins from oilseeds and pulses.[39]

Soy protein has developed steadily in providing a source of nutrition, replacing to some extent, some of the meat protein which is usually more expensive. In addition there is a growing trend towards vegetarianism, noticeably in the UK, but also in other developed countries. A typical process for the production of various soy products is shown in Fig. 1.

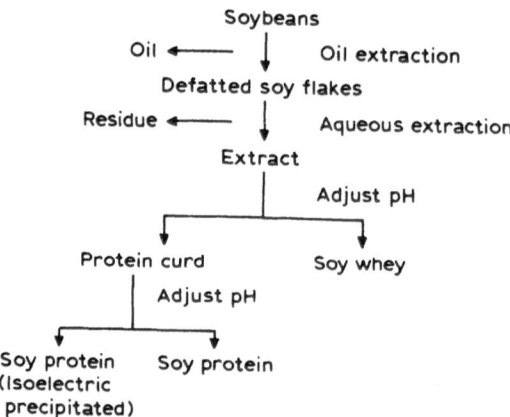

FIG. 1. Process for the preparation of Soy Proteins (Adapted from Ref. 40).

Whilst soy protein is generally regarded as a by-product of soya oil extraction, the shortages of protein and the relative surplus of oils and fats may change that perception. Oil extraction is a necessary part of soy protein production to ensure isolation of stable soy protein with desired properties. Soy protein isolates have functional and nutritional qualities which allow them to be of considerable use within the food industry.[40] Improvements in technology can improve the palatability of soya meal for use in India to overcome the scarcity of proteins in the diet.[41]

Soy proteins, however, often have a characteristic off-flavour note which has been described as 'beany' or 'cereal-like'.[42] The cause of bitterness in soy is due in part to phenolic acids and flavonoids, and is biochemically related to their common precursor L-phenylalanine.[43] Bitter compounds having an isoflavone structure have been isolated from soy by Chang et al.,[44] although oxidised lipid material also probably makes a contribution.[45] Undesirable flavour components may be removed by controlled contact with a mould of Rhizopus or Aspergillus.[46] These off-flavours may also be minimised by reducing the lipid content of the isolates before use, which confers increased stability on storage.[47] Extraction of soy protein with supercritical carbon dioxide is one method for increasing stability during storage.[48] Carbon dioxide is the preferred extractant but other liquids may be used.[49] The off-flavours from soy protein may be masked by tomato sauce, the combination improves the tomato flavour and substantially reduces the contribution from soy odour flavour and aftertaste.[50]

A cheese-like product can be produced from soy-milk curd coagulated by lactic fermentation and ripened for 3 weeks with inoculation by Penicillium caseicolum. The finished product showed a good flavour and texture although substantial hydrolysis of the protein had occurred.[51] The astringency of a yoghurt made from soy and whey was reduced by the addition of calcium sulphate.[52] It is, however, possible to produce dairy-like beverages from a soy protein concentrate by special extraction and heating processes to avoid noticeable off-flavours.[53]

The addition of soya to meat products as extenders (rather than replacements) is perhaps not surprisingly resisted by the meat industry, and indeed within the EEC, proposed legislation will attempt to limit the amount of non-meat protein to be allowed in traditional meat products. Notwithstanding this, the cost savings in adding soy to meat proteins is attractive and such addition is extensively used in catering operations, for example in the US military and school programmes. This can be successfully achieved with advantages to the quality of the meat products up to the limits where off-flavour and loss in qualities of some functionable properties become apparent.[54] Soya may be added in brines to a variety of whole muscle meats such as beef, poultry and seafood, although the level is again restricted to some extent by off-flavours.[55] There is a diminution in the quantities of thermally generated alkyl pyrazines in soy-meat mixtures which is directly proportional to the amount of soya added. This deficit in meaty flavour can be compensated for by the addition of flavouring substances.[56] The replacement of up to 40% of the ham by texturised soy proteins in ham patties may be successfully achieved, but at a 50% replacement level off-flavours became limiting.[57]

Similarly soy protein isolate and soy protein concentrate can be used to replace some of the meat in Bratwurst sausage. At a level above 3%, soy protein isolate adversely affected the flavour of the Bratwurst, but soy protein concentrate can be added at a higher level before the quality is affected.[58] Soy protein concentrate (with vital wheat gluten) when added to Frankfurters contributed some off-flavour notes which will need some alteration in spice formulation to improve the acceptability.[59] The determination of soy protein in meat mixtures may be achieved by SDS-PAGE electrophoresis[60] as well as by other techniques.

Meat-like products such as Tempeh can be made by fermenting cooked soybeans with a mould *Rhizopus digosporus*. The resulting firm cake can be used to make a vegetarian hamburger product. Tofu is a product made by coagulating soya protein with a calcium or magnesium salt. The texture of the resulting curds makes this suitable for a wide variety of culinary requirements from breakfast foods to desserts, in part due to its bland taste.[61]

Soya can be incorporated in a wide variety of other foods such as fish and pastries. However, the level of addition to fish gel, popular in Japan, seems limited to about 2% owing to the appearance of soy flavour in the fish product.[62] Soy protein concentrate can, however, be successfully incorporated into wheat flour used to make Egyptian pastries up to the 6% replacement level with no disadvantage in flavour terms and with some enhancement of protein efficiency ratio (PER) in rat studies.[63]

Whilst soya protein is by far the most studied and used of the alternative sources of protein, it is by no means the only vegetable protein source being used to augment traditional sources. For example, *Vicia faba* bean protein isolates usually have unacceptable off-flavours. This off-flavour may be reduced by a combination of defatting and isoelectric washing of the bean flour.[64] Field-pea flours have been investigated as protein supplements in foods particularly in cereal-based products. However, the level of utilisation seems to be limited by undesirable sensory attributes.[65] Peanut and Cowpea flours can replace up to 30% of the wheat flour in sugar cookies without loss of sensory attributes of the product.[66]

Extraction of protein from leaf sources is likely to be limited in the near future as the products are rather bitter and have a grassy flavour.[67] However, both undesirable green pigments and grassy flavour may be reduced from alfalfa protein concentrates by solvent extraction.[68] Perhaps surprisingly, tea has a significant amino acid content which presumably is derived from hydrolysis of protein material. The amino acids mainly have a bitter taste and are thought to contribute to the overall taste of tea.[69] The early deterioration of asparagus flavour in storage is thought to be at least partly explained by an increase in the protein content of the tip. Proteins are thought to contribute to the overall flavour of the spear.[70]

A new mycoprotein obtained from *Fusarium graminearum*, similar to the mushroom has been developed. Its flavour has been described as mildly wheaty.[71] A process for producing a microbial protein free from salty taste and other off-flavours has been developed.[72] A taste-modifying protein can be extracted from miracle fruit.[73]

Most of the proteins used in foods so far discussed have savoury, bitter or sour flavours. However, it is possible to extract intensely sweet proteins from African berries. These proteins, monellin and thaumatin, are approximately 100 000 times sweeter than sugar on a molecular basis, the sweetness being apparently due to conformation, but they do not apparently have any similarities in amino acid sequencing.[74] The sweetness properties of thaumatin can be altered by genetic engineering to produce differing sweet tasting proteins.[75]

Protein Interactions with Flavours

Although proteins themselves in general have little flavour, they can influence perceived flavour because they may contain bound off-flavours. They may modify flavour by selective binding. They may produce off-flavours, or provide the precursors of flavours, via such mechanisms as the Strecker degradation of amino acids.[76]

From studies on the binding of bovine serum albumin (BSA) with vanillin, L-tryptophan and N-acetyl-L-tryptophan, it was concluded that proteins are involved in two sorts of binding reaction. Ligand binding lowers the amount of flavour perception. On the other hand, slow release of ligand from proteins will contribute to the flavour persistence observed in foods.[77] With legume protein, the soluble proteins were able to bind alkyl acetates and ketones in line with the hydrophobicity of the protein. For insoluble proteins, there is physical binding as well as hydrophobic forces. Both mechanisms can cause lowering of the perceived intensity of added flavour by 70–80%.[78] Hydrophobic interactions with discrete binding zones is also offered as an explanation for the interactions of both non-volatile and volatile substances in food. The interactions are relatively weak and the binding properties change with denaturation of the protein.[79]

Chicken flavour appears to be bound in some way to soy protein in a formulated soup product, as the chicken flavour is increasingly suppressed by increasing amounts of soy protein.[80] Whey protein concentrate particularly, along with other food proteins, binds the model flavour compounds, 2-nonanone and nonanal. Whey protein concentrate is suggested as a carrier for volatile flavours in food systems.[81] Some volatile aldehydes can be bound to protein and thus retained over a long period of time, but such binding may lead to chemical reactions taking place which could generate unacceptable flavour compounds or non-volatile oligomers.[82]

At concentrations as low as 0·5% protein, BSA can bind diacetyl, used as a model flavour compound. Given the concentrations of proteins in foods it is suggested that these binding effects are very important for flavouring in processed and manufactured food.[83] When soy extenders are added to fresh meat products there is some loss of the meat flavour alkyl pyrazines. Flavourants can be added to foods to compensate for this loss.[56]

Binding sites on caseins may be reduced by glycosylation of $-NH_2$ groups. The modified protein retains less flavour volatiles than the original protein during air-drying of protein solutions.[84] Denaturation of fababean protein micellar mass leads to increased binding of vanillin compared with the native protein. Binding sites were also increased.[85] Sensory evaluation confirmed these findings.[86] Similar results were found with diacetyl bound to soluble pea protein. Physical treatment, heat or mechanical, increased the binding of diacetyl, although some protein rearrangements and lipid-protein reactions may interfere.[87]

Iron has been shown to bind to soya, especially soy isolate.[88] Presumably other transition metals will bind to sites which may therefore be unable, in competition, to bind flavour molecules.

Astringent flavours are caused by phenols in a wide range of foods such as beer, coffee and fruit, and it has been postulated that binding with proteins affects the astringency of these products.[89] Tea cream, the white insoluble precipitate which separates on cooling an infusion of tea, comprises a complex of caffeine, protein and the phenolics, theaflavins and thearubigins. This cream complex is thought to depress the flavour perception of flavoured teas through binding actions with the flavourants.[90]

During proteolytic breakdown of protein in the ageing of cheese, a flavourless fat free residue is formed. The loss of flavour is in part due to a reduction in size of the protein and hence less binding ability.[91]

The binding properties of protein in a protected porous structure are utilised in a patented process for manufacturing a flavour-retaining food product.[92] In another patented process, it is considered necessary to separate salt from a sweet protein to prevent a suppression of the flavour of the sweet protein. This has uses in chewing gum, it is claimed.[93]

The protein curuculin obtained from fruits has properties of modifying the taste perception of sour ingredients so that they taste sweeter. UHT treatment is claimed to stabilise the material so that it can be used in food products.[94]

The contribution of the knowledge of flavour chemistry to the food industry has been reviewed.[95]

PEPTIDES

Apart from the binding reactions just discussed, much of the foregoing chapter on the contribution to flavour by various proteins is almost wholly due to some reduction in molecular size of the protein, usually by hydrolysis, to form flavour-active components such as peptides. Peptides, for the purposes of this review, can be regarded as the breakdown products of proteins, which thus lose their functional properties. In terms of contribution to flavour they may range in molecular size from dipeptides (two amino acid fragments), the simplest peptides, to molecules comprising many amino acid residues and having molecular weights in the thousands.

Peptides may contribute a variety of flavours both desirable and undesirable off-flavours. For example, the formation of bitter peptides in the ripening of cheese is dependent on a number of factors. Whether they are detectable organoleptically depends on whether the quantities generated exceed the flavour thresholds and make a discernible contribution to overall flavour.[91]

The tastes of peptides can be sweet, bitter, salty or umami. All of these tastes have been obtained from peptides isolated from protein hydrolysates or by synthesis.[96] There are also some other types of tastes which have been identified in isolated peptides. A sour taste is one considered to be sufficiently characteristic to be included in the primary classifications. Astringency has also been attributed to peptides.[97]

Bitter Peptides

Peptides of molecular weight less than 5000 from 'cooked' and 'cooked-stored-recooked' products have been separated from red meat. The 'cooked' meat extract appeared to have equal quantities of hydrophilic (usually sweet) and hydrophobic (usually bitter or sour) peptides whereas the cooked-stored-recooked extract appeared to have mainly hydrophobic peptides.[21] From labelled sulphate addition to *Allium cepa* and other Allium species it was concluded that glutathione and γ-glutamyl peptides are intermediates in the biosynthetic pathway to flavour precursors in onions and garlic.[98] During the roasting of peanut (*Arachis hypogaea* L.) seed it has been shown that a methionine-rich polypeptide of molecular weight 70000 decreased. From this, it is suggested that methionine peptides may be involved in the formation of pyrazine compounds to give the roasted flavour of peanuts via Maillard (and other reactions) with sugars.[99]

Of the various tastes attributed to peptides perhaps the most limiting is the bitter taste which often makes the food unpalatable. In a review on bitterness in foods and beverages, it is concluded that peptides formed during proteolysis are primarily responsible for bitterness, though different peptides may be responsible for bitterness in different dairy products. For example, the peptides from milk largely pass into the whey during cheese making; hence different peptides may be responsible for bitterness in milk and in cheese. In enzymatic treatment of fish, the soluble portions were found to be bitter. Similarly enzymatic treatment of deboned chicken meat produced bitter components.[11] Bitter flavours in fermented dairy products often arise from the formation of bitter peptides by proteolysis. Bitter peptides are generally thought to be composed mainly of non-polar amino acids.[100] This view is expressed slightly differently as bitterness in peptides is ascribed to the peptides containing a high proportion of hydrophobic side chains.[9] The taste is also described as unclean.[5] Bitter peptides have been obtained from the digestion of α_{s1} and β-casein with alkaline protease from *Bacillus subtilis*.[101]

An investigation of the peptides formed from proteolytic breakdown of Cheddar cheese curd caused by *B. subtilis* proteinase showed that the bitter-tasting peptides are hydrophobic in nature.[102] Bitter peptides in meat flavour are associated with the bitterness caused by the hydrophobic action of amino acid side-chains.[17]

The structure of bitter peptides is, not surprisingly, given importance in assessing the acceptability of food. It is being investigated increasingly, particularly as the advanced techniques now available have allowed a better understanding of molecular structure and properties. For the purposes of this review, the standard abbreviations of amino acid residues used are listed in Table 1.

It has been suggested that the bitterness of synthesised oligopeptides containing various amino acids is correlated with the side-chain skeleton of the amino acid carbon chain having at least three carbon atoms. Only valine did not follow this general rule entirely. Of those amino acids studied, a somewhat intermediate taste perception was obtained from those oligopeptides which contained valine, the taste then being determined by structural characteristics.[103] In the case of proline-containing peptides, the most significant role for the proline residue in taste perception, would appear to be in a conformational alteration of the peptide

TABLE 1

ABBREVIATIONS USED FOR AMINO ACIDS

Alanine	Ala	Leucine	Leu
Arginine	Arg	Lysine	Lys
Asparagine	Asn	Methionine	Met
Aspartic acid	Asp	Phenylananine	Phe
Asn+Asp	Asx	Proline	Pro
Cysteine	Cys	Serine	Ser
Glutamine	Gln	Threonine	Thr
Glutamic acid	Glu	Tryptophan	Trp
Gln+Glu	Glx	Tyrosine	Tyr
Glycine	Gly	Valine	Val
Histidine	His	Anserine	Ans
Isoleucine	Ile	Carnosine	Car

molecule, in contrast to the bitterness conferred by hydrophobic amino acids.[104] The hydrophobicity of phenylalanine and tyrosine residues has been said to increase the bitter taste in peptides containing those amino acids.[105] Similarly the bitterness of leucine-containing peptides has been ascribed to the hydrophobicity of the leucine residues. The bitterness was stronger when the leucine was located at the C-terminus of such peptides.[106]

The bitterness of a tetrapeptide (Arg–Pro–Phe–Phe) and its derivatives was attributed to the maintenance of the Arg–Pro–Phe–Phe unit intact within peptide structures. If these units were concentrated together the bitterness increased. If two of the units (Arg–Pro) were used to make a tetrapeptide with glycine (Arg–Pro–Gly–Gly or Gly–Gly–Arg–Pro) the bitterness disappeared.[107] When the Gly–Gly residue is inserted into certain other peptides, the bitterness is again reduced. It is suggested that the Gly–Gly residue may prevent a hydrophobic group from binding to a taste receptor.[108]

In the hexapeptide Arg–Arg–Pro–Pro–Phe–Phe, if D-phenylalanine is substituted for L-phenylalanine, the bitterness is reduced. This is evidence to confirm that the importance of the configuration of the C-terminal hydrophobic amino acid makes a significant contribution to bitterness.[109]

The contribution to bitterness of various fragments of β-casein has been investigated. The tetradecapeptide H–Pro^{196}–Val^{209}–OH was shown to comprise two bitter peptides one of which was about twice as bitter as caffeine. The sequence H–Pro^{196}–Val^{209}–OH is much less bitter than $Arg^{202}-Val^{209}$ which is about 250 times as strong as caffeine.[110] The bitterness of the partial sequence (positions 61–67) of β-casein was also investigated through synthesis of two hydrophobic peptides ($Pro^{61}-Pro^{67}$ and $Tyr^{60}Ile^{66}$) which corresponded to common components of isolated bitter peptides. Compared with the previous examples, these peptides were still bitter, but much less so.[111] In investigations into the partial sequence (positions 82–88) of β-casein, a heptapeptide was synthesized and its bitterness assessed. It was concluded that the presence of a basic amino acid at the N-terminal was an important contributor to bitterness.[112]

Various studies have been undertaken to understand the origin of bitterness in peptides. The most important properties are believed to be the molar concentration of the most hydrophobic peptides and their average hydrophobicity and chain length. Using such parameters it is possible to establish a mathematical model which can predict the observed bitterness of 2-butanol extractable peptides.[113] Other workers have established a relationship between the free energy of amino acids and bitterness of peptides. When the sum of the free energies (F) divided by the number of amino acid residues exceeds 1350 then peptides are bitter provided the molecular weight is less than 6000.

$$Q = \Sigma AF \begin{cases} < 1350 \text{—not bitter} \\ > 1350 \text{—bitter (under 6000 Da)} \\ > 1350 \text{—not bitter (over 6000 Da)} \end{cases}$$

This rule can also be applied to the bitterness of theobromine in combination with diketopiperazines in Cocoa flavour.[114] A series of di- and tri-peptides have been synthesised and their bitterness evaluated. Compared with caffeine, the tri-peptide Leu–Glu–Leu was found to be the most bitter. The hydrophobicity of the amino acid residue was not apparently the only reason for the bitterness.[115] From studies on model derivatives of peptides it was concluded that bitterness required two reactive sites with the taste receptor, a 'binding site' and a 'stimulating unit'.[116] Through studies on derivatives of peptides the stereospecificity of bitterness has been proposed.[117] Assuming that bitterness is caused by the composition of hydrophobic amino acid residues, including the chain length, then it is possible to predict the bitterness of peptides using a software programme for an IBM (or compatible) computer.[118]

Various attempts have been made to remove, or mask, the bitterness of peptides, which otherwise limits the sensory acceptability of foods.[119] For example, the astringency of a soy-based yoghurt can be reduced by the addition of calcium cation in the form of calcium sulphate.[52] A systematic study of masking compounds to disguise the bitterness of quinine hydrochloride is of relevance mainly to the pharmaceutical industry,[120] but has applicability also to the food industry. Caffeine is generally used as the standard for bitter flavour in the food industry.

As previously observed, bitterness in milk protein is well known through hydrolysis of the protein. The bitter peptides contained in the hydrolysate of casein by protease may be further treated with aminopeptidase T either to reduce or eliminate the bitterness.[121] The aminopeptidase function, it is claimed, is to cleave single amino acids or pairs of amino acids from the N-terminal end of polypeptide chains. This prevents the formation of bitter peptides, greatly improving organoleptic and nutritional characteristics.[122] A variety of methods for debittering peptides, such as mixing with skimmed milk or adding gelatinised starch are said to be effective.[123] The use of wheat carboxypeptidase to debitter a fish protein concentrate has been described. The mechanism is thought to be through the cooperative effect of the selective release of hydrophobic amino acids and the formation of acidic dipeptides from the bitter peptides.[34] In a similar manner wheat carboxypeptidase is effective in reducing or eliminating bitterness from soybean protein.[124]

Sweet Peptides

There are perhaps few better examples of the complexity of the chemistry of taste, than studies on peptides, as not only do peptides have a bitter taste, many times stronger than that of caffeine, but there are also peptide sweeteners with a sweetness up to 50 000 times greater than sucrose. The research which followed the accidental discovery of Aspartame (L-aspartyl-L-phenylalanine methyl ester), which is about 180 times sweeter than sucrose, has led to the beginning of the understanding of the shape of the sweet receptor site.[125] Sweeteners based on L-aminocarboxylic acid esters have been patented.[126] Derivatives of Aspartame have been prepared with various sweetness characteristics and a model proposed to explain the differing degrees of sweetness.[127] A sweet peptide N-acetyl-phenylalanine-L-lysine has been synthesised from N-acetyl-L-phenylalanine ethyl ester using α-chymotrypsin as a catalyst.[128]

Salty and Other Tastes

To establish further the versatility of peptides in affecting taste reactions, other tastes such as 'salty', 'sour', 'umami' and 'delicious' have been described. One such example is a peptide with the taste of a beef soup.

The taste of the 'delicious' peptide is allegedly due to the interaction of basic and acidic fragments in the peptide.[129]

Peptides and their analogues have an effect on the perceived saltiness of sodium chloride.[130] Basic dipeptides such as L-ornithyl-taurine hydrochloride (Orn–Tau–HCl) and L-ornithyl-β-alanine hydrochloride (Orn–β-Ala–HCl) have a salty taste equal to or greater than salt, and also contain no sodium. Glycine methyl or ethyl esters have also a salty taste but weaker than that of the peptides previously mentioned.[131] Ornithyl-taurine and other dipeptides have the potential for replacing salt, or monosodium glutamate, for sodium-restricted diets.[132] A process for the preparation of salty peptides which are taurine derivatives has been described in a Patent.[133] The saltiness of the taste of L-ornithyl-taurine monohydrochloride has been disputed by a Swiss group who claimed that the saltiness previously reported was probably due to the presence of sodium chloride as an artifact of the method of preparation.[134] However, the claim was supported by the original authors who found out that the salty flavour is very much affected by pH and that ornithyl-β-alanine and sodium chloride enhance the saltiness of each other.[135]

The desire for salt has been found to be inversely proportional to the quantity of protein in the diet, according to studies in Japan.[136]

HYDROLYSED PROTEIN

The contribution to flavour of hydrolysed protein, in its many forms, has been covered in a recent review.[137]

Hydrolysed protein remains the largest contributor to flavourings derived from protein in terms of the quantity produced. The breakdown of protein in most hydrolysis reactions is virtually complete, amino acids being the main protein-derived product of the hydrolysis. The hydrolysis is accompanied by breakdown of amino acids with loss of ammonia, carbon dioxide and sulphur compounds. As the remaining molecules are highly active and as the hydrolysis of proteinaceous material is accompanied by hydrolytic breakdown of carbohydrate, lipids etc. in the various feedstocks, it is not surprising that there is the opportunity for considerable chemical change to take place in hydrolysis, or in the subsequent work-up of the hydrolysis liquor.

Hydrolysis by acids in particular gives rise to a large amount of a carbonaceous mass often called 'black residue'. The resulting hydrolysate is therefore usually dark brown in colour as well as having a strong savoury flavour. Enzymatic hydrolysis usually leads to products which are lighter in colour and have a much less pronounced meaty or savoury flavour. Hydrolysed protein from various sources is present in a very wide variety of savoury foods.

Acid Hydrolysis

Hydrolysis, using acid as a catalyst, remains a very well-used method of manufacturing hydrolysed protein.

The feedstock covers every normal protein; the acid used is most generally hydrochloric acid. For example it is claimed that egg (whole, yolk or white) may be hydrolysed with hydrochloric acid perhaps in combination with maize gluten to yield a chicken meat flavour hydrolysate for use in seasonings, vegetable products, bakery products etc.[138] Gelatin, and other feedstocks, may be hydrolysed by acid or alkali at high pressure (500 KPa). The temperature is only taken up to 80°C. The process has the advantage of short time and uses small quantities of acid. The product is used in seasonings.[139] Keratin is a preferred feedstock for acid hydrolysis. It is claimed that neutralisation may be carried out electrolytically, and the resulting product used as a precursor of meat aromas: it contains no mineral salts.[140]

It is also possible to produce an acid hydrolysate with a very low salt content, and a reduced content of non-polar amino acids, by gel filtration over a porous material, eluting the desired fraction.[141] Using mixed animal and plant protein (bonito meal and soybean meal); a hydrolysate can be obtained by hydrochloric acid which has a rich and distinctive flavour.[142] All-vegetable feedstocks may also be used, an example being the use of soy grits and dehydrated alfalfa flour which, when hydrolysed with hydrochloric acid, produces a dark hydrolysate with a specific flavour.[143]

Vegetable proteins may also be used for preparing a dietetic food seasoning, using hydrochloric acid.[144]

Hydrolysis of a feedstock with hydrochloric acid usually takes place in the presence of triglyceride fat material. It is known that chloro- and dichloro-propanols are formed during acid hydrolysis although they are found only in small quantities in the final product. Concern for the safety of the products of hydrochloric acid hydrolysis of proteins has led to investigations into the safety of the known chloro-compounds found in the hydrolysates. Thus under investigation are:

(a) Monols: Monochloropropanols (MCPs), e.g. 1-, 2- and 3-chloro-propanol.
Dichloropropanols (DCPs), e.g. 2,3-dichloropropan-1-ol, 1-3-dichloropropan-2-ol

(b) Diols: (MCDPs), e.g. 2-chloro-1,3-propandiol, 3-chloro-1, 2-propandiol

The amounts of these contaminants in hydrolysates is extremely small and it is considered that there should not be any potential hazard from this source.[145] Concern over these contaminants has led to limits being proposed in Germany and Holland for various chloro propanols. Other Governments are considering their position. A process has been described for the removal of chlorhydrins from the hydrochloric acid hydrolysate of vegetable proteins. The products may be used as seasonings.[146]

Enzymatic Hydrolysis

Whilst the production of hydrolysed protein by acids, particularly hydrochloric acid, remains the main process in terms of volume, there is a growing interest in the use of enzymes for producing flavoured hydrolysates for use in foods. They have the advantage of not producing chlorhydrins and they contain little salt.

One of the more limiting aspects of the production of protein hydrolysates is the off-flavours produced in part by the plastein reaction, i.e. the synthesis of peptides from peptide fragments and amino acids, which occurs at the same time as the hydrolytic breakdown reactions in most enzyme systems. As more information becomes available it is becoming feasible to produce protein hydrolysates free from bitter peptides, from a variety of feedstocks. Bitterness in the hydrolysates has previously tended to limit the acceptability of the various products.

For example, it is claimed that a new group of industrial enzymes, the aminopeptidases, in conjunction with specific endoproteases can remove the bitterness normally associated with protein hydrolysis.[141] The principle advantage would appear to be an extra degree of control over enzymatic processes which leads to more acceptable products, for example soy protein isolates of reduced viscosity.[148] Cheese or meat flavours may be produced by altering the conditions of the secondary hydrolysis using aminopeptidase. Thus flavour control of primary hydrolysates is effected.[149] A cascade system, using three enzyme systems, is proposed for control of the hydrolysis of commercial soya grits. The resultant hydrolysate is relatively free from bitterness.[150] A coffee flavour may be produced from an enzyme hydrolysed soya flour by utilising the water-soluble fraction from the hydrolysis for further reaction.[151]

Proteinaceous material may be recovered from the coarse and fine scrap-bone residues from the mechanical fleshing of beef and pig bones by enzymic processes. The hydrolysates may be used for seasonings which have the taste of meat.[152] It is claimed that a chicken-flavoured product may be prepared from heat-denatured comminuted chicken meat by means of a serine protease. The product has a high flavour intensity and may be used in soup, sauces, prepared meals etc.[153] Similarly, a partial hydrolysate with a high degree of hydrolysis may be obtained from a microbial biomass containing protein material. The fully soluble product is free from bitter taste and may be used in foods or pharmaceutical preparations.[154] It is possible to control the level of bitterness of a casein hydrolysed by rennet to produce products of high solubility, low bitterness at high values (e.g. 55) of degree of hydrolysis. Degree of hydrolysis is defined as number of peptide bonds cleaved divided by total number of peptide bonds expressed as a percentage.[155] An aminopeptidase may be used to hydrolyse further a casein hydrolysate to remove the bitter flavour.[121] A concentrated cheese flavouring can be produced by combining vegetable and casein hydrolysates, leucine and a fat containing aromatic substances.[156,157] Protein ingredients may be modified by enzyme treatment for example to produce cheese flavours. Some of the variability in performance will need to be resolved before full use may be made of the potential of enzymatic hydrolysis.[158]

Bacterial protease can be used to produce hydrolysates of high solubility and of nutritional value from fish protein.[159]

A combination of a protease from a fungus of the species *Aspergillus oryzae* and an enzymatic extract of porcine pancreas is said to be particularly suitable for the hydrolysis of soy protein isolates without causing bitter off-flavours to develop.[160] It is suggested that proteolysis in yoghurt preparation may be controlled by the addition of specified amounts of sucrose with a view to improving the quality of yoghurt.[161]

The demand for yeast extract has increased, following the expression of doubts over the safety of acid hydrolysed protein. In addition, with sodium labelling established in the USA and with Europeans under pressure to reduce salt (as the principal form of sodium in the diet) intake it is another advantage of yeast extract that the salt (or sodium) content can be minimised.

Indeed, yeast extracts are provided for flavours with the claim that they are able to replace salt in formulations.[162,163] A mixture of a primary yeast extract and a protein hydrolysate has uses as both a savoury flavour and as a flavour enhancer intended to replace MSG.[164] Dried yeast protein has been claimed to add flavour and nutritional value to a variety of foods.[165,166] Yeast and yeast derivatives may also be used in various meat products, especially sausages, with improved sensory qualities. The economics of partial meat replacement are said to be as favourable as those of substitution by soya protein.[167] A yeast culture *Candida utilis* used in a raw-dried Bulgarian sausage is said to improve the flavour of the product. In this case the major improvement was due more to a more pronounced lipolytic process caused by the yeast culture, rather than by the contribution to flavour by the yeast.[168] A suspension of *Terulopsis ethanolitolerans* cultivated on synthetic ethanol is autolysed to produce savoury flavourants.[169]

Apart from allowing reduction of sodium and MSG, it is claimed that yeast products can modify flavour favourably. The result of all these advantages is claimed to be wide applicability of the products.[170] This is aided by flavour of these products assessment through a descriptive flavour analysis panel for product applications.[171]

Soy sauce is much used in the far East, but it is also beginning to find wide applications in Western countries.[27] Soy sauce is one of a number of products suggested for indigenous fermented-food technologies for developing countries to assist in alleviating malnutrition.[172] At the other end of the scale of nutrition, to assist in the marketing of soy sauce for Western consumption, a low-sodium soy sauce has been developed.[147]

Ethyl carbamate, a known carcinogen, has been found in fermented products including some soy sauces. No ethyl carbamate has been detected in the non-fermented products. The level found indicates a minimal risk to health but rather more data are needed.[173]

Carboxypeptidase from *Saccharomyces cerevisiae* cell wall is used to debitter protein hydrolysates.[174] The flavour of protein hydrolysates with a high peptide content may be modified by heating with cysteine and a carbohydrate mixture. Intensive flavours ranging from roast beef to roast bread are said to be produced.[175]

Amino Acids

The main components resulting from the hydrolysis of protein are the amino acids. The twenty primary protein acids are those listed first in Table 3.

TABLE 2
ESSENTIAL AMINO ACIDS IN PROTEINS

Amino acid	Skim milk	Soy	Beef	Egg	Fish	Yeast	Pea
Lysine	8·6	6·8	8·3	6·3	6·6	6·8	7·8
Tryptophan	1·5	1·4	1·0	1·5	1·6	0·8	0·8
Phenylalanine	5·6	5·3	3·5	5·7	4·1	4·5	5·1
Methionine	3·2	1·7	2·8	3·2	3·0	2·6	1·0
Threonine	4·7	3·9	4·5	4·9	4·8	5·0	4·1
Leucine	11·0	8·0	7·2	9·0	10·5	8·3	8·2
Isoleucine	7·5	6·0	4·7	6·2	7·7	5·5	5·1
Valine	7·0	5·3	5·1	7·0	5·3	5·9	5·5

Data from Ref. 17.

Free amino acids are subject to a certain amount of degradation, through loss of the $-NH_2$ group and also decarboxylation, giving a variety of products. The nutritionally essential amino acid content of a variety of protein sources is given in Table 2.[17]

Amino acids are also very reactive, taking part in a variety of reactions leading to flavour generation. The best known of these is the Maillard series of reactions. These will not be discussed in this review as they are considered in depth in Chapter 4. Later here, however, some of the other reactions leading to flavour will be discussed.

Amino acids all make a contribution to taste. The basic tastes are usually exemplified by the common food components sodium chloride (salty), hydrochloric acid (sour), sucrose (sweet) and caffeine or quinine (bitter).[11] To this list must now be added the Umami taste, generally characterised by the taste of monosodium glutamate, stemming originally from workers in the far east but now gaining recognition in Europe and the U.S.A.[97,177]

The taste of individual amino acids have been characterised by several workers. The taste descriptors and threshold values are summarised here (Table 3).

Amino acids accumulate in fermented products such as cheese. Proline for example has been implicated in the sweet flavour of a Swiss variety of cheese. The addition of solid proline to a bland cheese imparts a sweet sensation, much more so than is expected from the same quantity of proline in solution.[100]

TABLE 3

TASTE DESCRIPTOR AND THRESHOLD VALUES OF AMINO
ACIDS

Amino acid	Taste	Threshold value (mg/dl)
Histidine	Bitter	20
Methionine	Bitter	30
Valine	Bitter	40
Arginine	Bitter	50
Isoleucine	Bitter	90
Phenylalanine	Bitter	90
Tryptophan	Bitter	90
Leucine	Bitter	190
Tyrosine	Bitter	No data
Alanine	Sweet	60
Glycine	Sweet	130
Serine	Sweet	150
Threonine	Sweet	260
Lysine	Sweet and bitter	50
Proline	Sweet and bitter	300
Aspartic Acid	Sour	3
Glutamic Acid	Sour	5
Asparagine	Sour	100
Glutamine	Flat	No data
Cysteine	—	No data
Glu Na (MSG)	Umami	30
Asp Na	Umami	100

Based on Ref. 96.

The deterioration of the flavour of salted chinese cabbage (at 27°C) is accompanied by an increase in free amino acids, not seen under storage at 7°C, or lower.[178] Changes in the flavour of Japanese pickles, 'Sunki', through the pickling process was attributed in part to an increase in aspartic acid content.[179]

In the preparation of kimchi, a traditional fermented vegetable food in Korea, it was found that in the sensory evaluation of the product the taste was closely related to the contents of non-volatile organic acid, free amino acids and pH.[180] The formation of free amino acids in sake brewed by rice germ addition is inhibited, the flavour and taste is thus improved. The factors isolated from rice germ which caused this reduction in free amino acids formation have been found to be magnesium, potassium and phosphate, magnesium being the most effective.[181]

During fermentation of cocoa seeds, the amino acids leucine, alanine, phenylalanine and tyrosine accumulate to a far greater extent than other amino acids present. These amino acids are thought to be especially reactive during roasting to produce the characteristic cocoa flavour.[182]

The palatability of tea is probably more dependent on favourable interaction of components than on single component groups. However, amino acids, present to about 6–7% on a dry weight basis, are amongst those bitter components which contribute to making tea a unique and attractive beverage.[69]

An association of glutamic and aspartic acids can be used as the basis for a salt replacer.[183]

The principle taste compounds in low salt fermentation of anchovy and yellow corvenia were thought to be free amino acids, with contributions from nucleotides and creatinine.[184] The amino acids glycine, glutamic acid, alanine and arginine are amongst the principal contributors to the flavour of scallop.[185] It is projected that if the free amino composition and other taste components could be controlled it would be possible to produce meat-like taste products from the waste in processing grass shrimp.[186] Not only human taste response to amino acids determines the acceptability of food, but fish also apparently display preferences related to the amino acids in feed preparation.[187,188]

The presence of certain amino acids and their esters enhances the saltiness taste of sodium chloride, although the acids are not themselves salty.[189]

There are many instances of amino acids undergoing changes as mentioned previously. The production of tyramine in beef, for example, is characteristic of spoilage. Tyramine and other similar amines, arise through the decarboxylation of amino acids by natural contaminating flora.[26] Pyrazines are formed by *Bacillus natto* from L-threonine and to a lesser degree from L-serine.[190]

The phenylalanine content of honeys is variable, and may be specific to a plant source. The decomposition of phenylalanine gives rise to a number of compounds implicated in flavour development.[45]

The thermal degradation of cystine in water at 160°C gives rise to a variety of sulphur compounds, including disulphides which have roasted or oniony flavours, under conditions of low pH.[191]

Apart from degradative changes, amino acids are reactive and will form flavour compounds with reactive aldehydes or ketones. For example serine and threonine react with sucrose under coffee roasting conditions to form pyrazines, which occur in all roasted foods.[192] Roasted or meaty flavours can also be produced by the reaction of cystine or cysteine with a furanone derivative. The quantity of flavour generated is dependent on pH: the lower the pH, the better the flavours produced.[193,194]

The contribution to flavour of the amino acids has been explained in various ways. Most of the hydrophobic L-amino acids have a bitter taste whereas all hydrophobic D-amino acids are sweet.[96] The lower molecular weight acids, except proline, are sweet irrespective of whether they are D- or L-enantiomers. For the higher molecular weight acids the L-enantiomers are predominantly bitter.[195] A number of D-amino acids, including Trp, His, Phe, Tyr, Leu, Gly and Apn are sweet, but their enantiomers, the L-amino acids, are known to be tasteless or bitter. The differences in sweetness in the enantiomeric form can be explained by the fact that the L-amino acids are not superimposable upon the receptor sites.[196] The apparent specific volume of the molecules can be used to predict the taste response. The taste properties of each amino acid encompass a range of tastes, sweet, bitter etc. However, the response may be dependent on size exclusion to certain receptor sites.[197,198] The taste properties of amino acids have been reviewed.[96,97]

FLAVOUR

Flavour, as has been noted, is defined as the organoleptic response to the parameters described as taste and aroma. Taste is largely due to the reaction with receptor sites on the tongue, or soft palate, from molecules which are mainly non-volatile. Aroma is the sensation obtained from interactions with olfactory sites, largely in nasal passages, and is largely due to the effects of volatile components.

Thus for the purposes of this review:

$$Flavour = Taste + Aroma$$

There are various other definitions. Taste in some other texts may be described as 'Flavour' and readers need to be careful to distinguish just what is meant by these terms. Aroma is also described as 'Odour' or 'Smell'. In this case there is little likelihood of confusion. In this review the definition of 'Flavour' above has been adopted for consistency.

Evaluation of Flavour

The evaluation of flavour is primarily by taste panels whose organisation and training is a large subject, covered comprehensively in the literature. An example of the use of panels, and other evaluation processes used for proteins, has been described.[199]

Taste panels are relatively expensive in resource requirements. Hence efforts are continually being made to replace panel evaluation by non-sensory methods. For example direct headspace gas chromatography, together with statistical evaluation of the chromatograms, has been used for assessing cheese quality.[200] The continuing development of analytical capability is widely used to understand flavour chemistry,[201] particularly in the use of computer techniques for both data collection and analysis.[202–204] Achievements and needs in flavour research have been reviewed.[202,205]

Taste

Taste is a combination of chemical sensations perceived by the papillae of the tongue. Most substances do not have a single taste: what is perceived is a complex sensation comprising one or more of four basic taste sensations: acidity, salinity, sweetness and bitterness.[206]

With sweet-tasting compounds some general rules have been established for predicting sweetness from structure. For bitterness, general rules do not appear to be established but observations on the relationship of structure with bitterness for amino acids and peptides have been documented.[207] Bitterness in low concentrations enhances the flavours of certain food preparations and in some cases can even serve as a measure of quality. Bitterness is an attribute which is also associated with a sense of efficacy in medicines.[208]

As has been previously mentioned, many authors regard taste as comprising four primary taste stimuli. Less attention has been given to the study of taste than has been given to understanding olfaction. Five primary tastes are generally assumed: sweetness, sourness, saltiness, bitterness and umami.[97] Using amino acids as models for studying the taste receptor mechanism it is proposed that glutamic acid imparts a 'umami' taste to food and that this should be considered as a primary taste.[209] However, in other studies on taste mechanisms, it has been shown that kainic acid selectively reduced the taste responses to glutamic acid, from which it is deduced that the taste response of glutamate does not follow the same route as the four primary tastes.[210]

Aroma

There are a very large and growing number of known volatile compounds from protein sources.[123]

Seven primary odours are recognised by some authors and there are several theories as to the nature of the interaction with the olfactory receptor sites. One of these theories is the site-fitting theory, where the response is governed by a fit between the odour molecule and the receptor site.[38] The response at the receptor sites is concentration of the aroma, especially as a function of time.[211]

The techniques used for aroma analysis place great emphasis on gas chromatographic separations, often aided by a mass detector. An example of the use of this combination has been seen in experiments to determine the source of undesirable flavours in gelatin.[33] As in the case of taste assessment from panels, the equivalent in aroma research employs the nose in sniffing aroma or through the mouth. For example, the effect of temperature, concentration and additive (NaCl and MSG) on the perception of a beef flavouring[212] has been so determined. An additional parameter of importance in determining the sensory response is time. A system to measure perceived intensity as a function of time has been developed.[213]

CHEMISTRY OF FLAVOUR DEVELOPMENT FROM PROTEINS

The complexity and variety of compounds contributing to flavour derived from protein sources is considerable, the biggest contribution to flavour coming from relatively small fragments of protein. By far the commonest form of fragmentation is hydrolysis. This can produce peptides, which may be bitter or sweet depending on the amino acids and their sequence in the protein. Ultimately amino acids are produced. The smaller molecules are chemically reactive and thus able to be involved in the many transformations which occur.

The main chemical reactions which take place during flavour development are outlined in a previous review.[137] A simplified schematic diagram showing these changes is shown in Fig. 2.

For example, nitrogen-containing aroma compounds found in meat may be derived from Fig. 2 in a series of reactions as:

(a) Reactions of reducing sugars and amino acids (Maillard).

(b) Thermal degradation of Amadori and Heyns rearrangement compounds (via Maillard second stage).

(c) Pyrolysis of amino acids (deamination, decarboxylation).

(d) Reaction of ammonia (from previous reaction) with α-dicarbonyl compounds (from Maillard).

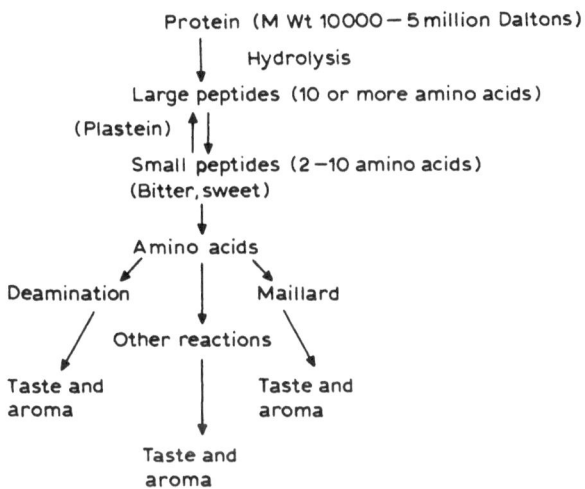

FIG. 2. Scheme showing routes to flavour from proteins.

Other heterocyclics found in meat contain oxygen and/or sulphur instead of, or as well as, nitrogen.[20] A comprehensive review of the chemistry of heterocyclic aroma compounds has been published.[214]

The reaction between ammonium sulphide and selected aldehydes gives rise to a large number of sulphur- and nitrogen-containing heterocyclic aroma compounds. Some of these compounds have been found in fried chicken and chips (french fries) potato flavour.[215]

The thermal degradation of cysteine and glutathione (α-Glu–Cys–Gly) in water gave rise to a large number of volatile compounds of potential importance in aroma. More came from cysteine than from glutathione. The different nature of the products suggested different routes, with nitrogen-sulphur heterocyclics from cysteine predominating, whereas glutathione gave largely a derivative of trithiolane, a sulphur-heterocyclic.[216] The thermal decomposition of the amino acid cystine

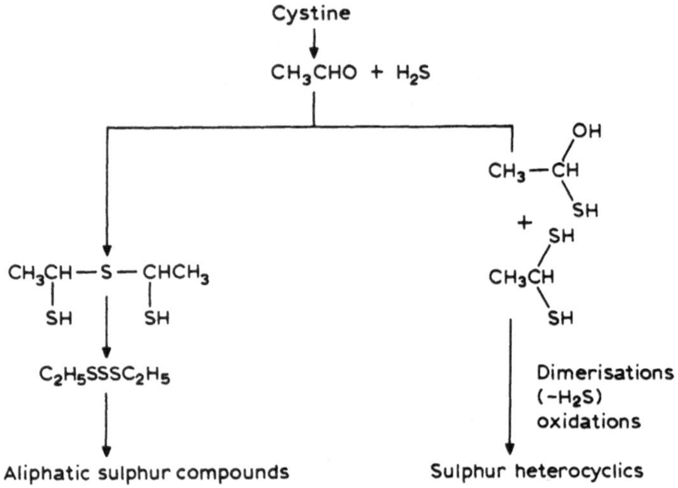

Cystine

$\downarrow$

$CH_3CHO + H_2S$

$$CH_3CH-S-CHCH_3$$
$$\quad\;|\qquad\qquad|$$
$$\quad SH\qquad\quad SH$$

$$C_2H_5SSSC_2H_5$$

$\downarrow$

Aliphatic sulphur compounds

$$\begin{array}{c} OH \\ | \\ CH_3-CH \\ | \\ SH \end{array}$$

$+$

$$\begin{array}{c} SH \\ | \\ CH_3CH \\ | \\ SH \end{array}$$

$\downarrow$

Dimerisations
($-H_2S$)
oxidations

Sulphur heterocyclics

FIG. 3. Sulphur compounds from thermal breakdown of cystine (based on Ref. 191).

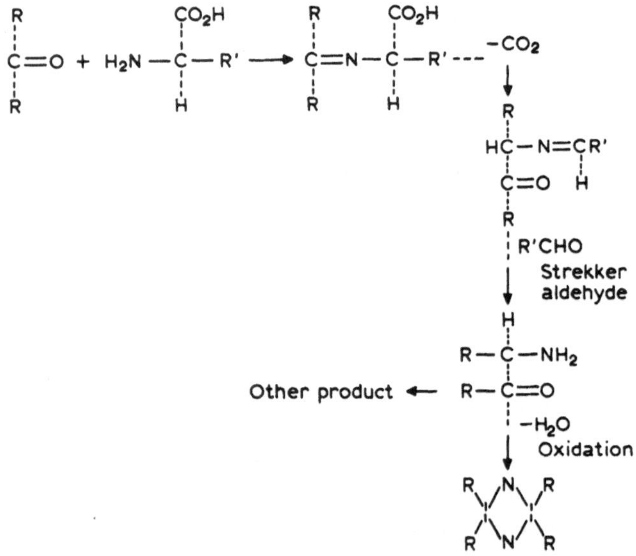

$$\begin{array}{c} R \\ | \\ C=O \\ | \\ R \end{array} + H_2N-\begin{array}{c} CO_2H \\ | \\ C \\ | \\ H \end{array}-R' \longrightarrow \begin{array}{c} R \\ | \\ C=N- \\ | \\ R \end{array}\begin{array}{c} CO_2H \\ | \\ C \\ | \\ H \end{array}-R' \cdots \overset{-CO_2}{\longrightarrow}$$

$$\begin{array}{c} R \\ | \\ HC-N=CR' \\ | \quad\;\; | \\ C=O \quad H \\ | \\ R \end{array}$$

$\downarrow$ R'CHO
Strekker aldehyde

$$\begin{array}{c} H \\ | \\ R-C-NH_2 \\ \\ R-C=O \end{array}$$

Other product $\longleftarrow$

$\downarrow$ $-H_2O$
Oxidation

$$\begin{array}{ccc} R & N & R \\ & \diagup \; \diagdown & \\ | & & | \\ & \diagdown \; \diagup & \\ R & N & R \end{array}$$

FIG. 4. Strekker degradation leading to pyrazines (based on Ref. 217).

gives rise to both aliphatic and heterocyclic sulphur compounds. The mechanism proposed for the transformations is summarised in Fig. 3. The reaction between both cysteine and cystine with 2,5 dimethyl-4-hydroxy-3(2H)-furanone in aqueous solution gives rise to the production of a wide range of heterocyclic compounds including thiazoles, thiazolines and pyrazines. The nature of the reaction products is, however, strongly dependent on pH.[193,194] In model systems, it has been proposed that a mechanism involving a Strekker degradation of the products of interaction of a dicarbonyl and an amino acid is involved (Fig. 4)[217]

Analytical Methods
The progress made in the field of flavour chemistry has to a great extent been dependent on the development of analytical and sensory procedures.[205] Developments in methodology have been reviewed.[218,219]

For the purposes of this review, the traditional stages of extraction, separation, detection and identification as described in a previous review[135] will be followed, although detection and identification are progressively coming together in one operation. Low temperature extraction of food constituents with supercritical carbon dioxide has many benefits, not the least being the preservation of sensitive flavour compounds. The liquid extracts may be separated by supercritical fluid chromatography (SFC).[220] The methods used for the isolation of food flavours (aroma) have been reviewed.[221]

The constituents may be detected by fourier transform infrared spectroscopy (FTIR), the data are useful in their own right, but are also complementary to that obtained by mass spectroscopy.[222] FTIR is also used in gas chromatography. The technique of freezing the eluate of a gas chromatograph on to a gold disc in an argon atmosphere before obtaining transmittance spectra is said to give an increased sensitivity of some 100 times over the use of FTIR in conventional GC/IR detection.[223]

A headspace sampling technique, utilising a porous polymer for trapping volatiles from a protein product followed by desorption in an injector part of a gas chromatograph, has been used to monitor the flavour deterioration of milk products.[224] Whilst the concentration may be more than adequate to obtain an olfactory response, the quantity needed to produce this response in those components of very low threshold value is below ready detection by even mass spectroscopy. In order to produce enough material for analysis it is possible to preconcentrate the volatile components. For example, in studies on herbs and spices the quantities of flavour volatiles are too low to be detected and a preconcentration step is used.

In a series of studies on the reactions of serine and threonine with sucrose under the conditions of coffee roasting, mass spectrometric identification is used to identify a variety of nitrogen-containing heterocyclic roast aroma compounds, notably pyrroles,[226] pyrazines,[192] pyridines and oxazoles.[227] Pyrroles and indoles can also be determined spectrophotometrically using p-dimethylaminobenzaldehyde in the presence of hydrochloric acid (the Ehrlich test), although many parameters affect the results.[228]

The mass of data generated during the analytical determinations is such that control of these data by computer is probably becoming essential. For example, in addition to controlling the parameters on GC analysis of headspace, the computer can be used to produce a library of retention indices.[229] It is suggested that the intensity of off-flavours in milk may be determined by multivariate analysis of gas chromatographic data.[230] It is claimed it is possible to use a general statistics package in the analysis of data in flavour research.[231] The status of data analysis in flavour research has been reviewed.[232]

USES OF PROTEIN-DERIVED FLAVOURS

Most studies on flavours are concerned with understanding the process which make and preserve flavours so that better use may be made of a food resource which is growing ever more scarce. However, a substantial body of work is also undertaken to provide flavours from protein sources which may be used for a variety of purposes, examples of which have already been encountered in this review.

The use of a cheese flavouring allows the reduction of up to 50% in the cheese solids content of a microwavable macaroni and cheese product.[233] Many of the flavours from proteins have a savoury taste and they find uses in meat products. Examples of this are proteins extracted or obtained as by-products of the meat industry.[19] Animal by-products from various sources may also be converted into meat flavours using technology derived from the pharmaceutical industry. These products have the advantage of being low in fat and low in salt.[29] A beef flavour can be made from the fermentation together of yeasts and vegetable materials. The flavour has claimed uses to enhance the flavour of reduced-fat and microwavable foods.[234] The utilisation of dark poultry meat may be improved by the addition of beef flavour.[235] Blood is an animal by-product for which edible uses are constantly being sought: one such investigation shows some promise.[18] It is now possible to produce a range of fish powders with many uses as flavourings.[38] Similarly technology has advanced sufficiently to allow the production of mycoproteins which have practical uses for many food products.[96]

The utilisation of vegetable proteins to produce meat substitutes has been extensively researched and various trial products have been produced. Products have also been launched on the market in the UK and in other countries. Certainly in the UK market the upsurge in interest in vegetarian products has rekindled interest in such products. Texturised vegetable protein, particularly soya, has a small but growing part of the meat-like protein market.

Using soya protein, for example, with meat products to extend or add to flavour is now a well-established technology. Several companies offer meat flavourings to ensure consumer acceptance.[236] Some of the flavours used for these products are described as derived from reaction or thermal process flavourings as they are derived from the controlled heating of precursors such as sulphur amino acids and sugars.[145] Soy proteins may be used in brines or injected into a range of whole muscle meats to advantage.[55] Beef bullock restructured steaks may be extended with soy protein or wheat gluten.[237] Up to 25% of ground meat is said to be replaceable by soy protein.[238] The economic advantages of the addition of soya protein in meat and meat products in catering markets for the military and schools in the US have been well-established.[54] It is suggested that toasted soya nuts may be used in snack foods and in other savoury products.[239]

The labelling of such products in Europe has not finally been agreed at the time of writing.

Non-meat protein additives have also been extensively used in poultry products.[240] It is even claimed that soya protein can replace milk protein to produce dairy-like beverages.[53] Other vegetable proteins such as peanut, cowpea and field-pea flours have uses to replace other scarcer or more costly proteins in bakery or snack products.[65,66]

Protein hydrolysates also have extensive use in flavourings, particularly the autolysates from soya or yeast. As well as flavouring uses, soy protein hydrolysates may also reduce the water activity in meat products, both to reduce the salt content and to extend the shelf-life.[241,242] Extracts of brewers' and bakers' yeasts can provide flavour enhancement of a wide variety of savoury products.[243] Collagen hydrolysates may also be used to reduce the salt content of meat products.[244,245] Another use for collagen hydrolysates is as a paste-like spread for bread.[92]

Apart from flavouring properties, a milk hydrolysate can be used to enhance the mouthfeel of low-fat foods.[246] A mix of soya protein hydrolysate with amino acids and emulsifier have been patented for use as flavour and mouthfeel character of beverages.[247,248] Hydrolysis of a mixture of animal and plant protein materials produces a new seasoning material.[142] Provision of flavouring with milk-fat may be provided by encapsulation of protein and peptides.[249]

The flavour enhancers, disodium 5'-inosinate (IMP) and disodium 6'-guanylate (GMP) can now be used with phosphatase-containing foods due to a coating which resists enzyme activity.[250]

REFERENCES

1. Aurand, L. W., Proteins. In *Food Composition and Analysis,*. ed. L. W. Aurand, A. E. Woods & M. R. Wells. Van Nostrand Reinhold Co., New York, 1987, pp. 232–282.

2. Schmidt, R. H., Bitter Components in Dairy Products. In *Bitterness in Foods and Beverages,* ed. R. L. Rouseff, Elsevier, Amsterdam, 1990, pp. 183–204.

3. Kilcast, E. A., Sensory Properties of Foods after Irradiation. In *Irradiation and Combination Treatments,* conference 1/2 March 1990, IBC Technical Services Ltd., London.

4. Harwalker, V. R., Boutin-Muma, B., Cholette, J., McKellar, R. C., Emmons, D. B., & Klassen, G., Isolation and partial purification of astringent compounds from ultra-high temperature sterilized milk. *J. Dairy Res.,* **56** (1989) 367–373.

5. Marshall, V. M. E., Flavour development in fermented milks. In *Advances in the Microbiology and Biochemistry of Cheese and Fermented Milk,* ed. F. Lyndon Davies & B. A. Law. Elsevier Applied Science Publishers, London, 1984, pp. 153–186.

6. Ashour, M. M., Abdel Baky, A. A. & El Neshawy, A. A., Improving the quality of domiati cheese made from recombined milk. *Food Chem.,* **20** (1986) 85–96.

7. Kumar, S. & Mathur, B. N., Proteolytic changes in raw buffalo milk preserved by LP-system. *Indian J. Dairy Sci.,* **41** (1988) 318–321.

8. Dousset, X., Demaimay, M., Ravaud, C., Levesque, A., Pinet, X. D. & Kergo, Y., Proteolysis and bitterness of UHT milk during storage as a consequence of refrigeration temperature of raw milk. *Le Lait,* **68** (1988) 143–156.

9. Law, B. A., Microbial proteolysis of milk proteins. In *Food Proteins,* ed. P. F. Fox & J. J. Condon. Proceedings of the Kellogg Foundation Inter-

national Symposium on Food Proteins, Cork, Eire, 1981. Applied Science Publishers, 1982, pp. 307–328.

10. Fox, P. F., Proteolysis during cheese manufacture and ripening. *J. Dairy Sci.*, **72** (1989) 1379–1400.

11. Rousseff, R., L., Bitterness in food products; an overview. In *Bitterness in Foods and Beverages*. Elsevier, Amsterdam, 1990, pp. 1–14.

12. Baldwin, K. A., Baer, R. J., Parsons, J. G., Seas, S. W., Spurgeon, K. R., & Torrey, G. S., Evaluation of yield and quality of cheddar cheese manufactured from milk with added whey protein concentrate. *J. Dairy Sci.*, **69** (1986) 2543–2550.

13. Banks, J. M. & Muir, D. D., Effect of incorporation of denatured whey protein on the yield and quality of Cheddar cheese. *J. Soc. Dairy Technol.*, **38** (1985) 27–32.

14. Banks, J. M., Elimination of the development of bitter flavour in Cheddar cheese made from milk containing heat-denatured whey protein. *J. Soc. Dairy Technol.*, **41** (1988) 37–41.

15. Anon., Say cheese for a smart award. *Lab. News*, (1988) (November 28) 3.

16. Hagberg, E. C., Haislip, J. R. & Johnson, B. R., Method for producing a highly flavoured cheese ingredient. US Patent No. 4 752 483, 1988.

17. Blenford, D., Functional proteins. *Food Flavour. Ingred. Process. Packag.*, **11** (1989) 55–9.

18. Oluski, V., Petrovic, M., Kelemen-Masic, D., Popov-Raljic, J. & Jodal, L., Properties of decolorized blood coagulate. *Technologija-Mesa*, **26** (1985) 335–338.

19. Anon, Proteins impart meaty flavours. *Food Process.*, **55** (1986) p. 8.

20. Bailey, M. E. & Einig, R. G., Reaction flavors of meat. In *Thermal Generation of Aromas*, ed. T. H. Parliment, R. J. McGorrin & C.-T. Ho. American Chemical Society, Washington D.C., 1989, pp. 421–432.

21. Spanier, A. M. & Edwards, J. V., Chromatographic isolation of presumptive peptide flavor principles from red meat. *J. Liquid Chrom.*, **10** (1987) 2745–2758.

22. Spanier, A. M., Edwards, J. V. & Dupuy, H. P., The warmed-over flavor process in beef; a study of meat proteins and peptides. *Food Technol.*, **42** (1988) 112–118.

23. Gray, J. I. & Pearson, A. M., Rancidity and warmed-over flavor. In *Advances in Meat Research*, Vol. 3, ed. A. M. Pearson & T. R. Dutson. Van Nostrand Reinhold, New York, 1987, pp. 221–270.

24. Bailey, M. E., Inhibition of warmed-over flavor, with emphasis on Maillard reaction products. *Food Technol.*, **42** (1988) 123–126.

25. St. Angelo, A. J., Koohmaraie, M., Crippen, K. L. & Crouse, J., Acceleration of tenderization/inhibition of warmed-over flavor by calcium chloride-antioxidant infusion into lamb carcasses. *J. Food Sci.*, **56** (1991) 359–362.

26. Edwards, R. A., Dainty, C. M., Hibbard, C. M. & Ramantanis, S. V., Amines in fresh beef of normal pH and the role of bacteria in changes in concentration observed during storage in vacuum packs at chill temperatures. *J. Appl. Bacter.*, **63** (1987) 427–434.

27. Blake, T., Trends in meat flavour technology. *Food Manuf.*, **62** (1987) 43, 45.

28. Nishimura, T., Rhue, M. R., Okitani, A. & Kato, H., Components contributing to the improvement of meat taste during storage. *Agric. Biol. Chem.*, **52** (1988) 2323–2330.

29. Andres, C., Meat flavours produced with pharmaceutical technology. *Food Process.*, *Chicago*, **46** (1985) 56.

30. Gorshkova, L. V., Kudryashov, L. S., Bol'shakov, A. S. & Goncharov, G. I., Effects of electrical and mechanical treatment during salting of beef on the content of free amino acids. *Izv. Vyssh. Uchebn. Zaved., Pisch. Teknol.*, **3** (1988) 47–49. (FSTA Abstract **21** (7) 7 S 58).

31. Johnson, L. P., Miller, M. F. & Reagan, J. O., The effect of various levels of added sodium chloride and potassium chloride on the chemical, physical and sensory characteristics of precooked, recombined beef chuck roasts. *J. Food Quality*, **12** (1988) 275–272.

32. Gillett, T., A., Adipose and connective tissue. In *Advances in Meat Research*, Vol. 3, *Restructured Meat and Poultry Products*, ed. A. M. Pearson & T. R. Dutson. Van Nostrand Reinhold, New York, 1987, pp. 73–124.

33. Kim, H. & Gilbert, S. G., Isolation and identification of volatile flavor compounds in commercial granular gelatin sample. In *Frontiers of Flavour*, ed. G. Charalambous. Conference 1–3 July 1987. Elsevier Science Publishers, Amsterdam, 1988.

34. Umetsu, H. & Ichishima, E., Mechanism of digestion of bitter peptide from fish protein concentrate by wheat carboxypeptidase. *Nippon Shok. Kog. Gakk.*, **32** (1985) 281–287.

35. Raksakulthai, N., Role of protein degradation in fermentation of fish sauce. *Diss. Abstr. Int. B*, **49** (1989) 2947 B.

36. Fonarev, N. A. & Rzhavskaya, F. M., Changes in protein components of canned mackerel. *Rybnoe Khozvaistvo*, **11** (1988) 85–88. (FSTA Abstract **22** (8) 8 R 9).

37. Cha, Y. J. & Ahn, C. B., Flavour components in sun-dried ray. *J. Korean Soc. Food Nutr.* **14** (1985) 370–374.

38. Lake, C., Powder potential. *Food Flav. Ingred. Packag. Process.*, **11** (1989) 41–43.

39. Lusas, E. W. & Rhee, K. C., Applications of vegetable food proteins in traditional foods. In *Plant Proteins: Applications, Biological Effects, and Chemistry*, ed. R. L. Ory. American Chemical Society, Washington D.C., 1986, pp. 32–45.

40. Arnoe, T., All-vegetable, isolated soy protein provides functional and nutritional benefits for today's food products. *Eur. Food Drink Rev.*, (Autumn) (1990) 51–52.

41. Naik, G. & Gleason, J. E., An improved soy-processing technology to help alleviate protein malnutrition in India. *Food Nutr. Bull.*, **10** (1988) 45–49.

42. Mittal, G. S. & Usborne, W. R., Meat emulsion extenders. *Food Technol.*, **39** (1985) 121–130.

43. Marshall, W. E., Bitterness in soy and methods for its removal. In *Bitterness in Foods and Beverages*, ed. R. L. Rousseff. Elsevier, Amsterdam, 1990, pp. 275–291.

44. Chang, S. S., Huang, A.-S. & Ho C.-T., Isolation and identification of bitter

compounds. In *Bitterness in Foods and Beverages*, ed. R. L. Rousseff. Elsevier, Amsterdam, 1990, pp. 267–274.

45. Speer, K. & Montag, A. Phenylalanine decomposition products as flavour compounds in honey. *Dtsch. Lebensm. Runds.*, **83** (1987) 103–107.

46. Friend, B. A., Gierhart, D. L. & O'Brien, J. K., Process for reducing the level of objectionable flavors in vegetable protein by microorganism contact. US patent 4 642 236 (1987).

47. Homma, S. & Fujimaki, M., Lipid in the soy protein isolate with beany flavour compounds. *Nutr. Sci. Soy Protein*, **6** (1985) 7–10.

48. Eldridge, A. C., Friedrich, J. P., Warner, K. & Kwolek, W. F., Preparation and evaluation of supercritical carbon dioxide defatted soybean flakes. *J. Food Sci.*, **51** (1986) 584–587.

49. Sevenants, M. R. Removal of textured vegetable product off-flavor by supercritical fluid or liquid extraction. US Patent 4 675 198 (1987).

50. McDaniel, M. R. & Chan, N., Masking of soy protein flavor by tomato sauce. *J Food Sci.*, **53** (1988) 93–96, 101.

51. Fuke, Y. & Matsuoka, H., Changes in proteins and protease activities during ripening of cheese-like products from soy milk using *Penicillium caseicolum*. *J Jpn. Soc. Food Sci. Tech. (Nipp. Shok. Kog. Gakk.)*, **34** (1987) 826–833.

52. Palielo, M. M. B., Reddy, K. V. & Da Silva, R. S. F., Calcium sulphate as organoleptic coadjuvant in the formulation of soy-whey yoghurt. *Lebensm.-Wiss.-Technol.*, **20** (1987) 155–157.

53. Moller, J. L., Dairylike beverages produced from soya protein concentrate. In *Proceedings of the 3rd European Conference on Ingredients and Additives*. Nov. 1988, Vol II, Food Ingredients Europe.

54. Kotula, A. W. & Berry, B. W., Addition of soy proteins to meat products. In *Plant Proteins: Applications, Biological Effects, and Chemistry*, ed. R. L. Ory. American Chemical Society, Washington D.C., 1986, pp. 74–89.

55. Young, L. S., Taylor, G. A. & Bonkowski, A., Use of soy protein products in injected and absorbed whole muscle meats. In *Plant Proteins: Applications, Biological Effects and Chemistry*, ed. R. L. Ory, American Chemical Society, Washington D.C., 1986, pp. 90–98.

56. Einig, R. G. & Bailey, M. E., Soy proteins and thermal generation of alkyl-pyrazines in meat flavor. In *Thermal Generation of Aromas*, ed. T. H. Parliment, R. J. McGorrin & C-T. Ho. American Chemical Society, Washington D.C., 1989, pp. 479–486.

57. Padda, G. S., Kesava Rao, V., Keshri, R. C., Sharma, N. & Sharma, B. D., Studies on physico-chemical & organoleptic properties of ham patties extended with texturised soy proteins. *J. Food Sci. Technol. Mysore*, **22** (1985) 362–365.

58. Stuhlberger, L. & Kotter, L. H. C., Use of soy isolates, soy concentrates and soy flours; soy protein in manufacture of Bruhwurst sausages. *Fleischerei*, **41** (1990) 41–44.

59. Keeton, J. T., Foegeding, E. A. & Patana-Anake, C., A comparison of non-meat proteins, sodium tripolyphosphate and processing temperature effects on physical and sensory properties of frankfurters. *J. Food Sci.*, **49** (1984) 1462–1465 & 1474.

60. Baltic, M. & Babic, L., Tracing of vegetal proteins in meat products by resorting to methods of SOS-Page electrophoresis. *Technol. Mesa*, **28** (1987) 34–37.

61. Wang, H. L., Uses of soybeans as foods in the west with emphasis on Tofu and Tempeh. In *Plant Proteins: Applications, Biological Effects, and Chemistry*, ed. R. L. Ory. American Chemical Society, Washington D.C., 1986, pp. 45–60.

62. Motohiro, T. & Taniguchi, H., Effect of soy protein isolate on food quality of fish gels. In *Recent Advances in Food Science & Technology*, Vol 2; *Oriental foods, meat and fishery products, food chemistry and engineering*, proceedings of a symposium, Taipei, 1980, ed. S. M. Chang. Hua Shiang Yuan Publ. Co., 1981, pp. 63–68.

63. Shehata, N. A., Ibrahim, A. A. & Ghali, N., Effect on protein quality of supplementing wheat flour with soy protein concentrate in making Egyptian pastries. *Nahrung*, **33** (1989) 753–759.

64. Schultz, M., Hoppe, K. & Schmandke, H., Off-flavour reduction in Vicia faba bean protein isolate. *Food Chem.*, **30** (1988) 129–135.

65. Klein, B. P. & Raidl, M. A., Use of field-pea flours as protein supplements in foods. In *Plant Proteins: Applications, Biological Effects, and Chemistry*, ed. R. L. Ory. American Chemical Society, Washington D.C., 1986, pp. 19–31.

66. McWatters, K. H., Use of peanut and cowpea flours in selected fried and baked foods. In *Plant Proteins: Applications, Biological Effects, and Chemistry*, ed. R. L. Ory. American Chemical Society, Washington D.C., 1986, pp. 8–19.

67. Vavreinova, S., Application possibilities of leaf protein in human nutrition. *Prum. Potravin*, **35** (1984) 321–322.

68. Favati, R., Fiorentini, R. & Galoppini, C., Pigment extraction from alfalfa protein concentrates, *Acta Aliment.*, **17** (1988) 239–244.

69. Yamanishi, T., Bitter compounds in tea. In *Bitterness in Foods and Beverages*, ed. R. L. Rousseff. Elsevier, Amsterdam, 1990, pp. 159–167.

70. King, G. A., Henderson, K. G., O'Donoghue, E. M., Martin, W. & Lill, R. E., Flavour and metabolic changes in asparagus during storage. *Sci. Hortic.*, **36** (1988) 183–190.

71. Byrne, M., Whatever happened to new protein? *Food Manuf.*, **63** (1988) 51–54 & 57.

72. Deger, H.-M. & Fricke, U. Process for obtaining a microbial protein isolate with particular properties. European Patent Appl. 0 207 423 A2 (DE) 29.6.85.

73. Kurihara, Y., Ookubo, K. & Halpern, B. P., Purification and chemical structure of taste modifiers: taste-modifying protein and ziziphin. ISOT IX/ AChemS VII Abstracts, p. 626.

74. Ogata, C., Hatada, M., Tomlinson, G., Shin, W-C. & Kim, S.-H., Crystal Structure of the intensely sweet protein monellin. *Nature*, **328** (1987) 739–742.

75. Weickmann, J. L., Lee, J-H., Blair, L. C., Ghosh-Dastidar, P. & Koduri, R. K., Exploitation of genetic engineering to produce novel protein

sweeteners. In *Progress in Sweeteners*, ed. T. H. Grenby. Elsevier Science Publishers Ltd., London, 1989, pp. 47–69.

76. Osnabrugge, W. van, How to flavor baked goods and snacks effectively. *Food Technol.*, **43** (1989) 74–82.

77. Dumont, J. P., Flavour-protein interactions: a key to aroma persistence. In *Flavour Science and Technology*, ed. M. Martens, G. A. Dalen & H. Russwurm Jr. Wiley, 1987, pp. 143–159.

78. Noar, S. R., Reversible protein: flavour interactions and their effect on sensory perception. *Diss. Abstr. Int. B*, **46** (1986) 2893.

79. Solms, J., Interactions of non-volatile and volatile substances in foods. In *Interactions of Food Components*, ed. G. G. Birch & M. G. Lindley. Elsevier Applied Science Publishers, London, 1986, pp. 189–210.

80. Malcolmson, L. J. & McDaniel, M. R. & Hoehn, E., Flavor protein interactions in a formulated soup containing flavored soy protein. *J. Can. Inst. Food Sci. Technol.*, **20** (1987) 229–235.

81. Jasinski, & Kilara, A., Flavor binding by whey proteins. *Milchwissenschaft*, **40** (1985) 596–599.

82. Kim, H. & Min, D. B., Interaction of flavor compounds with protein. In *Interactions of Food Components*, ed. G. G. Birch & M. G. Lindley, Elsevier Applied Science Publishers, London, 1986, pp. 404–420.

83. Land, D. G. & Reynolds, J., The influence of food components on the volatility of diacetyl. In *Flavour '81: Proceedings of the 3rd Weurman Symposium*, ed. P. Schreier. Walter de Gruyter, 1981, pp. 701–705.

84. Voilley, A., Fares, K. & Lorient, D., Aroma retention during air-drying of protein solutions. In *Advances in Food Technology*, Vol 2, *Proceedings of the 2nd World Congress of Food Technology, Barcelona, 1987*, ed. E. Primo Yufera & P. Fito Maupoey, pp. 1157–1167.

85. Ng, P. K. W., Hoehn, E. & Bushuk, W., Binding of vanillin by fababean proteins. *J. Food Sci.*, **54** (1989) 105–107.

86. Ng, P. K. W., Hoehn, E. & Bushuk, W., Sensory evaluation of binding of vanillin by fababean proteins. *J. Food Sci.*, **54** (1989) 324–325 & 346.

87. Dumont, J. P., Diacetyl retention by pea proteins: effect of physical treatments. In *Progress in Flavour Research 1984*, Proceedings of the 4th Weurman Flavour Research Symposium, Dourdan, France 1984, ed. J. Adda. Elsevier Science Publishers, Amsterdam, 1984, pp. 501–504.

88. Schnepf, M. I. & Satterlee, L. D. The interaction of iron with proteins and peptides. In *Interactions of Food Components*, ed. G. G. Birch & M. G. Lindley, Elsevier, London, 1986, pp. 43–63.

89. Clifford, M. N., Phenol-protein interactions and their possible significance for astringency. In *Interactions of Food Components*, ed. G. G. Birch & M. G. Lindley, Elsevier, London, 1968, pp. 143–163.

90. Seshadri, R. & Dhanraj, N., Flavour interactions in tea. In *Frontiers of Flavour*, Proceedings of the 5th International Flavour Conference, 1987, ed. G. Charalambous. Elsevier, Amsterdam, 1988, pp. 169–180.

91. Law, B. A., Flavour development in cheeses. In *Advances in the Microbiology and Biochemistry of Cheese and Fermented Milk*, ed. F. L. Davies, & B. A. Law. Elsevier Applied Science Pubs. Ltd., London, 1984, pp. 187–208.

92. Buckholz, L. & Stypula, R. J., Flavour-retaining food product. European patent application 0 306 000 A2 (1988).
93. Patel, M. M. & Dave, J. C., Chewing gum containing sweet protein and salt. International Patent no. WO 89/02703 A1 (1989).
94. Kurihara, Y., Method for stabilizing taste-modifier. European Patent Appl. 0 347 832 (1989).
95. Ensor, D. R., The contribution of flavor chemistry to the foods industry. In *Flavor Chemistry of Lipid Foods*, ed. D. B. Min & T. H. Smouse. The American Oil Chemists' Society, Champaign, Ill. 1989, pp. 1–12.
96. Kato, H., Rhue, M. R. & Nishimura, T., Role of free amino acids and peptides in food taste. In *Flavor chemistry: Trends and developments*, ed. R. Teranishi *et al.* American Chemical Society, Washington D.C., 1989, pp. 158–174.
97. Nishimura, T. & Kato, H., Taste of free amino acids and peptides. *Food Rev. Int.*, **4** (1988) 175–194.
98. Lancaster, J. E. & Shaw, M. L., γ-Glutamyl peptides in the biosynthesis of S-alk(en)yl-L-cysteine sulphoxides (flavour precursors) in Allium. *Phytochemistry*, **28** (1989) 455–460.
99. Basha, S. M. & Young, C. T., Changes in the polypeptide composition of peanut (Arachis hypogaea L.) seed during oil roasting. *J. Agric. Food Chem.*, **33** (1985) 350–354.
100. Hammond, E. G., The flavors of dairy products. in *Flavor Chemistry of Lipid Foods*, ed. D. B. Min & T. H. Smouse. The American Oil Chemists' Society, Champaign, Illinois, 1989, pp. 222–236.
101. Sohn, K.-H. & Lee, H.-J., Bitter peptides derived from alpha$_1$- and beta-casein digested with alkaline protease from *Bacillus subtilis*. *Korean J. Food Sci. Tech.*, **20** (1988) 659–665.
102. Cliffe, A. J. & Law, B. A., Peptide composition of enzyme-treated Cheddar cheese slurries, determined by reverse phase high performance liquid chromatography. *Food Chem.*, **36** (1990) 73–80.
103. Ishibashi, N., Ono, I., Kato, K., Shigenaga, T., Shinoda, I., Okai, H. & Fukui, S., Role of the hydrophobic amino acid residue in the bitterness of peptides. *Agric. Biol. Chem.*, **52** (1988) 91–94.
104. Ishibashi, N., Kubo, T., Chino, M., Fukui, H., Shinoda, I., Kikuchi, E., Okai, H. & Fukui S., Taste of proline-containing peptides. *Agric. Biol. Chem.*, **52** (1988) 95–98.
105. Ishibashi, N., Sadamori, K., Yamamoto, O., Kanehisa, H., Kouge, K., Kikuchi, E, Okai, H. & Fukui S., Bitterness of phenylalanine- and tyrosine-containing peptides. *Agric. Biol. Chem.*, **51** (1987) 3309–3313.
106. Ishibashi, N., Bitterness of leucine-containing peptides. *Agric. Biol. Chem.*, **51** (1987) 2389–2394.
107. Nosho, Y., Otagiri, K., Shinoda, I. & Okai, H., Studies on a model of bitter peptides including arginine, proline and phenylalanine residues. II. *Agric. Biol. Chem.*, **49** (1985) 1829–1837.
108. Shinoda, I., Nosho, Y, Kouge, Ishibashi, N, Okai, H., Tatsumi, K. & Kikuchi, E., Variation in bitterness potency when introducing Gly-Gly residue into bitter peptides. *Agric. Biol. Chem.*, **51** (1987) 2103–2110.

109. Shinoda, I., Nosho, Y., Otagiri, K., Okai, H. & Fukui, S., Bitterness of diastereometers of a hexapeptide *Agric. Biol. Chem.*, **50** (1986) 1785–1790.
110. Shinoda, I., Fushimi, A., Kato, H., Okai, H. & Fukui, S., Bitter taste of synthetic C-terminal tetradecapeptide of bovine beta-casein. *Agric. Biol. Chem.*, **49** (1985) 2587–2596.
111. Shinoda, I., Tada, M., Okai, H., & Fukui, S., Bitter taste of H-Pro–Phe–Pro–Gly–Pro–Ile–Pro–OH corresponding to the partial sequence. *Agric. Biol. Chem.*, **50** (1986) 1247–1254.
112. Shinoda, I., Okai, H. & Fukui, S., Bitter taste of H-Val–Val–Val–Pro–Pro–Phe–Leu–OH corresponding to the partial sequence. *Agric. Biol. Chem.*, **50** (1986) 1255–1260.
113. Adler-Nissen, J., Bitterness intensity of protein hydrolysates—chemical and organoleptic characterization. In *Frontiers of Flavor*, ed. G. Charalambous. Conference, Chalkidiki, Greece, 1–3 July 1987. Elsevier Science Publishers, Amsterdam, 1988, pp. 63–77.
114. Ney, K. H., Cocoa flavour—bitter compounds as its essential taste components. *Gordian*, **86** (1986) 84–88.
115. Ohayama, S., Ishibashi, N., Tamura, M., Nishizaki, H. & Okai, H., Synthesis of bitter peptides composed of aspartic acid and glutamic acid. *Agric. Biol. Chem.*, **52** (1988) 871–872.
116. Ishibashi, N., Kouge, K., Shinoda, I., Kanehisa, H. & Okai, H., A mechanism for bitter taste sensibility in peptides. *Agric. Biol. Chem.*, **52** (1988) 819–827.
117. Saroli, A., Structure-activity relationship of bitter compounds related to denatonium chloride and dipeptide methyl esters. *Z. Lebens. Unt. Forsch.*, **182** (1986) 118–120.
118. Ney, K. H., A program for IBM 8-compatible PC 'Γ' predicting the bitterness of peptides, especially in protein hydrolysates. *Alimenta*, **28** (1989) 9–13.
119. Behnke, U., Importance of peptides from enzymatically degraded proteins as components of foods. *Nahrung*, **29** (1985) 979–992.
120. Mogensen, L. & Adler-Nissen, J., Evaluating bitterness masking principles by taste panel studies. In *Frontiers of Flavor*. Proceedings of the 5th International Flavor Conference, Chalkidiki, Greece, 1987. Elsevier Science Publishers, Amsterdam, 1988, pp. 79–87.
121. Minagawa, E., Kaminogawa, S., Tsukasaki, F. & Yamauchi, K., Debittering mechanism in bitter peptides of enzymatic hydrolysates from milk casein by aminopeptidase T. *J. Food Sci.*, **54** (1989) 1225–1229.
122. Pawlett, D. & Fullbrook, P., Production of natural protein-based flavour and hydrolysates using amino peptidase enzymes. In *Proceedings of the 3rd European Conference on Ingredients and Additives Nov 1988*, Vol. I, ed. by Food Ingredients Europe.
123. Tamura, M., Mori, N., Miyoshi, T., Koyama, S., Kohri, H., & Okai, H., Practical debittering using model peptides and related compounds. *Agric. Biol. Chem.*, **54** (1990) 41–51.
124. Umetsu, H. & Ichishima, E., Mechanism of digestion of bitter peptides from soybean protein by wheat carboxypeptidase. *Nipp. Shok. Kog. Gakk.*, **35** (1988) 440–447.

125. Janusz, J. M., Peptide sweeteners beyond aspartame. In *Progress in Sweeteners*, ed. T. H. Grenby. Elsevier, Amsterdam, 1989, pp. 1–46.
126. Zanno, P. R., L-Aminodicarboxylic acid esters. US Patent 4 766 246 (1988).
127. Lok, S. M., I. Synthesis of high potency congener derivative of isoproterenol. II. Stereoisomeric approach to the molecular basis of taste. *Diss. Abstr. Int. B.*, **49** (1989) 3675 B.
128. Aso, K., Enzymatic approach to the synthesis of a lysine-containing sweet peptide, N-acetyl-L-phenylalanyl-L-lysine. *Agric. Biol. Chem.*, **53** (1989) 729–733.
129. Tamura, M., Nakatsuka, T., Tada, M., Kawasaki, Y., Kikuchi, E. & Okai, H. The relationship between taste and primary structure of 'delicious peptide' (Lys–Gly–Asp–Glu–Glu–Ser–Leu–Ala) from beef soup. *Agric. Biol. Chem.*, **53** (1989) 319–325.
130. Okai, H., An enhanced effect of peptides and their analogues on saltiness of sodium chloride. *J. Jpn. Soc. Food Sci. Tech. (Nipp. Shok. Kog. Gakk)*, **36** (1989) 769–776. (FSTA Abstract **22** (7) 7 T 27 (1990).)
131. Kawasaki, Y., Seki, T., Tamura, M., Kikuchi, E., Tada, M. & Okai, H., Glycine methyl or ethyl ester hydrochloride as the simplest examples of salty peptides and their derivatives. *Agric. Biol. Chem.*, **52** (1988) 2679–2681.
132. Anon. Salt-free salt. *Nutr. Rev.*, **43** (1985) 337–338.
133. Agnes, G. & Altamura, M. Taurine derivatives. British Patent Appl. 2 193 206 A (31.7.86) (1986).
134. Huynh-ba, T., Philippossian, G., Alleged salty taste of L-ornithyltaurine monohydrochloride. *J. Agric. Food Chem.*, **35** (1987) 165–168.
135. Seki, T., Kawasaki, Y., Tamura, M., Tada, M. & Okai, H., Further study on the salty peptide ornithyl-beta-alanine. Some effects of pH and additive ions on the saltiness. *J. Agric. Food Chem.*, **38** (1990) 25–29.
136. Kimura, S., Yokomukai, Y. & Komai, M., Salt consumption and nutritional state especially dietary protein level. *Am. J. Clin. Nutr.*, **45** (1987) 1271–1276.
137. Weir, G. S. D., Protein hydrolysates as flavourings. In *Developments in Food Proteins*, Vol. 4, ed. B. J. F. Hudson. Elsevier Applied Science Publishers, London, 1986, pp. 175–217.
138. Lieske, B. & Konrad, G., Process for manufacture of modified protein hydrolysates with a chicken meat flavour. GDR patent no DD 260 387 (1988).
139. Nitta Gelatin, Preparation of seasoning. Japanese patent 263783. (Unilever Patent Abstract 90-168330/22).
140. Kolbeck, W., Pyttlik, H. & Grasis, M. Product for Proteolysis. International patent WO 88/07822 (1988).
141. Rooij, J. F. J. de, & Meakins, S. E., Improved protein hydrolysate. European patent application 0 209 921 A1 (1987).
142. Miura, H., Nishiyama, K., Katsuragi, T. & Akatsuka, S., Preparation of new seasoning by hydrolysis of mixture of animal and plant protein materials. *Nipp. Shok. Kog. Gakk.*, **34** (1987) 98–101.
143. Dzanic, H., Protein hydrolysates from soy grits and dehydrated alfalfa flour. *J. Agric. Food Chem.*, **33** (1985) 683–685.

144. Belohlawek, L., Process and product of making a vegetable protein hydrolysate food seasoning. US patent 4 798 736 (1989).
145. May, C. G., Process flavourings. In *Food Flavourings*, ed. P. R. Ashurst, Blackie & Son Ltd., 1990, pp. 257–301.
146. Faesi, R., Werner, G. & Wolfensberger, U. Process for the production of seasoning. UK patent application 2 183 659 A (1987).
147. Duxbury, D. D. 40% less sodium in naturally brewed 'lite' soy sauce. *Food Process.*, **52** (1991) 96, 98.
148. Fullbrook, P., Pawlett, P. & Parker, D., Protein plus. *Food Process.*, **56** (1987) 11–13.
149. Parker, D. M. & Pawlett, D., Flavour control of protein hydrolysates. European patent application 0 223 560 (UK) (14.11.85) (1985).
150. Godfrey, T., Enzyme modifications for baked goods and flavour proteins. *Eur. Food Drink Rev.*, Autumn (1990) 43–44, 46, 48.
151. Krasnobajew, V., Mor, J.-R., Steiner, R. & Keller, A., Method for preparing new coffee aromas. International patent WO 86/03943 (1986).
152. Behnke, V., Ackermann, E. & Ruttloff, H., Production and utilisation of enzymic protein hydrolysate from meat residues from bones from mechanical deboning. I. Investigation of the enzymic hydrolysis of meat residues. *Nahrung*, **28** (1984) 397–407.
153. Lieske, B. & Konrad, G., Process for preparation of a flavour preparation with a chicken-like flavour. GDR patent DD 260 648 (1988).
154. Duwe, H., Kreuter, T., Lehwald, U. & Malter, M., Process for preparation of a non-bitter, fully water-soluble and non-acid precipitable partial hydrolysate of microbial protein. GDR patent DD 267 659 (1989).
155. Vegarud, G. E. & Langsrud, T., The level of bitterness and solubility of hydrolysates produced by controlled proteolysis of caseins. *J. Dairy Res.*, **56** (1989) 375–379.
156. Heyland, S., Fournet, G. & Bosch, H., Cheese flavoring product. US patent 4 544 568 (1985).
157. Heyland, S., Fournet, G. & Boesch, H., A flavouring product. UK patent application 2 151 897 A. (Switzerland) (29.12.85).
158. Kilara, A., Enzyme-modified food ingredients. *Process Biochem.*, **20** (1985) 149–157.
159. Rebeca, B. D., Pena-Vera, M. T. & Diaz-Castaneda, M., Production of fish protein hydrolysates with bacterial proteases; yield and nutritional value. *J. Food Sci.*, **56** (1991) 309–314.
160. Faigh, J. G., Stuart, M. J. & Talbott, L. L., Enzymatic hydrolysis of proteins. European Patent EP 0 325 986 A2 (1989).
161. Slocum, S. A., Jasinski, E. M., Anantheswaran, R. C. & Kilara, A., Effect of sucrose on proteolysis in yoghurt during incubation and storage. *J. Dairy Sci.*, **71** (1988) 589–595.
162. Anon. Salt booster brings no sodium burden. *Dairy Foods*, **89** (1988) 53.
163. Anon. Salt alternatives brew in marketplace. *Prep. Foods*, **157** (1988) 242.
164. Anon. YMR, a flavor enhancer (Ingredient shown at the New Orleans Food Expo). *Prep. Foods*, **157** (1988) 138.
165. Anon. Dried yeast protein enhances flavour. *Confect. Manuf.*, **23** (1986) 18.

166. Anon. Yeast protein enhances flavours and nutrition. *Food Process.* UK, **55** (1986) 13–14.

167. Mezeine Dudonisz, W., Functional properties of yeast derivatives and their application in the meat industry. *Husipar Minosegugyi Leanvvallata, Budapest,* (1988) (1) pp. 4–8. (FSTA Abstracts **22** (2) 2 S 117 1990).

168. Miteva, E., Kirova, E., Gadjeva, D. & Radeva, M., Sensory aroma and taste profiles of raw-dried sausages manufactured with a lipolytically active yeast culture. *Nahrung,* **30** (1986) 829–832.

169. Adamek, L., Rybarova, J., Rut, M., Karnet, J. & Stros, F., Manufacture of yeast autolysate. Czechoslovak Patent DS 256 863 (1988).

170. McCormick, R., The year of the yeast: yeast's sensory potential makes possible new applications. *Prep. Foods,* **156** (1987) 153–154.

171. Duxbury, D. D., Descriptive flavor analysis panel provides sophisticated service, product application concepts. *Food Process. US,* **51** (1990) 54, 56, 58.

172. Steinkraus, K. H., Indigenous fermented-food technologies for small-scale industries. *Food Nutr. Bull.,* **7** (1985) 21–27.

173. Hartman, T. G. & Rosen, R. T., Determination of ethyl carbamate in commercial protein based condiment sauces by gas chromatography-mass spectrometry. *J. Food Safety,* **9** (3) (1989) 173–182.

174. Thibault, P. A. & Monsan, P. F., Process for debittering of protein hydrolysates and the product so produced. French Patent Application FR 2 625 651 A1 (1989).

175. Lieske, B., Konrad, G. & Schulze, W., Manufacture of protein hydrolysates with modified flavour, GDR Patent DD 217 981 (1985).

176. Hardy, P. M., Protein amino acids. In *Chemistry and Biochemistry of Amino Acids,* ed. G. C. Barrett. Chapman & Hall, London, 1985, pp. 6–24.

177. Guion, P., Umami: an independent basic taste—a 20th century phenomenon. *Eur. Food Drink Rev.,* Summer (1989) 33–35.

178. Takama, R., Ishu, H. & Muraki, S., Quality changes of salted Chinese cabbages during storage and by freeze-drying. *J. Jpn. Soc. Food Sci. Tech. (Nipp. Shok. Kog. Gakk.),* **33** (1986) 701–707.

179. Itabashi, M., On the change of relative content of tasty and non-tasty amino acids during pickling process of Sunki-pickles. *J. Jap. Soc. Food Sci. Tech. (Nipp. Shok. Kog. Gakk.),* **35** (1988) 111–114.

180. Hawer, W. D. S., Ha, J. H., Seog, H. M., Nam, Y. J. & Shin, D. W., Changes in taste and flavour compounds of kimchi during fermentation. *Korean J. Food Sci. Tech.,* **20** (1988) 511–517.

181. Kitamoto, K., Miyake, M., Kohno, M., Watanabe, S., Takahashi, K., Totsuka, A. & Nakamura, K., Brewing of sake with excellent taste and flavour. II Inorganic components in rice germ affecting reduction of amino acid content in sake moroni mash. *J. Brewing Soc. Japan (Nipp. Jozo Kyokai Zasshe),* **90** (1985) 59–63.

182. Kirchhof, P.-M., Biehl, B. & Crone, G., Peculiarity of the accumulation of free amino acids during cocoa fermentation. *Food Chem.,* **31** (1989) 295–311.

183. Dubois, J. & Fabre, J., Composition having a salty flavour. French Patent Application FR 547 992 A1 (1985).

184. Cha, Y. J. & Lee, E. H., Studies on the processing of low salt fermented sea foods. VI Taste compounds of low salt fermented anchovy and yellow corvenia. *Bull. Korean Fisheries Soc.*, **18** (1985) 325–332.
185. Konosu-S., Watanabe, K., Koriyama, T., Shirai, T. & Yamaguchi, K., Extractive components of scallop and identifications of its taste-active components by omission test. *J. Jpn. Soc. Food Sci. Tech. (Nipp. Shok. Kog. Gakk.)*, **35** (1988) 252–258.
186. Tsen, H. Y. & Sun, S-T., Analysis of the taste extract of grass shrimp—the relationship between amino acid composition and proteases activity. *J. Chinese Agric. Chem. Soc.*, **25** (1987) 140–149.
187. Marui, T. & Kiyohara, S., Structure-activity relationships and response features for amino acids in fish taste. *Chem. Senses*, **12** (1987) 265–275.
188. Doving, K. B., Flavouring fish food. In *Flavour Science and Technology*, ed. M. Martens, G. A., Dalen & H. Russwurm Jr. Wiley, Chichester, 1987, pp. 253–258.
189. Tamura, M., Seki, T., Kawasaki, Y., Tada, M., Kikuchi, E. & Okai, H., An enhancing effect on the saltiness of sodium chloride of added amino acids and their esters. *Agric. Biol. Chem.*, **53** (1989) 1625–1633.
190. Ito, T., Sugawara, E., Miyanohara, J. I., Sakurai, Y. & Odagiri, S., Effect of amino acids as nitrogen sources on microbiological formation of pyrazines. *J. Jap. Soc. Food Sci. Tech (Nipp. Shok. Kog. Gakk.)*, **36** (1989) 762–764.
191. Shu, C.-K., Hagedorn, M. L., Mookherjee, B. D. & Ho, C.-T., Volatile components of the thermal degradation of cystine in water. *J. Agric. Food Chem.*, **33** (1985) 438–442.
192. Baltes, W. & Bochmann, G., Model reactions on roast aroma formation. IV Mass spectrometric identification of pyrazines IV. *Z. Lebensm. Unt. Forsch.*, **184** (1987) 485–493.
193. Shu, C.-K., Study of the reaction between cystine and 2,5-dimethyl-4-hydroxy-3(2H)-furanone. *Diss. Abstr. Int. B, 5* (1985).
194. Shu, C.-K. & Ho, C.-T., Effect of pH on the volatile formation from the reaction between cysteine and 2,5-dimethyl-4-hydroxy-3(2H)-furanone. *J Agric. Food Chem.*, **36** (1988) 801–803.
195. Kemp, S. E. & Birch, G. G., Tastes and solution properties of enantiomeric amino acids. *Chem. Senses*, **14** (1989) 214.
196. Wong, D. W. S., Sweeteners. In *Mechanism and Theory in Food Chemistry*, ed. D. W. S. Wong. Van Nostrand Reinhold, New York, 1989, pp. 264–282.
197. Kemp, S. E. & Birch, G. G., Structure, solution properties and sensory characteristics of amino acid molecules. *Chem. Senses*, **13** (1988) 703.
198. Birch, G. G. & Kemp, S. E., Apparent specific volumes and tastes of amino acids. *Chem. Senses*, **14** (1989) 249–258.
199. King, B. M., Flavour, the key to success for high-protein foods. In *Food Acceptance and Nutrition*, ed. J. Solms et al. Conference 7–10 April 1987, Academic Press, London, 1987, pp. 79–97.
200. Manning, D. J., Ridout, E. E. & Price, J. C., Non-sensory methods for cheese flavour assessment. In *Advances in the Microbiology and Biochemistry of Cheese and Fermented Milk*, ed. F. L. Davies & B. A. Law. Elsevier Applied Science Publishers, London, 1984. pp. 229–254.

201. Teranishi, R., New trends and developments in flavor chemistry; an overview. In *Flavor Chemistry Trends and Developments*, ed. R. Teranishi. ACS Symposium Series, American Chemical Society, 1989, pp. 1–6.

202. Pangborn, R. M., Sensory science in flavour research: achievements, needs and perspectives. In *Flavour Science and Technology*, Proceedings of the 5th Weurman Flavour Research Symposium, 1987, ed. M. Martens, G. A. Dalen & H. Russwurm Jr. Wiley, Chichester, 1987, pp. 275–289.

203. Montange, J.-Y., O'Halloran, S. & Strinsjo, H., Sensory data collection and evaluation with SENSTEC, a novel approach. In *Flavour Science & Technology*, Proceedings of the 5th Weurman Flavour Research Symposium, 1987, ed. M. Martens, G. A. Dalen & H. Russwurm Jr. Wiley, Chichester, 1987, pp. 543–548.

204. Roos, P. E., Dijksterhuis, G. B. & Punter, P. H. Automation of sensory analysis—new developments. In *Flavour Science and Technology*, ed. M. Martens, G. A. Dalen & H. Russwurm Jr. Wiley, Chichester, 1987, pp. 541–542.

205. Eriksson, C., Chemistry in flavour research, achievements, needs and perspectives. In *Flavour Science and Technology*, ed. M. Martens, G. A. Dalen & H. Russwurm Jr. Wiley, Chichester, 1987, pp. 5–21.

206. Wong, D. W. S., Flavors. In *Mechanism and Theory in Food Chemistry*, ed. D. W. S. Wong. Van Nostrand Reinhold, New York, 1989, pp. 231–263.

207. Maga, J. A., Compound structure versus bitter taste. In *Bitterness in Foods and Beverages*, ed. R. L. Rousseff. Elsevier, Amsterdam, 1990, pp. 35–48.

208. Brieskorn, C. H., Physiological and therapeutical aspects of bitter compounds. In *Bitterness in Foods and Beverages*, ed. R. L. Rousseff. Elsevier, Amsterdam, 1990, pp. 15–33.

209. Kurihara, K., Recent Progress in the taste receptor mechanism. In *Umami, A Basic Taste*, ed. Y. Kawamura, & D. M. R. Kare. Marcel Dekker Inc., 1987, pp. 3–39.

210. Schiffman, S. S., Recent insights into the mechanisms of taste transduction and modulation. *Food Chem.*, 21 (1986) 259–281.

211. Overbosch, P., & Soeting, W. J., Temporal aspects of flavoring. In *Flavor Chemistry: Trends and Developments*, ed. R. Teranishi, ACS Symposium Series, American Chemical Society, 1989, pp. 138–150.

212. Voirol, E. & Daget, N., Direct nasal and oronasal profiling of a meat flavouring: influence of temperature, concentration and additives. *Lebensmitt.-Wise.-u-Tech.*, 22 (1989) 399–405.

213. Overbosch, P., Flavour release and perception. In *Flavour Science and Technology*, ed. M. Martens, G. A. Dalen & H. Russwurm Jr. Wiley, Chichester, 1987, pp. 291–300.

214. Vernin, G. (ed.), *The Chemistry of Heterocyclic Flavouring and Aroma Compounds*. Ellis Horwood Ltd., Chichester, 1982.

215. Hwang, S.-S., A study of the reaction flavors. *Diss. Abstr. Int. B*, 47 (1987) 4363–4364.

216. Zhang, Y., Chien, M. & Ho, C. T., Comparison of the volatile compounds obtained from thermal degradation of cysteine and glutathione in water. *J. Agric. Food Chem.*, 36 (1988) 992–996.

217. Rizzi, G. P., New aspects on the mechanism of pyrazine formation in the Strecker degradation of amino acids. In *Flavour Science and Technology*, ed. M. Martens, G. A. Dalen & H. Russwurm Jr. Wiley, Chichester, 1987, pp. 23–28.
218. Shaath, N. A. & Griffin, P. M., Modern analytical techniques in the flavor industry. In *Frontiers of Flavor*, Proceedings of the 5th International Flavor Conference, Chalkidiki, Greece, 1–3 July, 1987. Elsevier, Amsterdam, 1988, pp. 89–108.
219. Teranishi, R., Development of methodology for flavor chemistry past, present and future. In *Flavor Chemistry of Lipid Foods*, ed. D. B. Min & T. H. Smouse. The American Oil Chemists' Society, Champaign, Ill., 1989, pp. 13–25.
220. Flament, I., Chevallier, C. & Keller, U., Extraction and Chromatography of food constituents with supercritical CO_2. In *Flavour Science and Technology*, ed. M. Martens, G. A. Dalen & H. Russwurm Jr. Wiley, Chichester, 1987, pp. 151–164.
221. Reineccius, G. A., Isolation of food flavours. In *Flavor Chemistry of Lipid Foods*, ed. D. B. Min & T. H. Smouse. The American Oil Chemists' Society, Champaign, Ill., 1989, pp. 26–34.
222. Morin, P., Caude, M., Richard, H. & Rosset, R., On-line carbon dioxide SFC-FTIR in aroma research, perspectives and limits. In *Flavour Science and Technology*, ed. M. Martens, G. A. Dalen & H. Russwurm Jr. Wiley, Chichester, 1987, pp. 165–174.
223. Williams, A. J., Tucknott, O. G., Lewis, M. J., May, H. V. & Wachter, L., Examples of cryogenic matrix isolation GC/IR in the analysis of flavour extracts. In *Flavour Science and Technology*, ed. M. Martens, G. A. Dalen & H. Russwurn Jr. Wiley, Chichester, 1987, pp. 259–271.
224. Mills, O. E., A headspace sampling method for monitoring flavour volatiles of protein products. *N. Z. J. Dairy Sci. Tech.*, 21 (1986) 49–56.
225. Koller, W. D., Preconcentration of volatiles by using static-head-space technique for mass spectrometric identification. In *Flavour Science & Technology*, ed. M. Martens, G. A. Dalen & H. Russwurm Jr. Wiley, Chichester, 1987, pp. 181–184.
226. Baltes, W. & Bochmann, G., Model reactions on roast aroma formation. III. Mass spectrometric identification of pyrroles IV. *Z. Lebensm. Unters. Forsch.*, 184 (1987) 478–484.
227. Baltes, W. & Bochmann, G. Model reactions on roast aroma formation. V. Mass spectrometric identification of pyridines. *Z. Lebensm. Unters. Forsch.*, 185 (1987) 5–9.
228. Al-Tamrah, S. A., Kinetic studies on the determination of pyrroles asnd tryptophan by Ehrlich reaction. *Anal. Lett.*, 22 (1989) 387–401.
229. Lendero, L., The use of capillary gas chromatography combined with headspace sampling technique and computerized retention, index data base for flavour analysis. In *Flavour Science and Technology*, ed. M. Martens, G. A. Dalen & H. Russwurm Jr. Wiley, Chichester, 1987, pp. 185–186.
230. Leland, J. V., Lahiff, M. & Reineccius, G. A., Predicting intensities of milk off-flavours by multivariate analysis of gas chromatographic data. In

Flavour Science and Technology, ed. M. Martens, G. A., Dalen & H. Russwurm Jr. Wiley, Chichester, 1987, pp. 453–468.

231. Persson, T., NMSP—a tool for data collection and data analysis in flavour research. In *Flavour Science and Technology*, ed. M. Martins, G. A. Dalen & H. Russwurm Jr. Wiley, Chichester, 1987, pp. 549–550.

232. MacFie, H. J. H., Data analysis in flavour research: Achievements, needs and perspectives. In *Flavour Science and Technology*, ed. M. Martens, G. A. Dalen & H. Russwurm Jr. Wiley, Chichester, 1987, pp. 423–438.

233. Duxbury, D. D., Natural flavor enhancers for microwavable foods. *Food Process.*, **66** (1991) 115–116.

234. Duxbury, D. D., Savory natural flavors enhance low-fat foods. *Food Process.*, **66** (1991) 100, 102.

235. Duxbury, D. D., Natural beef flavor-delicious in chicken. *Food Process.*, **66** (1991) 118, 120.

236. Ratz, W., Flavouring for the food industry. *Fleischerei*, **39** (1988) IV–V.

237. Miller, M. F., Davis, G. W., Seideman, S. C., Wheeler, T. L. & Ramsey, C. B., Extending beef bullock restructured steaks with soy protein, wheat gluten or mechanically separated beef. *J. Food Sci.*, **51** (1986) 1169–1172.

238. Duxbury, D. D., Isolated soy proteins add 'lite' image to meat products. *Food Process.* US, **49** (1988) 64, 66.

239. Tuley, L., Sunrise for soya. *Food Manuf.*, **66** (1991) 22–24.

240. Endres, J. G. & Monagel, C. W., Non-meat protein additives. In *Advances in Meat Research*, Vol. 3, ed. A. M. Pearson & T. R. Dutson. 1987, pp. 331–350.

241. Vallejo-Cordoba, B., Nakai, S., Powrie, W. D. & Beveridge, T., Protein hydrolysates for reducing water activity in meat products. *J. Food Sci.*, **51** (1986) 1156–1161.

242. Vallejo-Cordoba, B., Nakai, S., Powrie, W. D. & Beveridge, T., Extended shelf life of frankfurters and fish frankfurter-analogs with added soy protein hydrolysates. *J. Food Sci.*, **52** (1987) 1133–1136.

243. Duxbury, D. D., 'High tech' QFD and TQM programs produce all-natural ingredients. *Food Process. US*, **60** (1991) 46–50.

244. Hoffman, K. & Marggrander, K., Reducing the common salt content of meat products by the use of collagen hydrolysates. *Fleischwirtschaft*, **69** (1) (1989) 23–8, 65.

245. LaBell, F., Milk hydrolysate. *Food Process.*, **51** (1990) 102, 104.

246. Marggrander, K., Use of collagen protein hydrolysates. Paste-type spreads for bread. *Fleischerei*, **40** (1989) 229–231.

247. Schenz, Z. F. & Trumbetas, J., Flavour and mouthfeel character of beverages. US Patent 4 615 200 (1986).

248. Szczesniak, A. S. & Schenz, A. F., Improved fruit flavored beverages. European Patent EP 0 117 047 B1 (1987).

249. Braun, S. D. & Olson, N. F., Encapsulation of proteins and peptides in milkfat: encapsulation efficiency and temperature and freezing stabilities. *J. Microencapsulation*, **3** (1986) 125–126.

250. LaBell, F., Coated flavor enhancer resists enzyme activity. *Food Process.*, **60** (1991) 114–115.

INDEX